Geotechnologies and the Environment

Volume 18

The *Geotechnologies and the Environment* series is intended to provide specialists in the geotechnologies and academics who utilize these technologies, with an opportunity to share novel approaches, present interesting (sometimes counter-intuitive) case studies, and, most importantly, to situate GIS, remote sensing, GPS, the internet, new technologies, and methodological advances in a real world context. In doing so, the books in the series will be inherently applied and reflect the rich variety of research performed by geographers and allied professionals.

Beyond the applied nature of many of the papers and individual contributions, the series interrogates the dynamic relationship between nature and society. For this reason, many contributors focus on human-environment interactions. The series is not limited to an interpretation of the environment as nature per se. Rather, the series "places" people and social forces in context and thus explores the many socio-spatial environments humans construct for themselves as they settle the landscape. Consequently, contributions will use geotechnologies to examine both urban and rural landscapes.

More information about this series at http://www.springer.com/series/8088

Michael Leitner • Jamal Jokar Arsanjani
Editors

Citizen Empowered Mapping

Springer

Editors
Michael Leitner
Department of Geography and Anthropology
Louisiana State University
Baton Rouge, LA, USA

Jamal Jokar Arsanjani
Geoinformatics Research Group,
Department of Development and Planning
Aalborg University
Copenhagen, København, Denmark

ISSN 2365-0575 ISSN 2365-0583 (electronic)
Geotechnologies and the Environment
ISBN 978-3-319-51628-8 ISBN 978-3-319-51629-5 (eBook)
DOI 10.1007/978-3-319-51629-5

Library of Congress Control Number: 2017940639

Printed on acid-free paper

This Springer imprint is published by Springer Nature
The registered company is Springer International Publishing AG
The registered company address is: Gewerbestrasse 11, 6330 Cham, Switzerland

Contents

An Introduction to Citizen Empowered Mapping

Michael Leitner and Jamal Jokar Arsanjani

Introduction

This book on *Citizen Empowered Mapping* is a follow-up publication of the Special Content Issue (SCI) on *Crowdsourced Mapping* published in the *Cartography and Geographic Information Science (CaGIS)* journal in 2017 (Jokar Arsanjani et al. 2017). In order to understand how the idea to edit this book started, we have to go back more than 2 years ago. On 7 October 2014, a first solicitation of a manuscript proposal submission for a SCI on *Crowdsourced Mapping* to be published in the *CaGIS* journal was sent out. This solicitation was distributed to members of different list servers, including those from the University Consortium of Geographic Information Science (UCGIS), Cartography and Geographic Information Society (CaGIS), and four different Specialty Groups of the AAG (Applied Geography, Cartography, Geographic Information Science and Systems, and Spatial Analysis and Modeling). The solicitation welcomed submissions from the academic, public, and private sectors. Topics appropriate for the SCI were plenty and broadly defined. They included (1) volunteered geographic information (VGI); (2) citizen science; (3) humans as sensors (sensor-enabled humans); (4) mapping user-generated geographic information from the crowd, e.g., social media (Flickr, Instagram, Panoramio, etc.), telecommunication services, social networks (Facebook, Twitter, etc.), and sensor networks; (5) uncertainty mapping; (6) data/information quality visualization; (7) spatial/temporal visualization of crowd analysis; (8) indicators for volunteered geographic information and citizen science; (9) (re)mapping unmapped features from the crowd; (10) collaborative mapping and governments; (11) data mining and fusion of crowd-based data/information with authoritative data; (12) metadata production; (13) mapping costs and benefits; (14) crowd-based data/information for diverse applications, e.g., cities/environmental monitoring, disaster management, landscape analysis, land use/cover monitoring, social science, ecology, etc.; (15) emerging topics and datasets generated from/by the crowd; (16) big geo(data); (17) OpenStreetMap; (18) geospatial analysis and Internet of Things (IoT); and (19) digital divide and (geo)information dissemination.

By 30 November 2014, the deadline for manuscript proposal submission, a total of 54 proposals was submitted. One month later, the vast majority of corresponding authors of these 54 proposals were subsequently invited to submit a full manuscript. This decision was made solely by the guest editors. By the end of September 2015, a total of 33 manuscripts was received. Each manuscript was subsequently reviewed (double-blind) by at least two expert reviewers, following standard *CaGIS* review guidelines. Overall, the innovative aspect and the scientific quality of the research weighted heavily on the decision whether or not a manuscript was accepted or rejected. Altogether, 11 of the 33 manuscripts were finally accepted for publication in *CaGIS*. Of these 11 articles, six are included in the SCI on *Crowdsourced Mapping*, and the remaining five articles appear in the sixth volume of 2017 and in the first volume of 2018 of the *CaGIS* journal. The main reason is that similar to a regular issue of *CaGIS*, the SCI on *Crowdsourced Mapping* had 96 pages allocated to it. This page count was already exhausted with six of the 11 papers. We also contacted the corresponding authors from the 22 of the originally 33 received manuscripts that were not selected to be published in the *CaGIS* journal. We asked them whether they would be interested to have their research, pending necessary revisions, included as a book chapter in an edited book to be published by Springer entitled *Citizen Empowered Mapping*. We received a positive feedback from authors from 12 of the 22 manuscripts. These 12 manuscripts were revised according to the original reviewers' comments and suggestions by the book editors and represent the main chapters in this publication.

The book on *Citizen Empowered Mapping* organizes the 12 book chapters into three parts that include *Data Acquisition and Modeling*, *Data Quality and Reliability*, and *Environmental Monitoring and Perception*. This is completed by an *Introductory Chapter* written by the editors and an *Outlook Chapter* written by Giles Foody. Thirty-eight different authors contributed to the 12 main chapters of this book. This is an average of 3.2 authors per article and similar to the average number of authors (namely, 3.4) contributing to the *CaGIS* SCI of *Crowdsourced Mapping*, confirming the collaborative nature of this research topic. Contributing authors come from the USA (13), France (8), Mexico and the Czech Republic (4 each), Greece and Canada (3 each), Portugal (2), and New Zealand (1). Interestingly, while *Crowdsourced Mapping* seems to be a very popular research topic in Europe and in North America, with all but one contributing author coming from these two regions, it may not be of much interest in China and India (not a single author comes from these two populous countries). This supports a similar trend that was already observed for the SCI on *Crowdsourced Mapping*. It is also noticeable that almost 80% (30 out of 38) of all authors have an academic affiliation, with the remaining eight working in a governmental position. No author has a private sector affiliation. Eight of the 30 authors with an academic affiliation have their home department in geography, geographic analysis, geomatics, or geoinformatics. The second strongest group of authors from academia come from spatial planning or regional planning (five in total), followed by environmental science, computer/information science, and development studies (four each). Overall, it is clear that *Citizen Empowered Mapping* is a highly interdisciplinary topic that is of great research interest across a number of departments and colleges. Similar to the *CaGIS* SCI on *Crowdsourced Mapping*, geography, in general, and geographic information science, in particular, seem to play a leading role in this research effort.

In the following, the main research focus of each of the 12 chapters published in this book will be briefly introduced. Starting off the *Data Acquisition and Modeling* section is the chapter by **Touya and Baley**, in which the authors explore the automatic harmonization of OpenStreetMap data for large scale maps, with a process that transforms rough objects to make them consistent with detailed objects. In the second book chapter, **Gomez et al.** define new urban soundscape indicators to be depicted in noise maps that offer the possibility of several interpretations depending on the users' personal and cultural characteristics. The next chapter by **Lewis et al.** evaluates and compares Geospatial Software as a Service (GSaaS) platforms oriented toward providing basic mapping capabilities to non-GIS experts. The main goal of the book chapter by **Stylianidis et al.** is to initiate distributed and parallel techniques in mapping systems by implementing efficient map updating algorithms based on road network extraction methodology on satellite/aerial images. The fifth and final chapter in this section by **Sabino and Rodrigues** presents a productive network representation model, designed to discover indirect keywords and locations. The spatial dimension of the model enables indirect location discovery methods through the interpretation of the network as a graph, solely relying on keywords and locations that categorize or describe productive items.

The first book chapter in Part II titled *Data Quality and Reliability* by **Korris et al.** reports (1) on the implementation of a new crowdsourcing project to provide accurate and authoritative spatial data for The National Map (TNM) of the US Geological Survey (USGS) National Geospatial Program and explores (2) the quality of the volunteered geographic information (VGI) within the TNM by assessing horizontal positional errors, attribute errors, and errors of commission. The next article by **Shaker et al.** addresses the question whether routes computed using crowdsourced points of interests (POIs) are reliable between origin and destination locations. To address this question, the authors conducted experiments where routes (shortest and fastest) computed using crowdsourced POIs (e.g., through OpenStreetMap) were compared with routes computed using POIs obtained from professional and commercial sources. The final book chapter in Part II by **Craun and Hung** compares volunteer-provided data in OpenStreetMap (OSM) to Topologically Integrated Geographic Encoding and Referencing (TIGER) data from the US Census Bureau in order to evaluate the usability of these data as part of an authoritative dataset.

Part III in this book on *Environmental Monitoring and Perception* includes four chapters. The first chapter by **Leung et al.** critically assesses the use of social media as an observation network to identify rare phenomena, such as frostquakes, including the possibility of false positives and population bias. The next chapter by **Brandeis and Nyerges** tests the effectiveness of using a spatial decision unit (SDU) and compares the usability of free-form volunteered geographic information (VGI) and highly structured VGI. The main purpose of the next chapter in this section by **Martinez-Viveros et al.** is to help consolidate a Mexico City neighborhood

(Consejo Vecinal Roma) as a policy community by means of integrating a territorial vision of *The Roma*'s problems and opportunities, as perceived by the citizens involved in the process. The research presented in the final chapter by **Pánek et al.** involves the integration of Gould-style mental maps with Participatory GIS technologies. It describes the testing and implementation of the web-based crowd-sourcing tool PocitoveMapy.cz that is used for the collection of and visualization in maps of people's perceptions of environmental deprivation.

At the end of this introductory chapter, the editors would like to thank all 38 authors for contributing their research to this edited book, since without you this book would not have been possible.

Department of Geography and Anthropology
Louisiana State University
Baton Rouge, LA, USA
Michael Leitner

Geoinformatics Research Group
Department of Development and Planning
Aalborg University
Copenhagen, København, Denmark
Jamal Jokar Arsanjani

Reference

Jokar Arsanjani J, Leitner M, Zipf A (2017) Crowdsourced mapping. Cartogr Geogr Inf Sci (Special Content Issue) 44(2):95–184

Part I
Data Acquisition and Modeling

Chapter 1
Level of Details Harmonization Operations in OpenStreetMap Based Large Scale Maps

Guillaume Touya and Matthieu Baley

Abstract OpenStreetMap data comprise of very detailed (e.g. zebra crossing) and quite rough features (e.g. built-up area). But making large scale maps from data with inconsistent level of detail often blurs map comprehension. This paper explores the automatic harmonization of OpenStreetMap data for large scale maps, i.e. the process that transforms rough objects to make them consistent with detailed objects. A typology of the new operators that harmonization requires is presented and six algorithms that implement the operators are described. Experiments with these algorithms raise several research questions about automation, parametrization, or the level of abstraction of the transformation, which are discussed in the paper.

Keywords Level of detail • Volunteered geographic information • OpenStreetMap • Cartography • Legibility • Caricature

1.1 Introduction

As OpenStreetMap (OSM) is growing larger every day, practical applications based on OSM data are flourishing, but the initial goal of the project was to produce open topographical maps. A quick look at the default map output provided by OSM shows that it is difficult to create good legible maps out of the huge amount of data in OSM. One of the main obstacles to the creation of good legible maps from OSM data is the heterogeneity of its level of detail (LoD). For example, in the database, very detailed objects (e.g. zebra crossings) coexist with rough objects (e.g. shorelines extracted from Landsat imagery). This heterogeneity is troublesome for small scale maps, as detailed objects should be removed or simplified, but also for large scale maps, as rough objects are often inconsistent with detailed features of the map (Touya 2012a). Regarding small scale maps, it can be considered as a map generalization problem. Although there is little research effort on generalizing OSM data (Klammer 2013; Schmid and Janetzek 2013; Sester et al. 2014), this paper

G. Touya (✉) • M. Baley
IGN – COGIT, 73 avenue de Paris, 94165, Saint-Mandé, France
e-mail: guillaume.touya@ign.fr

M. Leitner, J. Jokar Arsanjani (eds.), *Citizen Empowered Mapping*,
Geotechnologies and the Environment 18, DOI 10.1007/978-3-319-51629-5_1

Fig. 1.1 Buildings on the *right* should be inside the built-up area (in *gray*) to make it a understandable spatial relation

focuses instead on legibility problems caused by the level of detail heterogeneity in large scale maps, i.e. scales larger than 1: 25,000. In such large scale maps, detailed objects can be displayed in the map without any generalization transformation, because the scale ratio makes map symbols close to their real extent on the ground. But rough objects are also included in the map, because it is often better to display rough information than display nothing all. For instance, if all forest polygons are rough, it is better to include them in the map rather than leaving them out. The first step to make such large scale maps is to find a way to infer the level of detail of OSM data in order to discriminate detailed objects from rough objects (Touya and Brando 2013; Touya and Reimer 2015). The understanding of a map is highly dependent on the way the reader grasps spatial relations between map objects. As a consequence, the level of detail inconsistencies are mainly damaging when occurring between spatially related objects. For instance, the gray built-up area in Fig. 1.1 does not include some of the town buildings on the right. The map reader may thus be troubled and may misinterpret what the built-up area is.

Dealing with LoD inconsistencies in large scale maps can be seen as a new automatic mapping process, namely to transform rough objects to make them consistent with detailed objects when both types of objects share a spatial relation that helps to understand the map. We call such a process harmonizing level of detail, implying that the harmonization increases LoD. It should be noted that generalizing heterogeneous data is also a kind of harmonization but it is not the focus of this paper. The automation of harmonization raises two questions. Is it possible to automatically harmonize OSM maps? Is it meaningful to transform data without any additional information from ground truth to make it more detailed? The work presented in this paper seeks to explore both questions by experimenting first attempts of automatic harmonization on OSM data.

After the introduction, the following second section of this paper, briefly discusses the notions of level of detail, scale and quality, and briefly describes a method to infer the level of detail, as a first step for harmonization. The third part precisely defines cartographic LoD harmonization and proposes a typology of possible harmonization operators. Section four is the core of the paper and describes six algorithms to harmonize different types of LoD inconsistencies. The fifth part discusses several issues on the automation and the meaning of harmonization for a map. The last section draws conclusions and explores further research.

1.2 Level of Detail and Data Quality

The scale of a map is the mathematical ratio between a distance measured in the map and the same distance measured on the ground. But the scale is not only a ratio, it is also closely related to the content of the map and its resolution (Mackaness 2007). Indeed, the scale limits the map to a certain extent, and the human perception limitations bound what can be displayed on the map. On the other hand, geographical information databases with vector data can be zoomed in and out, thus they cannot be defined by a single scale. So, we usually refer to the level of detail (LoD) when we want to define the resolution, or the granularity of a geographical database. Unlike scale, the level of detail is not a mathematically defined notion, and the fuzziness of its definition may make it hard to assess. In previous work, we defined the level of detail as a complex notion (Touya and Brando 2013) that encapsulates elements of:

- conceptual schema (a tree representation is more detailed than a forest),
- attribute resolution,
- geometric resolution, i.e. smallest length between two vertices (Fig. 1.2),
- geometric precision or accuracy,
- granularity (size of the smallest detail of geometries).

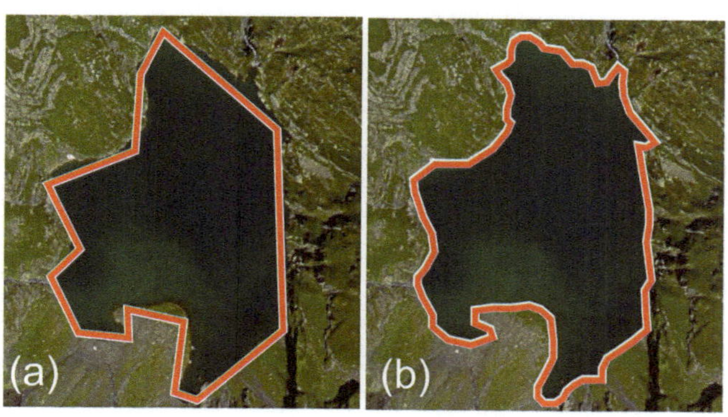

Fig. 1.2 Two captures of the same lake at different levels of detail

OSM can be considered as a large geographical database, where the LoD of features varies as the contributors with varying skills may use sources of varying scales or resolutions.

There is a substantial literature on all aspects of OSM data quality (Haklay 2010; Girres and Touya 2010; Mooney and Corcoran 2012). Although the level of detail comprises of some elements of data quality, the aim here is not exactly to assess data quality. A rough lake outline (i.e. with few vertices, see Fig. 1.2a) can be considered as a bad quality feature if it is expected to have a high level of detail, or conversely a good quality feature if the expected level of detail is not so high (if the aim is to make a map at a small scale, for instance). In this work, the focus is not on quality, precision or accuracy alone, so an inaccurate position will only be considered as a (major) factor for a low level of detail.

Scale can be inferred from the geometrical resolution and the analysis of similar features in existing maps (Reimer et al. 2014). For instance, Biljecki et al. (2014) proposed several metrics to infer LoD in 3D city models. Regarding OSM, the LoD inference is automatically possible using multiple criteria decision techniques (Touya and Brando 2013; Touya and Reimer 2015), where the used criteria correspond to different aspects of LoD (resolution, precision, etc.). Then, the LoD inconsistencies can be identified by searching for key anomalous, or improbable spatial relations between detailed and rough features. For instance, trees should not be located on roads, or land use parcels should not extend over coastlines (Touya and Brando 2013). Improvements and alternative methods are, of course, necessary to get a better inference of the individual level of detail of OSM data (Touya and Reimer 2015), but this is not the focus of the presented work. As a consequence, the results of the inference method from Touya and Reimer (2015), i.e. the classification of OSM features into one of five LoD categories from *street* LoD to *country* LoD, are used as inputs for the harmonization methods presented in the next sections.

The problem of LoD inconsistencies is not specific to the derivation of maps from OSM data, but may occur with any other volunteered geographical information. However, we choose to focus on OSM as the project is complete enough to derive large scale maps with a high density of information in areas with many contributors, such as, from Western Europe. OSM also contains data for every part of the world, making processes to automatically derive maps useful for many people.

1.3 Cartographic LoD Harmonization

1.3.1 Problem Statement

We define cartographic LoD harmonization as the mapmaking process that transforms features involved in a LoD inconsistency, in order to make the map more legible and comprehensible. When the target is a small scale map, or a map where the LoD of the rough feature of the inconsistency matches the scale, harmonization can be brought down to map generalization. But, in this paper, we only focus on large scale target maps, where simplification is not necessary and harmonization is a new problem. Moreover, the aim is not to provide quality control for the OSM dataset. When problematic LoD inconsistencies are identified, it is the readability that guides the transformation and not the quality control. In this case, the transformation should be a balance between position preservation and map legibility. As a result, caricature operations may be preferred to transformations guided by ground truth. OSM data are supposed to be more or less incomplete with some objects that could be in the data, but have not yet been captured. So, harmonization operations should take into account that the possible lack of detail may be due to incompleteness rather than LoD. For instance, when an object should be near a road and it is not, then the inconsistency may be caused by the inaccuracy of the object location, or by the absence of a road that exists in the real world. The key to harmonize LoD inconsistent spatial relations in the map is to transform the rough counterpart of the spatial relation while preserving the detailed member.

Research in cartographic generalization and multiple representation databases already focused on relations between geographical features at different LoDs. In multiple representation databases, spatial relations can be horizontal or vertical (Mustière and Moulin 2002; Burghardt et al. 2010). Horizontal relations involve features with a similar level of detail, such as a building located at the end of a dead end road, while vertical relations involve two features at different LoDs that represent the same real world feature, such as a city represented by a polygon or a point. Spatial relations involved in LoD inconsistencies are neither vertical, nor horizontal. Following the same vocabulary, we can consider them as diagonal relations.

There is no way to transform rough features into detailed and accurate abstractions of real world entities they represent, so, in a way, harmonization aims to make the map more readable. As stated by Monmonier (1996), “not only it is easy to lie with maps, it is ESSENTIAL.” Maps are systems of relationships (Mackaness et al. 2014), which means that most of the meaning of the map is conveyed by relations, and preserving relations by making them more legible improves the way maps are understood. However, transforming map features too much without safeguard can damage map readability more than it improves it. Then, our safeguard is the evaluation of harmonization to verify that map features are not too much transformed.

1.3.2 *Typology of Harmonization Operators*

In order to derive legible maps from LoD inconsistent information, harmonization requires operations that can be related to other automated cartography processes, such as cartographic generalization or text placement. Typologies of generalization operators (e.g. simplification, displacement, elimination) already exist, see for instance the ones from Foerster, Stoter and Kobben (2007) and Regnauld and McMaster (2007). In our case, we preferred referring to a broader typology of operators for multiscale maps from Roth et al. (2011). The following operators are introduced for harmonization as derivatives of Roth et al.'s operators:

- Merge/Dissolve
- Adjust Shape
- Displace and enhance
- Disambiguation

Merge is defined by Roth et al. (2011) as a "*replacement of a feature with a representative feature of equal dimensionality*" and illustrated by a group of small islands merged to the nearby big island. Regarding harmonization, the merge operation is useful to improve the geometry of a feature by merging detailed features that should be part of the rough feature. Geographical datasets often comprise high level objects that are aggregates of lower level objects of the dataset. For example, a city is an aggregate of buildings, roads, and parks. In OSM, such aggregate objects are very common and are generally less detailed than their components. This generates the most frequent LoD inconsistencies with obvious components that lie just outside the aggregate (see section 4.1). But merge is also coupled with **dissolve**, because the detailed features might sometimes be dissolved from the rough feature rather than merged, when they should not be part of the aggregate. For instance, a primary road cannot be part of a school site (see section 4.2).

Adjust shape is the adjustment of a least detailed shape without changing feature dimensionality, to avoid an improbable relation. It derives from the Roth et al. (2011) Adjust shape operator that was only dedicated to symbols and not to feature geometry. Modifying the shape of a lake in order to avoid intersections with roads is an example of the adjust shape operator (see section 4.4).

Displace and enhance is a displacement of a map feature in order to preserve or emphasize a relation. It is a mix of the "displace" and "enhance" operators from Roth et al. (2011). The displacement of trees along roads to enable a real alignment of tree symbols is an example of the "displace and enhance" operator (see sections 4.3 and 4.5).

Sometimes, the map reader does not know if the improbable relation is true or if there is a problem of data quality (e.g. missing features). **Disambiguation** is a new operator that aims at the removal of this ambiguity in map reading without consideration for the ground truth (that we do not know). It arbitrarily removes the improbable spatial relation. For instance, when a group of close buildings is inside a forest without clearing, the addition of the clearing is a disambiguation operation (see section 4.6). Finally, harmonization also has to make use of existing operators such as displacement to correct inaccurate positions.

1.4 Examples of Algorithms for Harmonization Operations

This section describes several algorithms (Table 1.1) that implement the operations introduced in the previous section.

1.4.1 *Built-Up Area Extension*

1.4.1.1 Algorithm Description

This algorithm seeks to extend the limit of built-up areas to include all buildings of the actual built-up area, i.e. buildings that lie just outside the limit, but also all buildings that are very close to these buildings. In this case, the only information available to draw a more detailed extent for the built-up area is that the components that caused the inconsistency should be inside the aggregate, and no longer outside. Several algorithms exist to compute the boundaries of built-up areas only using buildings (Boffet 2000; Chaudhry and Mackaness 2008; Walter 2008). We propose to use a similar strategy based on buffering the buildings, but in an iterative way. At each step, the algorithm searches buildings that are within a radius around the built-up area, but not yet inside the area. The radius used is 15 m, derived from Boffet (2000). Figure 1.3 shows the three steps of the algorithm, at each iteration: (i) buffers, with the same radius, are computed around the components to include, (ii) buffers are merged to the built-up area, and (iii) the outline is simplified to get a consistent resolution all along, as far as possible. In order to avoid too large extensions, the radius is cushioned at each iteration, with a 0.8 factor. The algorithm stops when there is no new building found within the search radius.

The boundary simplification algorithm is a simple Douglas and Peucker filter (1973), because the sharp shapes created by this simplification algorithm look like the shapes of initial rough built-up areas.

Table 1.1 Table summarizing algorithms described in this paper, with bold features in the inconsistency column represent the rough member of the relation

Algorithm	Operation	LoD inconsistency
Built-up area extension	Merge	Buildings that lie just outside a **built-up area**
Functional site adjustment	Merge	A **functional site** that excludes some of its component/ includes false components
Tree alignment along roads	Displace and enhance	**Trees** along a road but not aligned
Intersection removal	Adjust shape	Roads/paths intersecting a **lake**
Logical displacement	Displace and enhance	**Bus stops** too far from the nearest road
Clearing addition	Disambiguation	Building groups inside a **forest area**

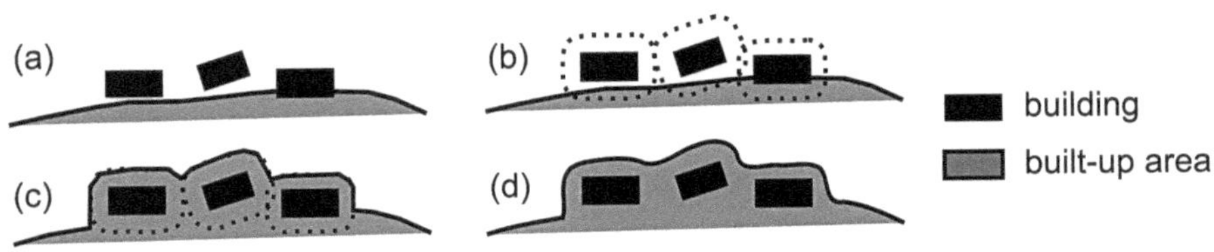

Fig. 1.3 Three steps of one iteration of the algorithm to extend built-up areas. (**a**) Initial state, (**b**) compute buffers around buildings, (**c**) merge the built-up area with buffers, (**d**) simplify to preserve the resolution

1.4.1.2 Evaluation

Constraint-based evaluation is commonly used to evaluate maps derived by automatic generalization (Stoter et al. 2009; Touya 2012b). Constraints, such as *building area should not be below 0.4 mm^2 on the map*, are defined according to map specifications and known eye perception limits (Salichtchev 1983). Such constraints have been defined to evaluate this harmonization algorithm, some for preserving the initial shape/granularity of features, and some for assuring that inconsistencies are really removed. Three constraints are used to evaluate harmonization:

- (C1) A shape preservation constraint that uses a surface distance (Girres and Touya 2010) to measure that the general shape has not been distorted too much.
- (C2) A granularity preservation constraint to measure that vertex density, used as a proxy for granularity by Reimer et al. (2014), has not increased too much after harmonization.
- (C3) A constraint that counts the number of dilated buildings intersecting the built-up area outline. The constraint is fully satisfied when there is no intersection.

An instance, or a monitor (Touya 2012b), is created for each constraint and each built-up area object. The monitors assess the constraint satisfaction given the current geometry of the built-up area. In this evaluation, the satisfaction of the constraint is retrieved before and after the harmonization to verify that C3 satisfaction has increased, while the C1 and the C2 satisfaction remained stable. The constraint satisfaction is expressed from 1 (not satisfied at all) to 8 (perfectly satisfied), similarly to the generalization constraints from Touya (2012b). A test area was chosen in the south west of France comprising around 50 built-up areas of small/medium towns with LoD inconsistencies. The built-up extension algorithm is triggered automatically with parameters given in the algorithm description. Results show a significant increase of C3 satisfaction, while C1 and C2 satisfactions slightly decrease but remain satisfied (Fig. 1.4). The mean of satisfactions only slightly increases, but the lack of an unsatisfied constraint after harmonization is a better indicator, and it proves the increased global quality of the harmonized map.

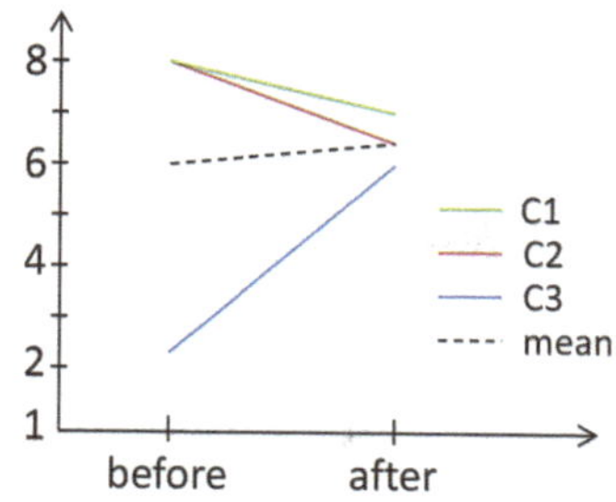

Fig. 1.4 Mean constraint satisfaction evolution during built-up extension harmonizations

1.4.2 Functional Site Adjustment

Regarding functional sites (Chaudhry et al. 2009; Mackaness and Chaudhry 2011), such as schools or hospitals, boundaries are often crisp and identified on the ground by buildings, walls, or barriers. So, in this case, the merge/dissolve operation should set more realistic bounds for the site, and as a consequence, an algorithm based on dilatation cannot be used. The principles of the proposed "functional site adjustment" algorithm is to infer the probability of features that are near the initial boundary, to be part of the functional site or not. Then, boundaries of the site are displaced around the included or excluded component, without introducing any gap.

First of all, a belonging function is defined for each type of site. The function uses semantics and topology to infer if an object is a component of the site, or not (Chaudhry et al. 2009). We decomposed the belonging function into two functions that are specialized for each type of functional site: A *functional belonging* that uses semantics, e.g. a library or a football field is more prone to be part of a school while a primary road is not; and a *spatial belonging* that uses topological and metrical measures, e.g. a building that is 99% inside the site is more likely to be a part of the site. Both belonging functions give negative values for unlikely partonomy and positive values for probable partonomy. These functions were implemented for two types of sites, namely schools and hospitals. For instance, schools functional belonging value is 3 for objects tagged as churches or sports fields, 10 for objects tagged as libraries, and −10 for buildings tagged as commercial, and −15 for roads tagged as highways (Table 1.2). The spatial belonging computes the percentage of overlap between both geometries with a positive value over 60% and a negative value below 30%. Then, functional and spatial belongings are summed and compared to two thresholds: A belonging threshold (2 was used for schools and hospitals) and an exclusion threshold (−2 was used). Between the thresholds, the boundary is not modified by the object tested.

Finally, the boundary is adjusted to include or exclude the intersecting objects. When the object to include (or to exclude) has a polygonal geometry, a polygon union (or difference) is used with a smoothing of intersections, i.e. the vertices just before the intersections are removed from the geometry (see the zoomed image

Table 1.2 Bonuses and maluses affected by the function belonging of school components for specific tags of OSM

Tag key	Tag value	Functional belonging
Building	*Any value*	1
Building	Dormitory	5
Building	Church	3
Building	Chapel	3
Building	Civic	3
Building	Commercial	−10
Building	Industrial	−10
Building	Residential	−15
Amenity	School	10
Amenity	University	10
Aenity	College	10
Amenity	Library	10
Leisure	Pitch	5
Sports	*Any value*	5
Highway	Primary	−15
Highway	Secondary	−15
Hghway	Tertiary	−10
Highway	Residential	−10

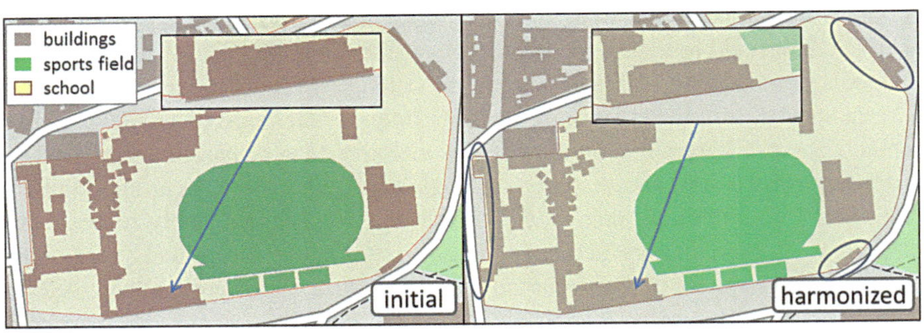

Fig. 1.5 High school with rough limits harmonized to be consistent with the detailed buildings and sports grounds

in Fig. 1.5). When the object has a linear geometry, the vertices of the boundary between the intersections with the line are replaced with the vertices of the line between the intersections.

Figure 1.5 shows the example of a high school whose initial boundaries intersect with several buildings, with some buildings (at the bottom and on the left) that are included in the high school and one (on the right) that is excluded.

1.4.3 Tree Alignment Along Roads

When the "displace and enhance" operator is required, we have to choose which feature is displaced to enhance the relation. In the case of trees along roads, trees are often poorly detailed because they are hard to capture precisely, which makes tree alignments often overlap with road symbols (Fig. 1.6). Here, we propose to displace trees because their location is already inaccurate, and because their displacement does not cause many repercussions in terms of symbol overlap in the map. On the contrary, moving or distorting the road would require some propagation of the transformation to connected roads.

The algorithm displaces trees in order to reduce the overlap between symbols and enhances the relation by forcing the alignment to the road symbol (Fig. 1.6). Tree alignments on the right and on the left of a road are first identified by computing left and right buffers on roads and counting the trees inside each buffer. Then, right and left offset lines (like buffers without the caps) are computed with a distance that is the addition of the road symbol width and half the tree symbol radius (to maintain some overlap that mimics the fact that tree branches might overlap with road pavement). Finally, trees are projected on the offset and moved apart a little, in case projections are too close to each other (the tree symbol diameter is used as a minimal distance). When a tree is along two roads (i.e. at a crossroad), which is detected by the fact that a tree belongs to the left and right side of two different roads, the projected position is the intersection of both offset lines.

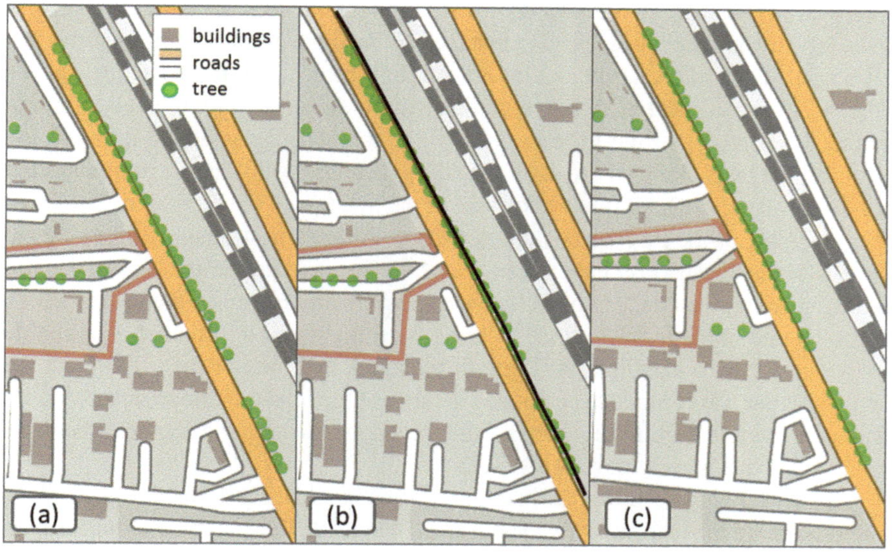

Fig. 1.6 (**a**) Inaccurate trees overlapping a road. (**b**) computation of an offset on the side of the road with trees. (**c**) trees aligned on the offset line

Figure 1.6 shows some automatic results obtained at the 1: 5000 scale of the French city of Bordeaux, which is quite a large city where trees have been captured extensively by OSM contributors. Hundreds of overlapping alignments have been automatically identified and successfully displaced. Some remaining problems have appeared when there are several rows of trees, because the row along the road sometimes overlaps with other rows inside a park or a square. This problem would require some displacement propagation, and this issue is discussed in section 5.2.

1.4.4 Intersection Removal

Intersection removal is the modification of a linear or areal object that is rough, with a more detailed boundary, using other detailed objects that are in relation with the rough object, in order to figure out how the more detailed outline needs to be drawn. For instance, when objects like paths or buildings intersect the outline of a lake, a new outline is drawn avoiding such inconsistencies. Depending on the type of objects involved in the inconsistency, the intersection removal has to be made differently, and different implementations are possible, e.g. if space is required between the intersecting objects or if objects might be adjacent. In this paper, we describe an algorithm for paths or roads that cross lakes, so a space is required between the path and the lake boundary.

Lakes have bona fide boundaries (Smith and Varzi 2000), i.e. there is a physical discontinuity that marks the boundary on the ground. So, more information on the intersection removal can be deduced from the geographical characteristics of objects involved in the inconsistency. For instance, a detailed bicycle path (captured by GPS tracks) can intersect the outline of an rough lake captured on satellite images (Fig. 1.7). The bicycle path has a certain width, so the harmonized lake outline cannot be adjacent to the bicycle path. Then, a small gap is added between the path and the lake, using an offset on the lake side of the path. The algorithm has the following steps: (i) Determine which side of the line (given the order of the vertices) the main part of the lake is (i.e., the portion with the largest area); (ii) remove the portion of the lake that is on the other side of the line; (iii) define an offset of the line (i.e. a buffer without cap) only on the side where the main part of the lake is with a width related to the real width of the line (use some width information if available in the tags), plus a small gap (0.1 mm in the map), and remove the offset from the lake polygon; (iv) compute a cushioned version of the offset on the segments before and after the intersecting segments of the line to avoid sharp differences where the intersection has been removed (Fig. 1.7).

However, Fig. 1.7 shows that, sometimes, paths crossing a lake are just ridges and there is no inconsistency that needs to be resolved. To avoid bad harmonization in such cases (Fig. 1.7c), the intersection removal algorithm is improved with a pre-step that automatically identifies parts of a path that probably belong to a bridge. First, the algorithm checks the following semantics: If there is a tag "bridge" or "man_made"='bridge' on the line, it is considered to be a bridge.

Fig. 1.7 (**a**) A bicycle path intersecting a lake; (**b**) harmonized lake outline; (**c**) an example of bridges with bad harmonization; (**d**) harmonization with automatic detection of bridges

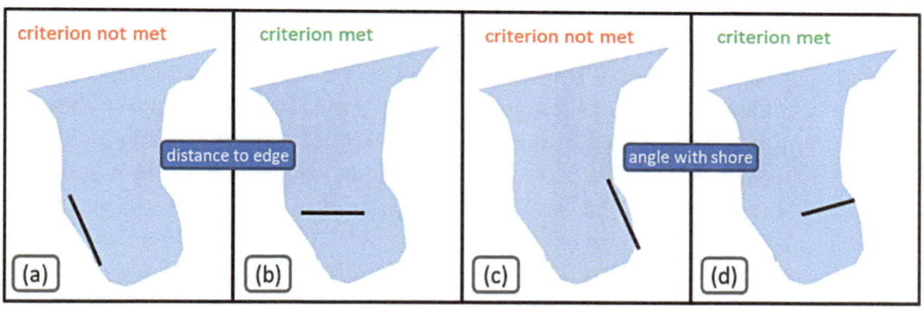

Fig. 1.8 (**a**) a segment whose middle point is as close to lake boundaries as ending points (criterion not met), (**b**) a segment that meets distance to edge criterion, (**c**) a segment not orthogonal to lake boundary (criterion not met), (**d**) a segment orthogonal to lake boundary (angle shore criterion met).

Then, the two following characteristics of bridge sections are used for the automatic identification:

1. The middle of a bridge section is more "inside" the lake than either of its endpoints (is this what you want to say?) (Fig. 1.8a, b),
2. The angle between the bridge section and the nearest lake shore is close to 90° (Fig. 1.8c, d).

Each segment of intersecting lines is inspected for both criteria. In order to check criterion 1, the shortest distance to the boundary of the lake is computed for both end vertices and for the middle point of the segment. Only the minimum distance for the end vertices is kept and compared to the middle point distance. If the middle point distance is significantly longer (a factor of 1.1 was used in the tests), the criterion is met (Fig. 1.8a, b). In order to check criterion 2, the orientation of the tested intersecting segment is computed and compared to the mean of the orientations of the shore segments around the tested segment. If the angle difference is close to 90° (in fact bigger than 60°), the criterion is met (Fig. 1.8c, d). When both criteria are met, the segment is considered as part of a bridge. Figure 1.7d shows that the identification of bridges greatly improves the automatic harmonization.

1.4.5 Bus Stop Displacement

Bus stop displacement is an instance of logical displacement of bus stops that cannot be located where they are captured. This has two possible causes: Either the bus stop is misplaced by lack of precision, or the road that serves the bus stop is missing. The proposed algorithm has three steps: (i) Verify the context to decide if displacement is required; (ii) find the most probable serving road; (iii) compute the displacement vector for the most probable serving road.

The main criteria to decide whether displacement is required or not is the distance to roads: If it is over 12 m, the bus stop is considered misplaced unless a public transport area is also defined. If the bus stop is inside a building or other features, like sports fields, it strengthens the probability of a misplaced bus stop. In the case of Fig. 1.9, the ambiguous bus stop is in the middle of a large building, which is unlikely a bus station (i.e. the only case where a bus stop should be in a building) considering the semantics and the shape of the building. So the bus stop is most likely misplaced.

The most probable serving road is first computed by finding the nearest roads within a specified radius (roads further than 120 m are not considered in our experiments). The nearest road is the default serving road, but obstacles can discard this choice. The segment between the bus stop and its projection on the road is buffered and intersections with buildings, hydrological features or barriers are

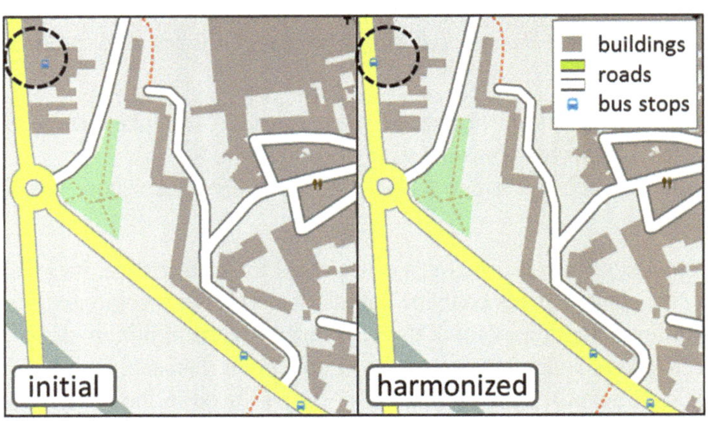

Fig. 1.9 One bus stop is far from the road and harmonized by displacement

searched. If one obstacle is found, the nearest other road without an obstacle is then considered as the most probable serving road. But if the bus stop already intersects a building (like in Fig. 1.9), this building is not considered as an obstacle.

Finally, the displacement is computed in the direction of the most probable serving road, to be close to the road (the width of the road is estimated according to the semantics attached to the road), and in a location with space. This location with space is computed the same way as in Duchêne et al. (2012), by removing the spaces already occupied by other objects (the same objects used to find obstacles). Figure 1.9 shows an example of a bus stop automatically displaced.

1.4.6 Clearing Addition

1.4.6.1 Algorithm Description

In the case of buildings grouped inside a forest, the proposed algorithm computes the extent of the building group and then computes a probable clearing geometry from the group extent and the surrounding objects. First, buildings are grouped using a process similar to Boffet (2000) and Chaudhry and Mackaness (2008), and similar to the built-up extension algorithm, merging buffers computed around each building inside a forest. The buffer used here is larger, 25 m, in order to create a larger group of buildings. However, the choice of a unique buffer parameter value, effective on a large area, is a complex problem. See section 5.1 for further discussions. Then, when groups are identified, the buffer of the convex hull of the group is computed, as a gap between buildings and trees is quite probable (Fig. 1.10b). Finally, the network helps to refine the clearing extent (Fig. 1.10c). The clearing polygon is cut into several parts using intersecting roads and paths, and the small parts that contain no building are removed from the clearing geometry. As roads and paths are often "natural" boundaries to clearings/forests, this last step results in a more realistic clearing boundary.

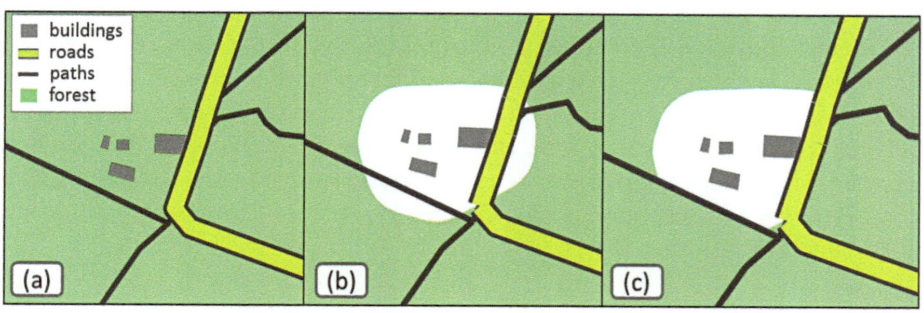

Fig. 1.10 (**a**) an initial building group inside a forest, (**b**) a first clearing geometry is computed by merging the dilated buildings, (**c**) removal of clearing parts that lie over a path or a road

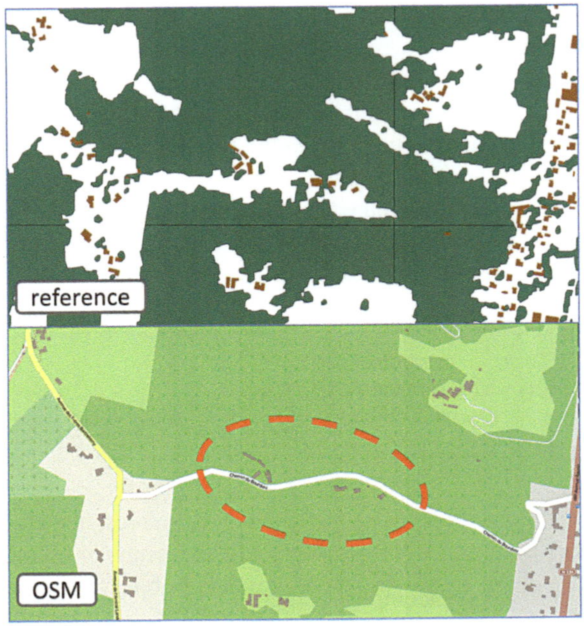

Fig. 1.11 Obstacles to conduct comparative evaluation: The clearing is not a clearing in the reference data, but only a recess in the forest

1.4.6.2 Comparative Evaluation

Reference data from IGN, the French national mapping agency, were used to compare some harmonized results to existing high quality maps. Comparing clearings automatically obtained by this algorithm to reference clearings is a complex task because the reference is quite different from the OSM forest data (Fig. 1.11). In particular, the difference is illustrated by clearings in the harmonized map that are not clearings in the reference, but just a recess in the forest (Fig. 1.11).

In order to compare clearings as two polygons, we have to close the clearings in the reference data when they actually are not holes. This is done by intersecting reference forests with an expanded envelope of the harmonized clearings. Then, shapes of clearings are compared using the surface distance between two polygons (Girres and Touya 2010) described in equation 1. Comparisons were carried out on 33 clearings automatically created in three different areas in France. The average surface distance is 0.73, which is quite a large value as 1 stands for disjoint polygons and 0 for equal polygons. However, we consider this as a validation that harmonized clearings approximately occupy the same space as actual clearings. The aim never was to create more realistic clearings, which is impossible only using OSM buildings.

$$surfacedistance\,(A, B) = 1 - \frac{area\,(A \cap B)}{area\,(A \cup B)} \tag{1.1}$$

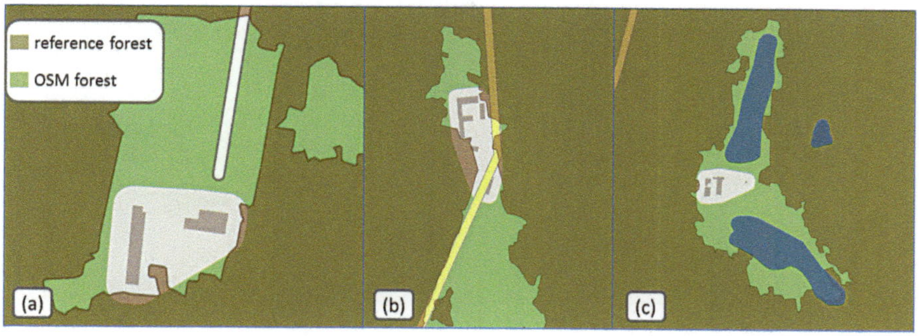

Fig. 1.12 Comparison of harmonized and reference clearings. (**a**) and (**b**) the actual clearings are bigger than the harmonized ones. (**c**) The lack of ponds (to be included in the clearing) in the OSM makes shapes quite different

Figure 1.12 shows some clearing harmonizations used in this comparative evaluation with the reference forest data. Figure 1.12c shows that other objects (in this example, ponds) should be included in the computation of clearings and also that the incompleteness of OSM greatly penalizes harmonization, since ponds displayed in Fig. 1.12c are not captured in OSM, but are extracted from the reference data.

1.5 Discussion

1.5.1 The Importance of Parameterization

Experiments on large datasets from different countries and landscapes confirmed our assumption on the difficulty to find the best parameter values for the proposed harmonization algorithms. First, harmonization algorithms are hard to parameterize as parameter values are hard to correlate with visual results. It is a classical problem in automatic mapping processes, such as map generalization algorithms (Weibel et al. 1995), or label placement. For instance, there is no obvious value for defining how far a building can be considered to be "just outside" a built-up area.

Moreover, it appears that harmonization algorithms parameters are context-dependent insofar as parameter values are adapted to some situations and other situations require different parameter values. For instance, Fig. 1.13a clearing uses the standard set of parameters empirically defined (25 m buffer around buildings to cluster the buildings), but does not look like the real clearing drawn in the IGN map (Fig. 1.13c). A specific set of parameters (75 m buffer radius to cluster the buildings), which fails for most other cases, gives, here, much better results. This example suggests that trying different set of parameters and keeping the best result,

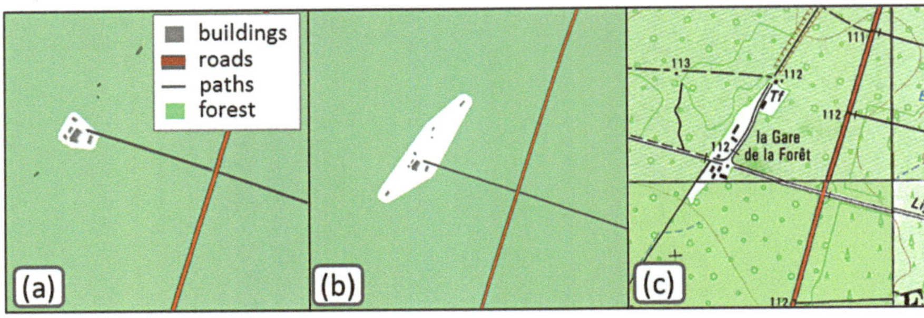

Fig. 1.13 The clearing created with standard parameters (**a**) does not look like the clearing in the IGN map (**c**); however, different parameter values give a closer result to the IGN map (**b**), but would fail in most other cases

and/or defining situation-specific parameters, might be a more robust solution than finding the best parameters and applying them everywhere.

1.5.2 Required Cartographic Knowledge and Degree of Automation

The automation of cartographic processes, such as label placement, style definition, or generalization often requires some acquisition and formalization of the knowledge of cartographers (Buttenfield and McMaster 1991). Automatic harmonization of large scale OSM maps does not avoid this bottleneck, and some kind of cartographic knowledge base is necessary for several steps of the process.

The first type of cartographic knowledge to formalize to enable harmonization is the identification of the key anomalous spatial relations. Some have been identified in this paper, but the list is not exhaustive, and additional relations may be of interest in other parts of the world (only French OSM datasets are used in this work) where landscapes are different and the detailed and rough features may not be the same. Then, there is what Taillandier et al. (2011) call *control knowledge*, which allows the definition of good parameters for automatic operations, and also allows the guidance of processes that chain several operations. Control knowledge requires experimenting with harmonization techniques to find out what leads to better maps.

In order to acquire this knowledge base in a more generic way, learning and artificial intelligence techniques could be used (Weibel et al. 1995). Formalizing the knowledge to share it would also be beneficial, and the collective project of building on ontology for on-demand mapping processes (Gould et al. 2014) could be used in this way.

Then, this knowledge could be used in processes able to chain harmonization operations, to adapt parameters to specific geographical situations, and to handle propagations of transformations. It could be interesting to adapt optimization (e.g.,

Harrie and Sarjakoski 2002; Sester 2005) or multi-agent techniques (e.g. Duchêne et al. 2012) used in automated map generalization to enable such harmonization processes. Nevertheless, problems should be less complex than map generalization, as large scale map symbols allow more free space in the map than small scale map symbols.

1.5.3 Abstraction Versus Realism

Harmonization operations try to guess what the consistent detailed information would be from the detailed objects of the dataset. Thus, information that does not correspond to ground truth is introduced into the map, in order to make the map readable. For the same reason, map generalization also distorts ground truth, by moving or simplifying objects. Thus, it is necessary to wonder if harmonization should aim at realistic harmonized representations, which mislead the reader and making him/her believe that the map is a realistic view of ground truth. Or, aim at abstract harmonized representations that show to the map reader that the information is not exactly as it is represented on the map. Figure 1.14 illustrates both strategies for the clearing creation around building groups. None of the representations is close to ground truth, or even to its representation in the IGN map. But Fig. 1.14b is clearly a more realistic representation of the clearing than Fig. 1.14c, which is more abstract or sketchier.

Research in computer graphics and non-photorealistic rendering (like maps) show that blurring an object or making it sketchier may convey information on data quality (Wood et al. 2012). Dashes also proved to convey uncertainty information (Boukhelifa et al. 2012). So, a sketchy or dashed clearing outline could also be an efficient alternative to our proposed realistic or schematic harmonization. In the case of bus stop disambiguation, some instances may remain unsolved (Fig. 1.15), and blurring or sketching the bus stop symbol could convey the uncertainty and avoid misinterpretation for the map reader.

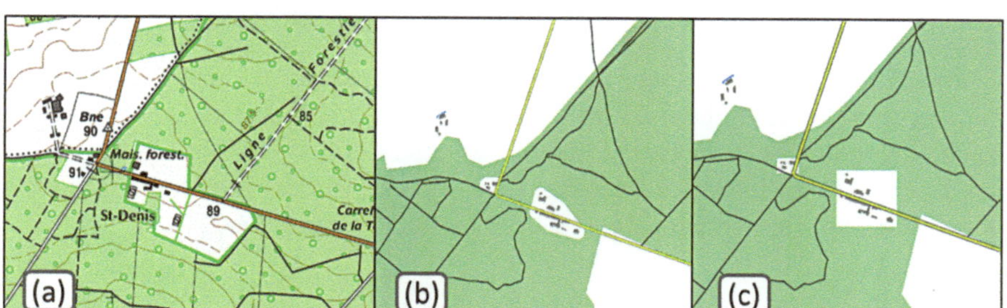

Fig. 1.14 (**a**) The actual shape of a clearing in the IGN map; (**b**) a computed realistic shape; (**c**) a computed abstract/schematic shape

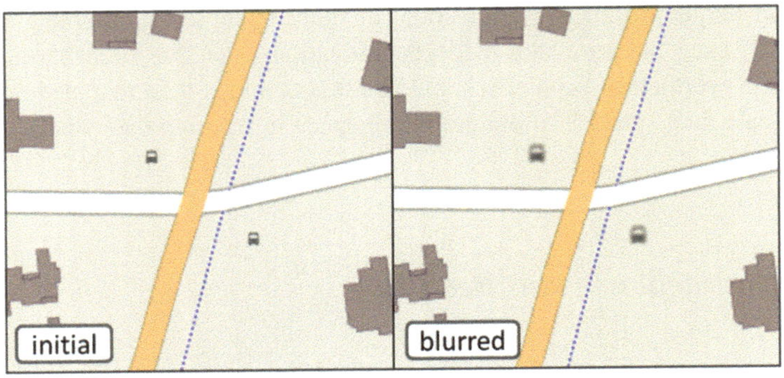

Fig. 1.15 Unsolved disambiguation (which road serves the stops?) could be overcome by blurring symbols to convey uncertainty

1.5.4 Update OpenStreetMap?

In a certain way, harmonization operations improve OSM data quality and correct some mistakes in the database. So it may be tempting to use the harmonized operations to push updates in the OSM database. However, harmonization is a process dedicated to cartography, so some transformations carried out may be irrelevant for the OSM database. Operations that include some kind of caricature should not be used to improve the OSM database. For instance, the alignment of trees along roads caricatures the tree alignment to make it straight, in order to improve the map clarity. But positional accuracy is then lost for some trees. However, occurrences of trees badly aligned are consistency problems of the OSM database that should be corrected, but with different operations that focus more on placement accuracy than map legibility.

Some other operations could be pushed in the database as it improves the level of detail of some rough objects, but it should be done carefully as it breaks a general rule of OSM to rely on ground truth to contribute to the project. For instance, the extension of built-up area to nearby buildings improves the level of detail of the built-up area but there is no checking in the field or with images that the new extent is close to ground truth.

Finally, we believe that some operations could be included without further checking, because the modifications are sure. For instance, "Adjust shape" operations avoid situations that cannot exist, like a path intersecting a lake or a forest in the sea. Existing OSM tools like KeepRight[1] or Osmose[2] already search for such kind of problems, for further manual corrections by OSM contributors.

[1] http://keepright.ipax.at/

[2] http://osmose.openstreetmap.fr/fr/map/

1.6 Conclusion

To conclude, this paper tackles a new cartography problem raised by OpenStreetMap data, namely the level of detail harmonization that improves the level of detail of some rough objects in large scale maps. Several types of harmonization operations are proposed and experimented on OSM datasets. Some problems raised by the experiments are discussed, such as the need for realism or abstraction and the automation of the proposed algorithms in a process to harmonize a complete map. The first results show that it is a promising topic to explore in automated cartography.

Further research should clearly focus on harmonization processes, to be able to automatically chain harmonization operations, and solve complex problems that involve many objects and require optimization techniques. The processes should tackle the dependency of parameters on the geographic context of features. Of course, each operation presented in the paper could be improved and new operations have to be designed for the LoD inconsistencies that are not mentioned in the paper. Furthermore, as harmonization operations transform data into something realistic but false, abstract harmonization should be investigated with user tests, to know if map user better understand realistic or abstract harmonization. Finally, one of the main characteristics of OSM is that contribution patterns change all over the world (Jokar Arsanjani et al. 2015) and the inconsistencies encountered in the datasets tested in this paper might not occur elsewhere. Recent research showed that there is some relation between social aspects and contribution patterns in a region of the world (Mashhadi et al. 2015), so it can be inferred that LoD might differ according to these varying patterns. As a consequence, experiments should be carried out to analyze the influence of varying contribution patterns on LoD harmonization problems.

References

Biljecki F, Ledoux H, Stoter J, Zhao J (2014) Formalisation of the level of detail in 3D city modelling. Comput Environ Urban Syst 48:1–15

Boffet A (2000) Creating urban information for cartographic generalisation. In: International symposium on Spatial Data Handling, Beijing, China, 10–12 August 2000

Boukhelifa N, Fekete JD, Bezerianos A, Isenberg T (2012) Evaluating sketchiness as a visual variable for the depiction of qualitative uncertainty. IEEE Trans Visual Comput Graph 18(12):2769–2778

Bucher B, Brasebin M, Buard E, Grosso E, Mustière S, Perret J (2012) GeOxygene: built on top of the expertness of the french NMA to host and share advanced GI science research results. In: Bocher E, Neteler M (eds) Geospatial free and open source software in the 21st century. Springer, Berlin/Heidelberg, pp 21–33

Burghardt D, Petzold I, Bobzien M (2010) Relation modelling within multiple representation databases and generalisation services. Cartogr J 47(3):238–249

Burghardt D, Duchêne C, Mackaness WA (eds) (2014) Abstracting geographic information in a data rich world: methodologies and applications of map generalisation. Springer, Heidelberg

Buttenfield B, McMaster R (eds) (1991) Map generalization. Making rules for knowledge representation. Longman, London
Chaudhry OZ, Mackaness WA (2008) Automatic identification of urban settlement boundaries for multiple representation databases. Comput Environ Urban Syst 32(2):95–109
Chaudhry OZ, Mackaness WA, Regnauld N (2009) A functional perspective on map generalisation. Comput Environ Urban Syst 33(5):349–362
Douglas DH, Peucker TK (1973) Algorithms for the reduction of the number of points required to represent a digitized line or its caricature. Cartographica Int J Geogr Inf Geovisual 10(2): 112–122
Duchêne C, Ruas A, Cambier C (2012) The CartACom model: transforming cartographic features into communicating agents for cartographic generalization. Int J Geogr Inf Sci 26(9): 1533–1562
Foerster T, Stoter J, Köbben B (2007) Towards a formal classification of generalization operators. In: Proceedings of 23rd international cartographic conference. Moscow, Russia, 2007
Girres JF, Touya G (2010) Quality assessment of the french OpenStreetMap dataset. Trans GIS 14(4):435–459
Gould NM, Mackaness WA, Touya G, Hart G (2014) Collaboration on an ontology for generalisation. In: Proceedings of 17th ICA workshop on generalisation and multiple representation. Vienna, Austria, 2014
Haklay M (2010) How good is volunteered geographical information? A comparative study of OpenStreetMap and Ordnance Survey datasets. Environ Plan B Plan Des 37(4):682–703
Harrie LE, Sarjakoski T (2002) Simultaneous graphic generalization of vector data sets. GeoInformatica 6(3):233–261
Jokar Arsanjani J, Zipf A, Mooney P, Helbich M (2015) An introduction to OpenStreetMap in geographic information science: experiences, research, and applications. In: Jokar Arsanjani J, Zipf A, Mooney P, Helbich M (eds) OpenStreetMap in GIScience. Lecture notes in geoinformation and cartography. Springer International Publishing, New York, p 1–15
Klammer R (2013) TileGen – an open source software for applying cartographic generalisation to tile-based mapping. In: Proceedings of 26th international cartographic conference. Dresden, Germany, 2013
Mackaness WA (2007) Understanding geographic space. In: Mackaness WA, Ruas A, Sarjakoski T (eds) The generalisation of geographic information: models and applications. Elsevier, Amsterdam, pp 1–10
Mackaness WA, Chaudhry OZ (2011) Automatic classification of retail spaces from a large scale topographic data-base. Trans GIS 15(3):291–307
Mackaness WA, Burghardt D, Duchêne C (2014) Map generalisation: fundamental to the modelling and understanding of geographic space. In: Burghardt D, Duchêne C, Mackaness WA (eds) Abstracting geographic information in a data rich world. Springer, Heidelberg, pp 1–15
Mashhadi A, Quattrone G, Capra L (2015) The impact of society on volunteered geographic information: The case of OpenStreetMap. In: J JA, Zipf A, Mooney P, Helbich M (eds) OpenStreetMap in GIScience. Lecture Notes in Geoinformation and Cartography. Springer International Publishing, New York, pp 125–141
McMaster RB, Shea KS (1988) Cartographic generalization in digital environment: a framework for implementation in a GIS. In: Proceedings of GIS/LIS'88, San Antonio, Texas, USA, 1988, pp 240–249
Monmonier M (1996) How to Lie with maps, 2nd edn. University Of Chicago Press, Chicago
Mooney P, Corcoran P (2012) The annotation process in OpenStreetMap. Trans GIS 16(4): 561–579
Mustière S, Moulin B (2002) What is spatial context in cartographic generalisation? In: Proceedings of joint international symposium and exhibition on geospatial theory, processing and applications, Ottawa, Canada, 2002, pp 274–278
Regnauld N, McMaster RC (2007) A synoptic view of generalisation operators. In: Mackaness WA, Ruas A, Sarjakoski LT (eds) Generalisation of geographic information. Elsevier, Amsterdam, pp 37–66

Reimer A, Kempf C, Rylov M, Neis P (2014) Assigning scale equivalencies to OpenStreetMap polygons. In: Proceedings of AutoCarto 2014, Pittsburgh, USA, 2014

Roth RE, Brewer CA, Stryker MS (2011) A typology of operators for maintaining legible map designs at multiple scales. Cartogr Perspect 68:29–64

Salichtchev KA (1983) Cartographic communication: a theoretical survey. In: Taylor DRF (ed) Graphic communication and design in contemporary cartography, vol 2. Wiley, New-York, pp 11–36

Schmid F, Janetzek H (2013) A method for high-level road network extraction of OpenStreetMap data. In: Proceedings of the International Cartographic Conference 2013 (ICC 2013), Dresden, Germany, 2013

Sester M (2005) Optimization approaches for generalization and data abstraction. Int J Geogr Inf Sci 19(8):871–897

Sester M, Jokar Arsanjani J, Klammer R, Burghardt D, Haunert JH (2014) Integrating and generalising volunteered geographic information. In: Burghardt D, Duchêne C, Mackaness WA (eds) Abstracting geographic information in a data rich world. Springer, Heidelberg, pp 119–155

Smith B, Varzi AC (2000) Fiat and bona fide boundaries. Philos Phenomenol Res 60(2):401–420

Stoter J, Burghardt D, Duchêne C, Baella B, Bakker N, Blok C, Pla M, Regnauld N, Touya G, Schmid S (2009) Methodology for evaluating automated map generalization in commercial software. Comput Environ Urban Syst 33(5):311–324

Taillandier P, Duchêne C, Drogoul A (2011) Automatic revision of the control knowledge used by trial and error methods: application to cartographic generalisation. Appl Soft Comput 11(2):2818–2832

Touya G (2012a) What is the level of detail of OpenStreetMap? In: Workshop on Role of Volunteered Geographic Information in Advancing Science: Quality and Credibility, Columbus, Ohio, USA, 2012

Touya G (2012b) Social welfare to assess the global legibility of a generalized map. In: Xiao N, Kwan MP, Goodchild MF, Shekhar S (eds) Geographic information science 7th international conference, GIScience 2012. Springer, Berlin, Heidelberg, pp 198–211

Touya G, Brando-Escobar C (2013) Detecting Level-of-Detail inconsistencies in volunteered geographic information data sets. Cartographica 48(2):34–143

Touya G, Reimer A (2015) Inferring the scale of OpenStreetMap features. In: Jokar Arsanjani J, Zipf A, Mooney P, Helbich M (eds) OpenStreetMap in GIScience. Springer International Publishing, Switzerland, pp 81–99

Walter V (2008) Automatic interpretation of vector databases with a Raster-Based algorithm. In: Proceedings of ISPRS Congress XXXVII, Commission II, WG II/4, Beijing, China, 2008

Weibel R, Keller S, Reichenbacher T (1995) Overcoming the knowledge acquisition bottleneck in map generalization: The role of interactive systems and computational intelligence. In: Frank AU, Kuhn W (eds) Spatial information theory a theoretical basis for GIS. Springer, Berlin, Heidelberg, pp 139–156

Wood J, Isenberg P, Isenberg T, Dykes J, Boukhelifa N, Slingsby (2012) Sketchy rendering for information visualization. IEEE Trans Vis Comput Graph 18(12):2749–2758

Chapter 2
Cartographic Representation of Soundscape: Proposals and Assessment

Saúl Gomez, Catherine Domingués, Pierre Aumond, Catherine Lavandier, Gaëtan Palka, and Kamal Serrhini

Abstract Environmental noise is a major concern for city dwellers, however, actual noise maps are not adapted to them. The work described here is developed in the context of the CartASUR project which specifically addresses these deficiencies of noise maps. CartASUR collected perceptual data in several places of Paris using a survey and aiming to define new urban soundscape indicators to offer the possibility of several interpretations depending on the users' personal and cultural characteristics, and to show them on maps. In this chapter, the indicators *sound pleasantness* and *global loudness* are presented. Several cartographic proposals are made to portray indicators during one, two, or three periods (day, evening, night) on maps, and their characteristic features (symbol design, visual variables) are discussed. Cartographic proposals are assessed through a survey which addresses the understanding of cartographic symbols and global properties of maps. The survey concludes that map readers prefer to view complete (both sound pleasantness and global loudness) and precise (three measurement periods) information, even when the amount of information leads to complex maps, which are considered the most attractive and useful. Nevertheless, the survey shows that these complex maps are not well interpreted by a large part of map readers. Use of visual variables *color* and *quantity* is discussed and proposals are made to improve symbol understanding.

Keywords Soundscape • Sound pleasantness • Perceptive map • Semiology of graphics • Survey

---

S. Gomez • C. Domingués (✉)
Université Paris-Est, IGN/LaSTIG, COGIT, 73 avenue de Paris, 94160, Saint-Mandé, France
e-mail: catherine.domingues@ign.fr

P. Aumond • C. Lavandier
Université de Cergy-Pontoise, Laboratoire MRTE, 33 boulevard du Port, 95011, Cergy-Pontoise Cedex, France

G. Palka
Swiss Federal Institute WSL, Zürcherstrasse 111, 8903, Birmensdorf, Switzerland

K. Serrhini
Université François-Rabelais de Tours, UMR CNRS 7324 CITERES, équipe IPAPE, 33 allée Ferdinand de Lesseps, 37200, Tours, France

M. Leitner, J. Jokar Arsanjani (eds.), *Citizen Empowered Mapping*, Geotechnologies and the Environment 18, DOI 10.1007/978-3-319-51629-5_2

2.1 Introduction

Environmental noise is a major concern for city dwellers. A survey conducted in (2014) by IFOP, the French Institute of Public Opinion, found that 32–40% of the French population were rather troubled by noise. Such disturbance has a real negative impact on health. For example, Basner et al. (2013) described the effects of noise exposure on the sequence and duration of various stages of sleep. According to the World Health Organization (WHO 2011), environmental noise should not only be considered as a nuisance, but also as a public health problem because of "the relationship between environmental noise and specific health effects, including cardiovascular disease, cognitive impairment, sleep disturbance, and tinnitus".

In 2002, the European Parliament and the Council of the European Union adopted a directive (2002/49/EC) relating to the assessment and management of environmental noise. Consequently, cities are required to manage direct action on noise reduction (speed control, noise barriers, etc.) and to disseminate strategic noise maps. However, despite the positive aspects that current noise maps offered in several realms (mainly in physical aspects of sound as shown by Miedema and Oudshoorn (2001)), these noise maps have deficiencies regarding aspects of understanding and interpretation of noise (Schiewe and Weninger 2013). Usually, noise map data come from measurements directly taken from the city or from traffic noise and the calculations are made using mathematical models. The resulting maps are purely based on physical indicators. Particularly, noise is measured in decibels which follow a logarithmic scale that is difficult for most users to understand. Furthermore, the addition of noises depends on the volume of each individual noise and on the difference between them. Noise additions are shown in Fig. 2.1.

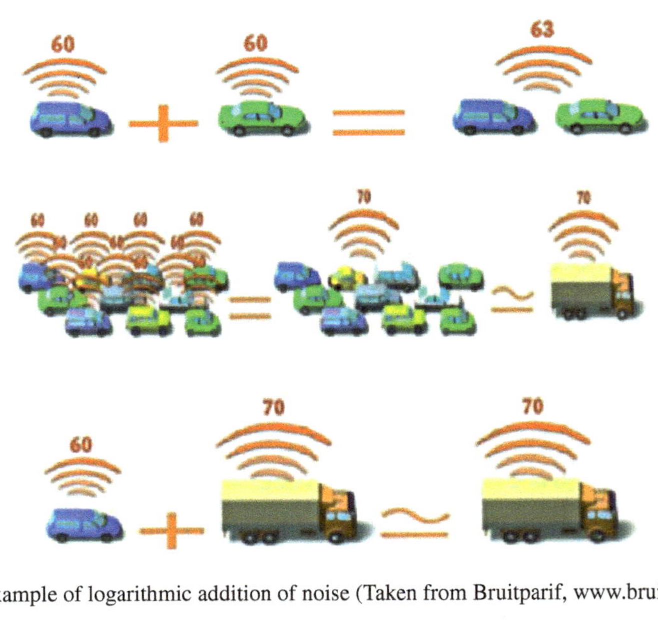

Fig. 2.1 Example of logarithmic addition of noise (Taken from Bruitparif, www.bruitparif.fr)

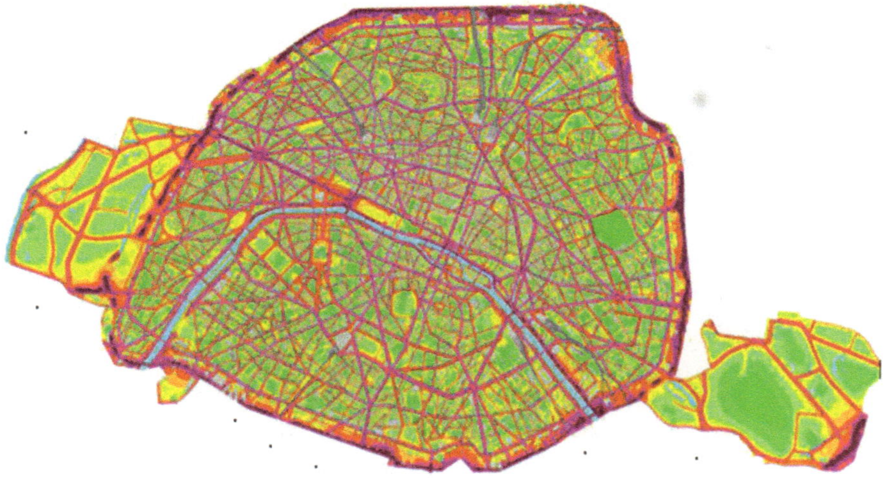

Fig. 2.2 Traffic noise map (L_{DEN} indicator) of Paris in 2007 (http://api-site-cdn.paris.fr/images/154512.pdf, pg. 24)

For example, if one car produces 60 decibels, two cars would produce 63 decibels. Or, if a 60-decibel car is behind a 70-decibel truck, since the noise difference is ten decibels or more, the higher noise conceals the weaker and the resulting loudness would be 70 decibels, as if there is just one truck. These examples seem to refute city dwellers' daily experience of sound (Schiewe and Weninger 2013).

Figure 2.2, the noise map of Paris in 2007, shows an example of an indicator recommended by the European Directive. It is the L_{DEN} indicator (a weighted average level for day, evening, and night sounds) which is also based on a logarithmic scale and which represents a weighted average sound level calculated over 24 h. It contradicts the natural human and physiological interpretations of sound and noise annoyance, because it is one single figure which is not able to describe the overall quality or nature of each sound that is heard by city dwellers at each location on the map (Haberle et al. 1984). Instead, over the last decade, soundscape researchers put forward soundscape descriptors that are closer to the perception of users (see (Aletta et al. 2016) for a review)). Guski et al. (1999) studied the personal and social aspects of the description of noise annoyance and showed that the same sound is interpreted differently depending on personal and social connotations. Morel (2012, pg. 3) concluded that "*it is necessary to improve the noise maps by the definition of complementary relevant indicators from the point of view of the individual to characterize the perceived noise annoyance*".

The CartASUR project (Cartographic representation of urban sound quality) specifically addresses these deficiencies of noise maps. This project, involving several universities, research laboratories, and public organizations, focuses on urban dwellers' feelings. It aims to define new urban soundscape indicators to offer the possibility of several interpretations depending on the users' personal and cultural characteristics, and to show them on maps. Section 2.2 explains the definition of these perceptual indicators by the CartASUR project. Section 2.3 introduces the cartographical proposals made to show indicators on maps. These maps have been assessed by a survey detailed in Sect. 2.4. Section 2.5 displays the results of the survey and comments on them. Conclusions and perspectives are laid out in Sect. 2.6.

2.2 Definition of Soundscape Indicators in the CartAsur Project

Guastavino (2007) showed that sound and noise are two different cognitive objects. Sound is an isolated phenomenon, independent of the source, which can be described in terms of physical properties. By contrast, noise is a sensitive phenomenon, inseparable from the source and interpreted according to the environment in which it is heard. In seeking a common approach to explaining soundscape (Aletta et al. 2016), soundscape researchers defined it as the "*acoustic environment as perceived or experienced and/or understood by a person or people in context*" (ISO 2014). Lavandier et al. (2013) proposed indicators of sound quality which depend not only on perceived loudness, but also on the presence of sound sources, such as traffic, birds, or voices (Guastavino 2007). Ricciardi et al. (2015) developed a survey in Paris regarding the global sound environment characterization, the perceived loudness of some emergent sources, and the perceived presence (in duration) of identified source that do not emerge from the background. Thus, they proposed indicators of urban sound quality based on linear regressions with perceptive variables. The authors showed that soundscape assessment depends on different perceptual data, such as sound pleasantness, global loudness, presence of birds, voices, etc. Based on these studies, CartASUR aimed to show that subjective indicators could complement a physical description of sound, and assist in creating their cartographic representations. In this study, the sound quality addresses the overall perception of the acoustic environment, measuring whether a soundscape is pleasant or unpleasant.

CartASUR aimed to define the soundscape in several places of Paris using a survey. Data were collected by city dwellers to record various perceptual and acoustic data in 29 locations across two districts of Paris (see Fig. 2.3). The locations had been selected to represent the diversity of places in Paris (Ricciardi et al. 2015) such as thoroughfares, crossroads, small streets, schools, parks, gardens, etc. Different periods had been also chosen in order to respect the rhythm of the sound environment in a city (Brocolini et al. 2013). In each survey location, participants were asked to answer a questionnaire about the various sound sources making up the landscape and contributing to their soundscape appraisal. There were 18 items on the questionnaire, such as sound pleasantness, global loudness, presence of trucks or buses, cars, voices, footsteps, birds, water, and wind. This perceptual data was rated on a scale of 1 (*weak*) to 11 (*strong*). Finally, the locations were perceptively

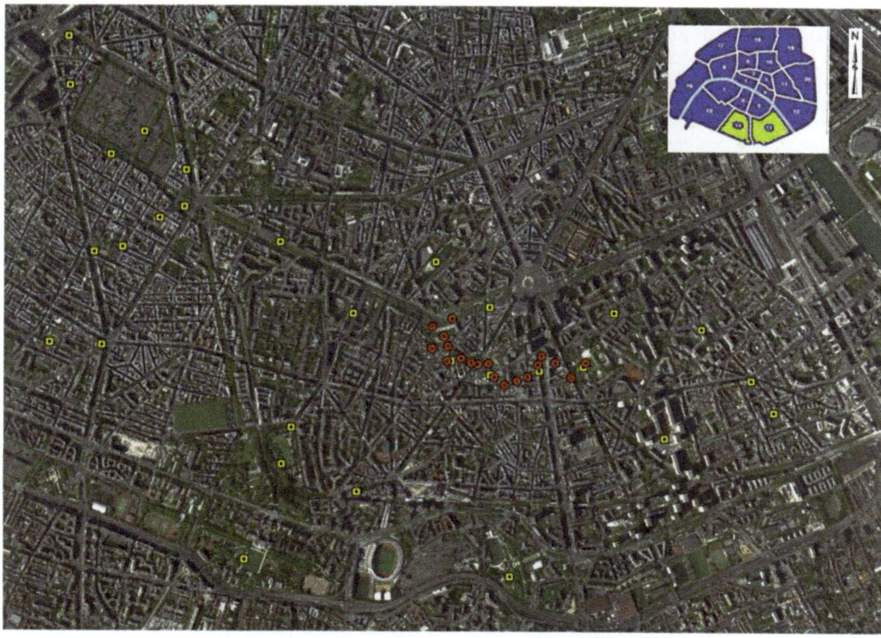

Fig. 2.3 Survey locations in the 13th and 14th districts of Paris. Fixed survey locations are portrayed by *yellow dots*; *red dots* show survey locations along an urban walking trip, during a field experiment (Aumond et al. 2016)

assessed between twice and five times a day, in summer and in winter. In total, 204 situations have been perceptively evaluated by at least 15 persons, being 3409 perceptive measurements.

The CartASUR approach is described in Fig. 2.4. Perceptual survey data (step 1 in Fig. 2.4) made it possible to understand how each noise source (cars, voices, birds, etc.) influences the overall opinion of sound pleasantness (Delaître et al. 2014). One of the purposes was to define formulas based on the survey to predict indicators across the entire city. A global sound quality indicator, modelled from the 3409 perceptive measurements was thus proposed on a scale from 1 (*unpleasant*) to 11 (*pleasant*) with a linear regression model, according to Lavandier et al. (2015):

$$Pleasantness = 8.11 - 0.38 * (Overall\ Loudness) + 0.20 * V + 0.15 * B - 0.14 * T, \quad (2.1)$$

where V is perceived presence of voices, B perceived presence of birds, and T perceived presence of traffic.

This model explains 34% of the individual variance of participant answers in this experimentation (correlation of 0.58 between the 3409 individual real sound pleasantness and the pleasantness predicted by the model). This correlation reaches a value of 0.89 if the average values of the sound pleasantness for each of the 204

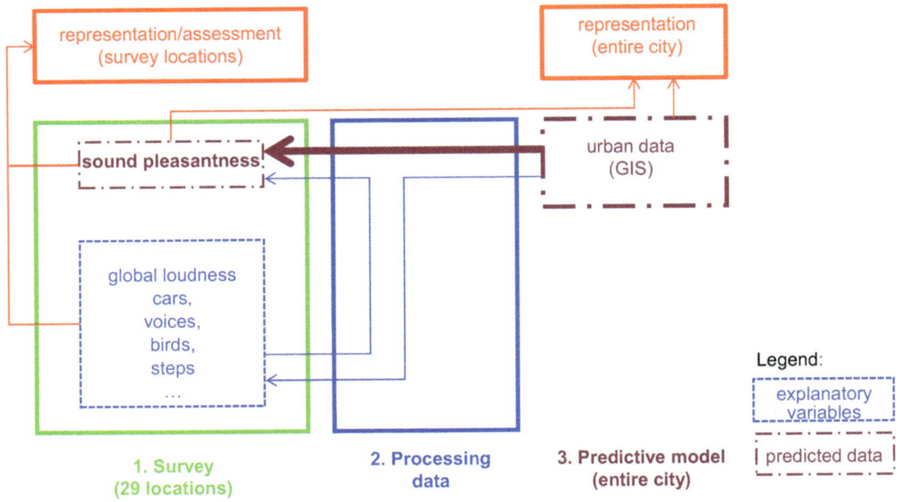

Fig. 2.4 CartASUR diagram

urban assessed situations are compared with the proposed model values, which are constructed from the averages of the independent perceptive variables (step 2 in Fig. 2.4). On the other hand, these variables can also be linked to urban data, which are contained in geographic and urban databases (step 3 in Fig. 2.4), which are part of the geographical information systems (GIS) of the city. It therefore becomes possible to predict perceptual data such as sound pleasantness, from the city GIS data (purple arrow in Fig. 2.4). This chapter presents several proposals for the representation of perceptual data collected in survey locations, including their assessment.

2.3 Soundscape Representation

The analysis of perceptual and acoustic survey data showed that sound pleasantness depends on the surrounding context. Consequently, one purpose of a representation is to provide map readers with the possibility of understanding and imagining the sound environment context (global loudness, presence of traffic, voices, and birds) and to interpret it according to personal criteria. Indeed, sound pleasantness and global loudness in a city are generally correlated, but not always. City dwellers have experimented with this situation. It may therefore be difficult for them to distinguish between both indicators. For example, it is not obvious for some people that a location may be loud and pleasant, or very quiet and very unpleasant. Consequently, maps have to show both indicators and the cartographic representation of the explanatory variables of soundscape. To simplify a comparison between both indicators, it would be useful that both are portrayed on the same symbol, the drawback being that the symbol becomes rather complicated. Thus, some cartographic representations are being suggested to portray sound pleasantness and global loudness in the survey locations, and to show urban features explaining the items which play a role in the formulas.

2.3.1 Features of the Cartographic Symbology

Cartographic symbols are designed to show two indicators, including sound pleasantness and global loudness. The symbols also need to allude to the explanatory variables, such as presence of traffic, voices, and birds. Indicators are portrayed by symbols and explanatory variables must be inferred from the map base.

Cartographic symbols need to portray sound pleasantness and global loudness, indicators that vary according to the time of the day. The temporal variable is something that does not need to be portrayed by a visual variable. Instead, it could be combined with the indicators, which would make it more difficult for the map to convey the message. Sound pleasantness is a perceptual ordinal variable. In this survey, it can vary between 1 (*unpleasant*) and 11 (*very pleasant*). In order to simplify the representation, these eleven values were reclassified into the following four categories: *unpleasant, rather unpleasant, rather pleasant, and pleasant*, thus eliminating the neutral assessment class. Global loudness is a perceptual ordinal variable too, which is rated on a scale of 1 (*weak*) to 11 (*strong*) in the survey. As for sound pleasantness, the eleven classes were reclassified into four categories, also: *quiet, rather quiet, loud,* and *very loud*. Mapping both sound pleasantness and global loudness is made possible by using a bivariate map. The difficult aspect is then to create the legend. In this case, it is a rectangular box with four categories a side, resulting into 16 (4 × 4) smaller boxes, with each box representing a unique relationship of the pleasantness and loudness variables. As Jeong and Gluck (2002) explained, "t*hose maps are extremely difficult to understand because the users need to refer to the arbitrary legend all the time*". Leonowicz (2006) confirmed these findings by a survey and concluded that "*one-variable choropleth maps are more effective while reading the spatial distribution, and well designed two-variable choropleth maps are more effective in reading the spatial relationship*". As there is no spatial relationship between sound pleasantness and global loudness, one of the aims of CartASUR was to distinguish between these variables. As a result, bivariate maps were abandoned.

Another cartographic solution was chosen to represent both indicators. It is based on the use of two different graphical variables (Bertin 1983) with the same symbol. In this case, color (i.e. variation of hue, lightness and saturation) was chosen to portray pleasantness and the number of colored bands to portray loudness, with the color of the band showing the level of pleasantness. Alberts and Rubio Alferez (2012) studied the use of colors according to the European Directive. Kornfeld et al. (2011) attempted a first recommendation base for noise mapping and Weninger (2013) concluded that "*to decide on a specific range of colours, the specific case of application has to be analysed with regard to the user group, the user tasks, and the map makers' aim*" and proposed a new scale of colors which is better adapted to noise representation. The cartographic proposals took this research into consideration. The point was to facilitate map users' lecture so that the number of levels was reduced and highly contrasted colors were chosen. In the ColorBrewer (Brewer et al. 2003), several diverging color schemes are offered and one example was chosen to design the symbols.

To show indicator variations according to the time of the day, symbols were divided depending on periods that they show. The number and duration of periods are specified by the L_{DEN} indicator. At most, symbols were divided into three parts for the three periods shown, including the day (from 8am to 6pm), evening (from 6pm to 9pm), and night (from 9pm to 8am). The size of each part of the symbol may or may not be proportionate to the corresponding period duration.

The map base was compiled using urban data from the GIS of the city. Different cartographic proposals were put forward to portray data connected to the explanatory variables (for example, vegetation to suggest the presence of birds) and to translate them into point symbols in order to facilitate the data interpretation of survey locations. These proposals of the explanatory variables were assessed but the results are too long to be discussed in this chapter, which focuses on the representation of indicators.

2.3.2 *Symbology Proposals*

Four proposals of cartographic symbology were made (Figs. 2.5, 2.6, 2.9, and 2.10) to portray indicators for sound pleasantness and global loudness. Proposals differ according to the information the symbols display (sound pleasantness and/or global loudness) and by their design. Symbols that correspond to legends are portrayed in Figs. 2.7, 2.8, 2.11, and 2.12. They all have the same size (the visual variable *size* is not charged with a meaning to portray the indicators).

Legends 1 and 2 (Figs. 2.5 and 2.6) just provide information on sound pleasantness. Only one visual variable is used to portray pleasantness levels, namely variations in color hue (with variations in color saturation and lightness). Compared

Fig. 2.5 Legend 1

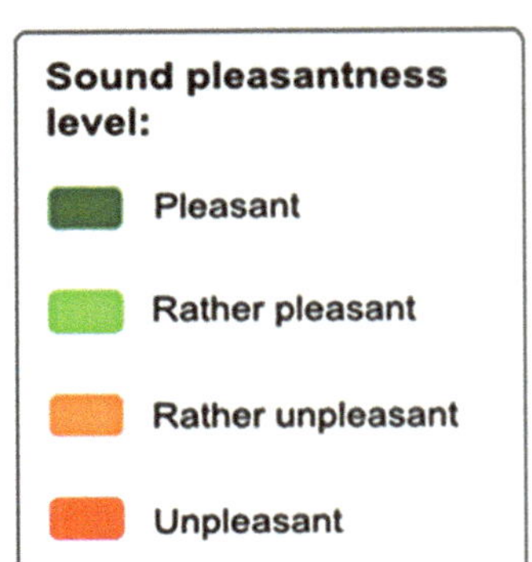

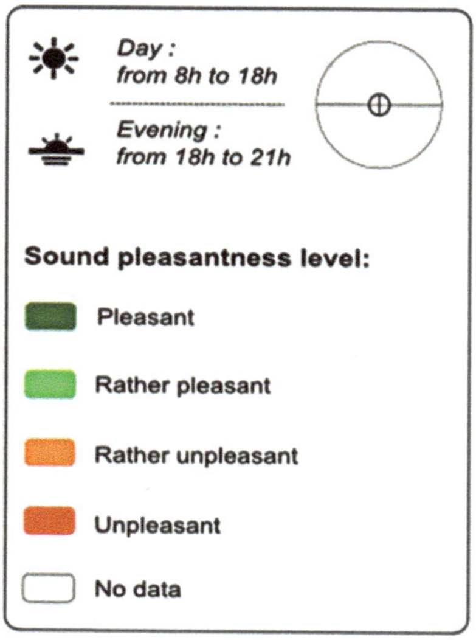

Fig. 2.6 Legend 2

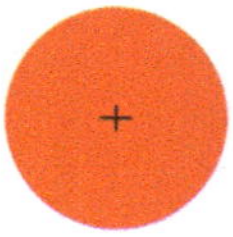

Fig. 2.7 Symbol example of Legend 1

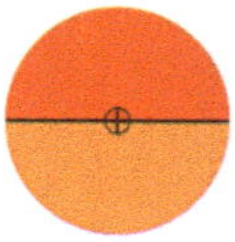

Fig. 2.8 Symbol example of Legend 2

to symbol 1, symbol 2 distinguishes between two periods, namely day and evening. For the latter, the legend specifies that the upper part of the symbol refers to day, and the lower part of the symbol refers to evening (with this symbol, the period size is not proportional to period duration). The symbol in Fig. 2.7 shows that the soundscape is *unpleasant*, whereas in Fig. 2.8, the soundscape is *unpleasant* during the day and *rather unpleasant* at night.

Proposals 3 (Fig. 2.9) and 4 (Fig. 2.10) are more complex. They show both sound pleasantness and global loudness. They are a combination of two graphical variables, including color hue (with variations in color saturation and lightness) to show pleasantness, and the number of bands to show loudness. They also distinguish between several periods. Proposal 3 shows two periods (day and evening) and can be interpreted similar to proposal 2 (upper half of symbol for day, and lower half for evening). Proposal 4 shows three periods (day, evening, and night). In this case, the symbol is divided into three parts (the size of each part matches the number of hours). In the example of Legend 3 (Fig. 2.11), the symbol shows that it is *unpleasant & loud* during the day, and *rather unpleasant & rather quiet* in the evening. The example of Legend 4 (Fig. 2.12) shows a symbol indicating that it

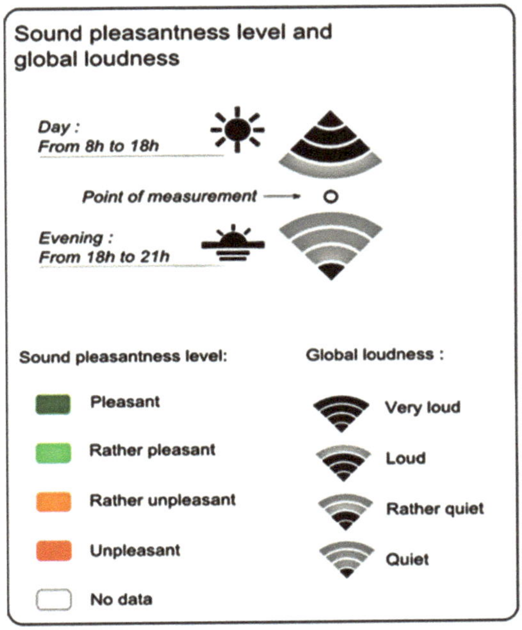

Fig. 2.9 Legend 3

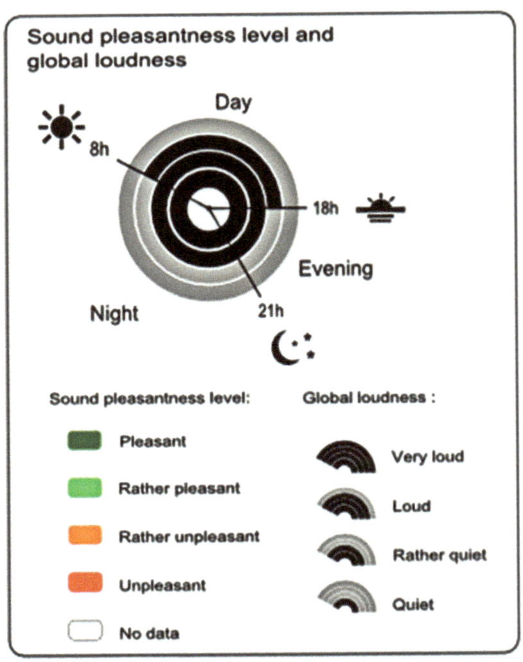

Fig. 2.10 Legend 4

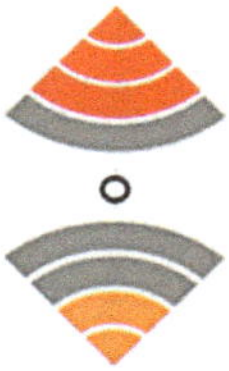

Fig. 2.11 Symbol example of proposal 3

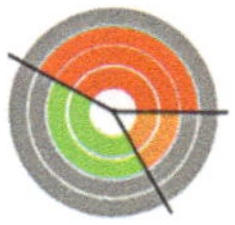

Fig. 2.12 Symbol example of proposal 4

is *unpleasant* & *loud* during the day, *rather unpleasant* & *rather quiet* during the evening, and *rather pleasant* & *rather quiet* at night.

2.4 Assessment Survey of Cartographic Proposals

Four cartographic proposals were made. The purpose was to select the best symbology to help city dwellers distinguish between the two indicators and to imagine soundscape from urban features (in the same way as formulas compute the two indicators through urban features). Consequently, an evaluation survey of the cartographic proposals was developed. Its purpose was to test the interpretation and understanding of various map elements relating to the design of soundscape. The survey was quantitative and based on a close-ended questionnaire which was split into different sections. The aim of each section was to assess one cartographic proposal by identifying the understanding and users' preference of the indicator(s) portrayed.

2.4.1 Participants of the Assessment Survey

The final mapping was designed to be used by any city dweller, regardless of his/her experience, academic background, or training. Consequently, survey participants were not asked whether they possessed specific work-related, social, or cultural characteristics. In order to receive as many answers as possible, the questionnaire was launched on the Limesurvey[1] online platform. This enabled a speedy and easy distribution and facilitated the storage of answers. Calls were made in professional sites and mailing lists; the questionnaire was provided in two languages (French and Spanish) to include responses from remote geographic locations and it was available from March 2015 to July 2015.

[1] www.limesurvey.org

In total, 174 individuals, mainly between 19 and 39 years of age, responded to the questionnaire. In Table 2.1, a column *Survey* shows the percentage share of participants by age, whereas a column *Paris* shows the share of Paris population for the same age ranges. Both groups vary from each other; this can be explained by the method of data collection, namely the advertising in specialized electronic reviews or blogs and direct solicitation of researchers or doctoral students. The under-18 category is under-represented because this category is difficult to reach through calls in professional sites or mailing lists. Similarly, the over-60 age group is under-represented because very few people were interested in the survey whereas there are numerous in this category in Paris.

Most participants have a university education (compare Table 2.2) and often use maps (compare Table 2.3) and even use noise maps (compare Table 2.4).

In order to analyze what participants initially thought of the information displayed, the questionnaire did not allow them to return to modify any of their original answers. The average response duration observed to complete the entire questionnaire was between 20 and 25 min.

2.4.2 The Medium of the Assessment Survey

The page layout was the same for every proposal (see Fig. 2.13). It was divided into several parts. Every map showed the same area with the same five points of measurement (from *Point 1* to *Point 5*). The proposal name was on the top left corner, the legend in the rectangle on the bottom left corner. Questions, which aimed to verify whether the information was understood by map readers, depended on the cartographic proposal and were placed into the top right corner. Participants had to check the box(es) reflecting their choice. It should be noted that it was mentioned that more than one answer was possible to check.

Table 2.1 Age ranges of participants

Age ranges	Survey	Paris
From 0 to 18	1%	24,6%
From 19 to 29	39%	11,6%
From 30 to 39	29%	12,4%
From 40 to 49	9%	13,4%
From 50 to 59	7%	13,1%
From 60 to 90	1%	24,9%
Average age	29	41

Table 2.2 Socio-economic rankings of participants

Socio-economic rankings	
Artisan, shopkeeper or businessman	1%
Employee with a university education	67%
Employee without a university education	6%
Student	22%
Unemployed	3%
Other	1%

Table 2.3 Map use frequency of participants

Maps use frequency	
Never	1%
Rarely	6%
Once a year	10%
Once a month	16%
Once a week	17%
Several times a week	50%

Table 2.4 Noise map use of participants

Noise map use frequency	
Never	33%
At least one time	67%

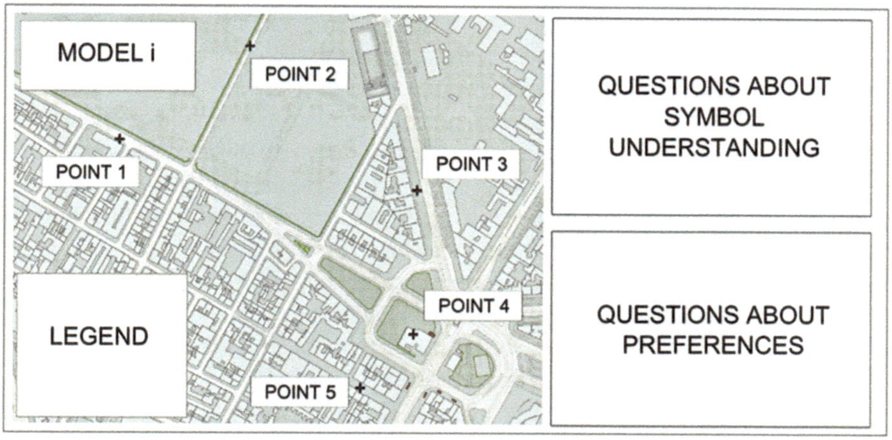

Fig. 2.13 Pattern of the page layout

2.4.3 Questions of the Assessment Survey

The questionnaire was quantitative and close-ended. It included a total of 72 questions that were divided into eight sections. It aimed to assess the participants' understanding of cartographic symbols and their preferences.

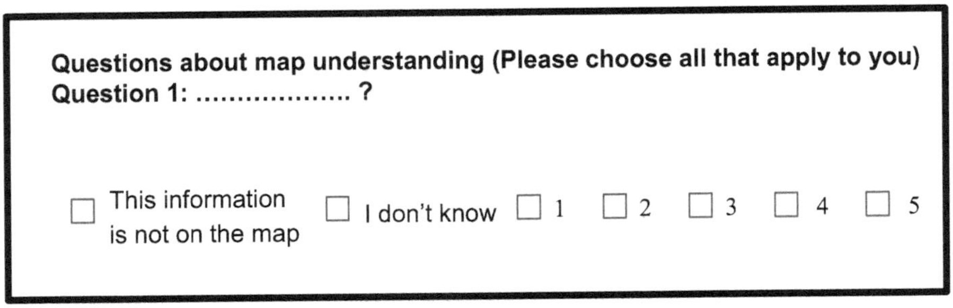

Fig. 2.14 Type of questions for each proposal (*Top right* corner of page layout)

2.4.3.1 Questions About Understanding of Cartographic Symbols

CartASUR maps aim to help city dwellers build an image of soundscape and distinguish sound pleasantness from global loudness. Part of the survey was therefore intended to assess the mapping proposals by finding out whether the message delivered was understood by the map reader. The message is based on both sound pleasantness and global loudness, so the questions aimed to check that readers distinguished between both notions and that they did not answer about loudness, when the symbol showed information about pleasantness. Questions *Q.2* for proposals 1 and 2 were formulated to assess this and may be seen as being tricky (see below). Four maps were made (one for each cartographic proposal) and readers were asked to interpret symbols and to compare them.

Figure 2.14 shows the type of questions about the map understanding. There were two questions in proposals 1 and 2, and four questions in proposals 3 and 4 with this type of questions. Figure 2.15 shows the complete example of proposal 3.

Questions 1 and 2 aimed to check whether readers understood that the symbol was about sound pleasantness and that they could distinguish it from loudness. The formulation of both questions was very similar for the four proposals. For proposals 1 and 2 which only show sound pleasantness, Question 1 verified that map readers could understand information shown with the symbol and Question 2 aimed to distinguish between sound pleasantness and global loudness. Since symbols do not show global loudness, the right answer was: "*The information is not on the map*".

For proposals 3 and 4, two questions (Questions 3 and 4) were added to assess periods and measurement understanding. Both questions were supposed to verify that readers understood representations of periods with the chosen symbols. Proposal 3 only gives information for the day and the evening. So, the right answer for Question 4 was: "*The information is not on the map*". For Question 3 and Question 4, there was only one right answer. Questions (in italics) of each proposal are listed below.

Proposal 1:

- *Q. 1: What is the most pleasant point according to noise (pleasantness)?*
- *Q. 2: What is the loudest point according to noise (loudness)?*

a

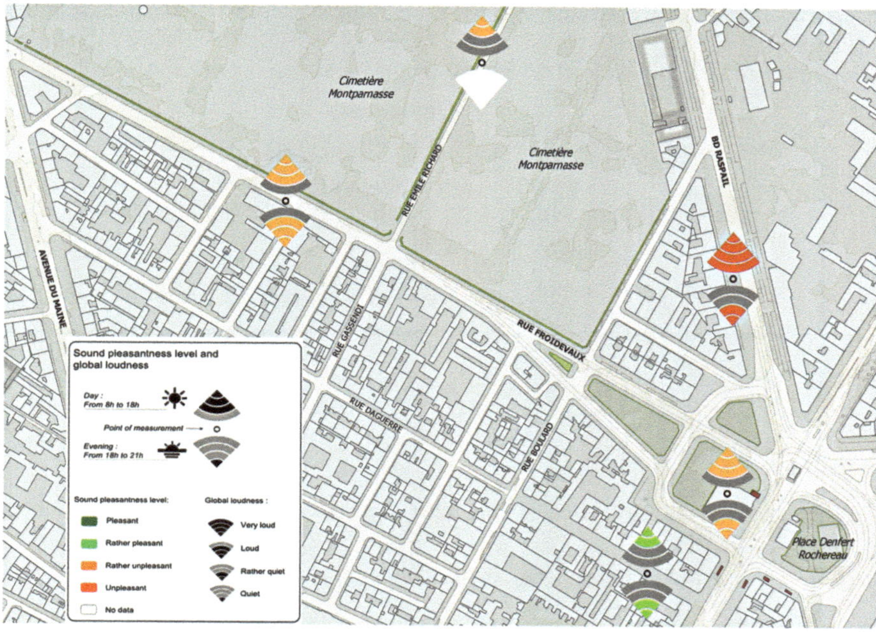

b

1. Questions about understanding (please choose all that apply to you)

Q. 1: During the evening, what is the most unpleasant point according to noise (pleasantness)?

☐ This information is not on the map ☐ I don't know ☐ 1 ☐ 2 ☐ 3 ☐ 4 ☐ 5

Q. 2: During the evening, what is the loudest point according to noise (loudness)?

☐ This information is not on the map ☐ I don't know ☐ 1 ☐ 2 ☐ 3 ☐ 4 ☐ 5

Q. 3: In what point the noise has not been measured in the evening?

☐ This information is not on the map ☐ I don't know ☐ 1 ☐ 2 ☐ 3 ☐ 4 ☐ 5

Q. 4: At 23h, what is the loudest point (loudness)?

☐ This information is not on the map ☐ I don't know ☐ 1 ☐ 2 ☐ 3 ☐ 4 ☐ 5

Fig. 2.15 Questionnaire about proposal 3, including the map (**a**), questions about understanding (**b**), and questions about global map properties (**c**)

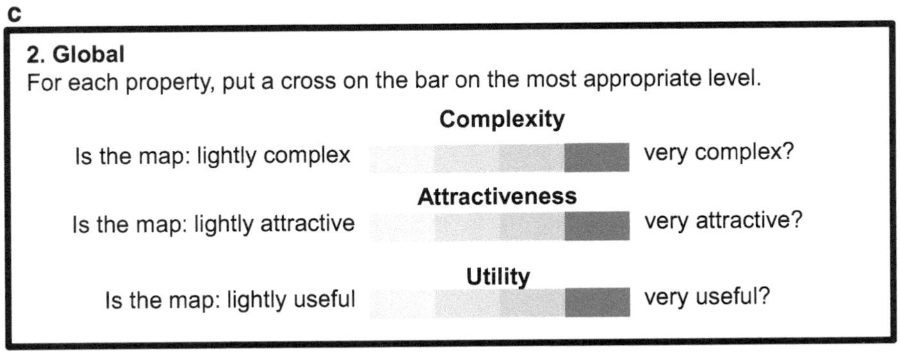

Fig. 2.15 (continued)

Proposal 2:

- *Q. 1: During the day, what is the most unpleasant point according to noise (pleasantness)?*
- *Q. 2: During the evening, what is the quietest point according to noise (loudness)?*

Proposal 3:

- *Q. 1: During the evening, what is the most unpleasant point according to noise (pleasantness)?*
- *Q. 2: During the evening, what is the loudest point according to noise (loudness)?*
- *Q. 3: In what point the noise has not been measured in the evening?*
- *Q. 4: At 23h, what is the loudest point (loudness)?*

Proposal 4:

- *Q. 1: During the day, what is the most unpleasant point according to noise (pleasantness)?*
- *Q. 2: During the day, what is the loudest point according to noise (loudness)?*
- *Q. 3: In what point is the loudness different during the day and during the evening (loudness)?*
- *Q. 4: In what point is the loudness the same during the day and during the evening (loudness)?*

2.4.3.2 Questions About Readers' Preferences

The survey aimed to assess readers' preferences about maps, in terms of their understanding, where appropriate. The aim was to examine the map readers' preferences about global map properties and the amount of information. Consequently, three questions were asked for each mapping proposal, using close-ended questions.

In your opinion, how many periods in the symbol do you prefer (only one answer)?

☐ one period (day) ☐ two periods (day and evening) ☐ three periods (day, evening, and night)

In your opinion, what is more relevant information about noise (only one answer)?

☐ sound pleasantness ☐ sound pleasantness and global loudness

Fig. 2.16 Questionnaire about amount of information

These questions were located in the rectangle at the bottom right corner in Fig. 2.15 shows the complete example of proposal 3.

Global Map Properties

Global map properties may be complexity, attractiveness, utility, range of colors, information density, etc. They may relate to the map base as they affect the whole map and not just one, singular point. This set of questions addressed the map's complexity, attractiveness, and utility. There were four levels for each property, for example: "*Is the map: lightly complex, moderately complex, complex, very complex?*". Participants had to put a cross on the bar at the level which seemed to be the most appropriate (see Fig. 2.15). The same set of questions was repeated for each proposal.

Amount of Information

The last set of questions (Fig. 2.16) was not related to a specific cartographic proposal; it focused on the amount of information which can be portrayed on maps according to map readers' preferences. The information relates to the number of periods and indicators. Periods may be portrayed in three ways, including one period (day) as in proposal 1, two periods (day and evening) as in proposals 2 and 3, or three periods (day, evening, and night) as in proposal 4. With regards to indicators, sound pleasantness may be portrayed on its own as in proposals 1 and 2, or with global loudness as in proposals 3 and 4. It should be noted that the representation of global loudness was not studied on its own because the main indicator is sound pleasantness.

2.5 Results and Interpretation

The survey addressed the understanding of cartographic symbols and global properties of maps. The questionnaire answers were grouped into these two themes which are explored in the next sections.

Table 2.5 Distribution of answers for proposals 1, 2, 3, and 4 with boxes about sound pleasantness being highlighted in grey

Proposals:	1		2		3				4			
	Q.1	Q.2	Q.1	Q.2	Q.1	Q.2	Q.3	Q.4	Q.1	Q.2	Q.3	Q.4
Correct answers	93%	49%	89%	52%	89%	56%	94%	68%	95%	68%	56%	57%
Incorrect answers	6%	1%	9%	12%	11%	41%	4%	31%	4%	30%	42%	41%
"I don't know"	1%	3%	2%	2%	1%	3%	2%	1%	1%	2%	2%	2%
Wrong answers		47%		34%		(26%)						

2.5.1 *Symbol Understanding*

Questions 1 to 4 addressed the understanding of cartographic symbols, portraying sound pleasantness and global loudness. Table 2.5 shows the distribution of answers by questions. Answers are grouped into categories which distinguish between "incorrect answers" and "wrong answers" concerning Question 2; differences are explained in the section "Confusion between pleasantness and loudness" below.

2.5.1.1 Understanding of Pleasantness Symbol

The understanding of the pleasantness symbol is measured by *Q.1* questions and results are mostly correct. On average across all *Q.1* questions, over 91% participants understood the information about pleasantness, regardless of the number of periods shown in the symbol.

2.5.1.2 Understanding of Loudness Symbol

In terms of loudness, the symbol understanding is measured by *Q.2*, *Q.3* and *Q.4* questions. Participants got the majority of the answers correct. On average, there are 63% correct answers. There are slightly fewer correct answers for question *Q.3* of proposal 3, where participants were asked to identify what information was missing/excluded. On average, the proportion of correct answers declines to 58%. When the answers for *Q.2* in proposals 1 and 2 are examined, it is clear that participants did not realize that only pleasantness was portrayed. They answered the questions about loudness instead and consequently, the answers were wrong.

2.5.1.3 Confusion Between Pleasantness and Loudness

The comparison between *Q.1* and *Q.2* answers enables us to verify whether the two concepts, sound pleasantness and global loudness are understood and differentiated by participants. When interpreting *Q.2* answers, the distinction is made between “incorrect answers” and “wrong answers”. When participants replied about loudness looking at maps that only portrayed sound pleasantness information (Q.2 in proposals 1 and 2), their answers showed that they confused the two concepts. The answers are recorded as “wrong answers”. There are 47% “wrong answers” for proposal 1 and 34% for proposal 2. For *Q.2* in proposals 3 and 4, both pleasantness and loudness were portrayed, so the answers which are incorrect are recorded as “incorrect answers”. This analysis can be examined further. Indeed, amongst the 41% participants who gave an incorrect answer in proposal 3, 26% selected the point which was the more unpleasant instead of the loudest. This seems to show ongoing confusion between the two indicators, even when both are portrayed. Lastly, for proposal 4, the point which was the loudest (*Q.2*) was also the more pleasant point. So, in this case, answers which do not indicate the right point are really incorrect and they are recorded as “incorrect answers”. They represent 30% of the answers. This means the legend was simply not understood by the reader. The findings could be that symbols proposed do not help map readers to differentiate between concepts of sound pleasantness and global loudness, even when legends point out differences between such representations. Lastly, the number of correct answers on loudness interpretation (*Q.2* for all proposals) increases from the first proposal (49%) to the 4th proposal (68%), suggesting that map readers have learned to distinguish the two perceptual dimensions as the survey did progress.

2.5.1.4 Understanding of Representing Time Periods

Symbols in proposals 2, 3, and 4 were divided into several parts to show different time periods of the day or during a 24 h period. Proposals 2 and 3 divide the day into two periods (day and evening) and proposal 4 into three periods (day, evening, and night). The size of each part of the symbol may be proportionate to the corresponding period duration (as in proposal 4) or not (as in proposals 2 and 3). Survey participants mostly understood the suggested time period representation, when the question was simple, such as in Question *Q.3* of proposal 3, when no information of the time period was asked. When the question addressed something more specific, the number of correct answers decreased. This can be seen in *Q.4* of proposal 3, which asked about the time period representation of symbols (the representation of the time period may not depend on the indicator). In this case the number of correct answers is 68%. This shows that the time period information is not easy to read with the symbol in proposal 3. The symbol chosen in proposal 4, which portrays more time periods, would even be more difficult to comprehend.

2.5.1.5 Comparison Between Locations

Questions *Q.3* and *Q.4* in proposal 4 asked readers to make comparisons between locations. This task was the most complex of the questionnaire. In order to answer this question correctly, readers had to interpret symbols and summarize pieces of information about both loudness and time periods. Less than 58% of participants answered both questions correctly. However, one goal of reading maps is to compare locations according to the indicators represented (sound pleasantness and/or global loudness) and a comparison must be made possible. In this case, just over half of the readers (56 and 57%) were able to correctly make this comparison.

2.5.1.6 Use of Color Hue (and Color Lightness and Color Saturation)

Questions *Q.2* of proposals 3 and 4 about loudness (*What is the loudest point?*) were similar to question *Q.1* about pleasantness (*What is the most pleasant point?*). Nevertheless, the numbers of correct answers were very different. 62% of participants correctly identified the loudest point, but, interestingly 91% correctly identified the most pleasant point. An explanation may be found in the visual variables used for the various representations. It seems that color that was used to portray sound pleasantness maybe much easier to comprehend and to interpret than quantity variation in the form of several bands used to portray global loudness. The prevailing influence of color has been described by Bertin (1983) and our results provide an additional confirmation. A significant part of this misunderstanding may be due to the difficulties in interpreting the visual variable chosen (quantity variation in the form of bands) to visualize global loudness variations. In these symbols, color variations seem to conceal quantity variations.

2.5.1.7 "I Don't Know" Answers

Lastly, the results included a few "I don't know" answers (less than 2%). It seems that map readers did not hesitate to interpret symbols, but they often made a mistake when symbols were complex.

Some findings can be derived from this section regarding symbol understanding. Even though sound pleasantness and time period representations are well understood, loudness representation is less understood. This is true for simple tasks. When tasks are more complex and require a comparison between locations and an understanding is required about both indicators and time periods, just over half of all survey participants were able to interpret symbols correctly. Interestingly, participants did not check the "I don't know" answers. They did not realize that they might make a mistake and they proceeded with an answer that sometimes resulted in making mistakes. Two alternative hypotheses can be expressed: (i) A large part of map readers does not distinguish between the two concepts of sound pleasantness and global loudness, even when definitions about both concepts are available at the beginning of the questionnaire, including the theme of questions, or (ii) The use of quantity variations to show global loudness variations is so complicated to understand that it hampers all other understanding tasks. In addition, the visual variable *color* seems to capture map readers' attention.

2.5.2 *Preferences*

2.5.2.1 Global Properties

Legends and symbols are shown in Figs. 2.5, 2.6, 2.7, 2.8, 2.9, 2.10, 2.11, and 2.12. As can be seen, they gradually provide more information and become more complex. Table 2.6 shows participants' answers about map complexity. Not surprisingly, proposals 1 and 2 are considered to be lightly to moderately complex with both categories adding up to over 90% each for all answers. Assessments of proposals 3 and 4 are almost equally distributed between the categories *lightly* and *moderately complex* (47% and 43%) and *complex* and *very complex* (53% and 57%).

Table 2.7 shows the participants' answers about map attractiveness. An assessment of map complexity does not seem to influence the evaluation on attractiveness, quite the opposite. Attractiveness increases as complexity does and, paradoxically, the map with the most complex symbology (proposal 4) is considered to be the most attractive.

Likewise, Table 2.8 shows that utility, like attractiveness, increases as complexity does. The most useful maps are therefore the most complex ones. 78% and 77% of participants rated proposals 3 and 4 to be useful or very useful. However, "only" 63% rated proposals 1 and 2 with the two same categories.

To summarize the results about the symbols' overall properties, it can be concluded that the most attractive and the most useful symbols are also the

Table 2.6 Map complexity

COMPLEXITY	Proposals:			
	1	2	3	4
Lightly complex	79%	34%	7%	6%
Moderately complex	17%	57%	40%	37%
Complex	3%	7%	45%	41%
Very complex	1%	2%	8%	16%

Table 2.7 Map attractiveness

ATTRACTIVENESS	Proposals:			
	1	2	3	4
Lightly attractive	17%	15%	15%	9%
Moderately attractive	39%	44%	38%	35%
Attractive	43%	40%	46%	51%
Very attractive	1%	1%	1%	5%

Table 2.8 Map utility

UTILITY	Proposals:			
	1	2	3	4
Lightly useful	13%	7%	4%	5%
Moderately useful	24%	30%	18%	18%
Useful	55%	56%	65%	59%
Very useful	8%	7%	13%	18%

Table 2.9 Participants' preferred number of time periods

One period (day)	9%
Two periods (day and evening)	26%
Three periods (day, evening and night)	65%

Table 2.10 Participants' preferred number of indicators

One indicator (sound pleasantness)	10%
Two indicators (sound pleasantness and loudness)	90%

most complex ones. Participants are aware of the importance of both pieces of information (sound pleasantness and global loudness) offered in proposals 3 and 4 to understand soundscape. Nevertheless, by comparing these results with those on symbol understanding, complex maps are preferred even though map readers are not always able to make complex operations with them, such as comparing locations.

2.5.2.2 Amount of Information

Tables 2.9 and 2.10 show that map readers largely prefer to have a lot of information displayed with symbols. This preference is valid for both time periods (see Table 2.9) and indicators (see Table 2.10). Such results are consistent with findings on global map properties. The most attractive and useful proposal is proposal 4, which also includes the most complex map symbols with three time periods and two indicators portrayed.

This answer about preferred number of indicators does not depend on the noise map use of participants (Table 2.4).

As a conclusion, it can be said that map readers prefer to view complete (both sound pleasantness and loudness) and precise information (three time periods), even when the amount of information leads to complex maps, which are then considered to be the most attractive and useful by map readers.

2.6 Conclusions and Perspectives

The European Directive's definition of "noise mapping" takes into account the need for more information about noise pollution in cities. Strategic noise maps are based on the L_{DEN} indicator, which is not easy to interpret by the general population. "*A positive attitude towards maps [...] also produces a positive judgment about user-oriented cartography as a decision support tool*" (Reinermann-Matatko 2013). Noise representation is a complicated task due to concept complexity. A new approach is to directly map the sound pleasantness instead of the L_{DEN} exposure level. In previous studies, a sound pleasantness indicator based on loudness, presence of traffic, presence of voices, and presence of birds was put forward. However, the strong link between sound level and sound pleasantness still introduces some difficulties for the general population to understand both dimensions.

In addition, the specific and personal nature of map readers interferes with the interpretation of urban sounds and this is an element that can hardly be predicted. Therefore, the CartASUR project develops maps that can also be interpreted in a particular way by each user. CartASUR aims to portray the soundscape of Paris based on two indicators, including sound pleasantness and global loudness which are directly measured in 70 places and predicted across the entire city. To portray them, four symbology proposals have been designed and assessed. This chapter focuses on cartographic proposals and the assessment of symbol understanding and map readers' preferences which were measured in the survey. Other proposals about representation of traffic and presence of voices and birds are not discussed.

The findings are that map readers are aware of soundscape complexity and they express their preferences for maps which portray both indicators according to different periods. Even if these maps are complex, map readers consider them to be the most attractive and the most useful. Nevertheless, the survey showed that most map readers do not interpret these complex maps well and such maps do not provide a suitable medium to perform complex tasks. In this respect, this result agrees with the main conclusion of Hegarty et al. (2009) on the users' preference for complex information. Besides, survey participants were used to handling maps. In contrast, the results of map readers, who are more representative of the Parisian population is expected to be worse. To address this difficulty, a proposal could be to portray each indicator on a different map. Then, the main drawback would be that the comparison between sound pleasantness and global loudness would require to handle two maps at the same time.

An explanation for the wrong interpretation of indicator symbols may be the chosen visual variables. Pleasantness variations are associated with color (actually color hue, lightness and saturation vary in the same time) variations, whereas loudness variations are shown with a number of bands. The *color* (hue, lightness, and saturation) visual variable seems to capture map readers' attention best. So, color variations are easy to read and push variations of the other visual variable into the background, especially in this case where sound pleasantness seems to be closely tied to global loudness. To confirm this hypothesis, an additional test could have been done that may swap the visual variables used for pleasantness and loudness symbols, to verify whether a loudness symbol using color variations would be more correctly interpreted.

Another hypothesis is that map readers do not read legends carefully. Hegarty et al. (2009) pointed out how knowledge about represented data was important to interpret symbols and understand maps. The effect explained in Sect. 2.5.1.3.

"Confusion between pleasantness and loudness" shows that for an audience that includes not only specialists when doing sound research, training would be more significant than merely providing information required to distinguish between the two concepts, pleasantness and loudness. Besides, the authors are currently planning an additional test related to this research, based on an eye tracker. The objective is to study how readers look at different map indicators, especially at the duration and the number of times they return to the legend, and the order of reading operations. This analysis should indicate whether participants really try to understand legend information or, by contrast, answer quickly without much thought.

The findings in this book chapter show that the need for understanding noise maps is still of high interest to the general population. The prediction formulas of sound pleasantness have shown that this indicator depends on presence of traffic, birds, and voices. In order to help map readers to imagine sound pleasantness, it would be useful to portray these perceptual variables, i.e. to design cartographic objects which allude to traffic, birds, and voices and to assess them for map readers. The prediction formulas are able to determine sound pleasantness and global loudness in each location of the city. Thus, the representation of predicted indicators would be continuous across the entire city of Paris, whereas the survey representation in this chapter only shows indicators at specific survey locations.

Acknowledgment This CartASUR project is funded by ADEME, the French Environment and Energy Management Agency (Agreement n° 1217C0035).

References

Alberts W, Rubio Alferez J (2012) The use of colors in END noise mapping for major roads, Prague. In: EURONOISE 2012, Czech Republic, 5 p. http://www.carreteros.org/explotacion/2012/6.pdf. Accessed 22 Oct 2015

Aletta F, Kang J, Axelsson Ö (2016) Soundscape descriptors and a conceptual framework for developing predictive soundscape models. Landsc Urban Plan 149:65–74

Aumond P, Can A, De Coensel B, Botteldooren D, Ribeiro C, Lavandier C (2016) Sound pleasantness evaluation of pedestrian walks in urban sound environments. In: Proceedings of 22nd international congress on Acoustics, Buenos Aires [to be published]

Basner M, Babisch W, Davis A, Brink M, Clark C, Janssen S, Stansfeld S (2013) Auditory and non-auditory effects of noise on health, Philadelphia, Unit for Experimental Psychiatry, Division of Sleep and Chronobiology, Department of Psychiatry, University of Pennsylvania Perelman School of Medicine. http://www.ncbi.nlm.nih.gov/pmc/articles/PMC3988259/ [visited on: 2015/10/22]

Bertin J (1983) Semiology of graphics: diagrams, networks, maps. University of Wisconsin Press, (first published in French in 1967, translated to English by Berg W.J. in 1983)

Brewer CA, Hatchard GW, Harrower MA (2003) ColorBrewer in print: a catalog of color schemes for maps. Cartograph Geograph Inform Sci 30(1):5–32. www.ColorBrewer.org

Brocolini L, Lavandier C, Quoy M, Ribeiro C (2013) Measurements of acoustic environments for urban soundscapes: choice of homogeneous periods, optimization of durations and selection of indicators. J Acoustic Soc Am 134(1, Pt. 2):813–821

Delaitre P, Lavandier C, Ribeiro C, Quoy M, D'Hondt E, Gonzalez E, Kambona K (2014) Influence of loudness of noise events on perceived sound quality in urban context. Melbourne. In: Inter Noise 2014, 10p

Guski R, Felscher-Suhr U, Schuemer R (1999) The concept of noise annoyance: how international experts see it. J Sound Vibr 223(4):513–527

Guastavino C (2007) Categorization of environmental sounds. Canad J Exp Psychol 61

Haberle M, Dovener D, Schmid D (1984) Inquiry on noise causing complaints in residential areas near chemical plants. Appl Acoust 17:329–344

Hegarty M, Smallman HS, Stull AT, Canham MS (2009) Naïve cartography: how intuitions about display configuration can hurt performance. Cartographica: Int J Geogr Inf Geovisualization 44(3):171–186

IFOP (2014) http://www.ifop.com/?option=com_publication&type=poll&id=2799

ISO (2014) Acoustics-Soundscape-Part 1: definition and conceptual framework. International Organization for Standardization, ISO 12913-1 TC 43/SC 1. pp 1559–1564

Jeong W, Gluck M (2002) Bivariate thematic maps with auditory and haptic display. Proceedings of the 2002 international conference on Auditory Display, Kyoto, Japan, July 2–5. http://onlinelibrary.wiley.com/doi/10.1002/meet.1450390130/full

Kornfeld AL, Schiewe J, Dykes J (2011) Audio cartography: visual encoding of acoustic parameters. In Lecture notes in geoinformation and cartography, London, 18 p

Lavandier C, Delaitre P, D'Hondt E, Gonzalez E, Kambona K (2013) Urban sound quality assessment with mobile technology: The Cart_ASUR project, Proceedings of Acoustics 2013, New Delhi, India

Lavandier C, Delaitre P, Ribeiro C (2015) Global and local sound quality indicators for urban context based on perceptive and acoustic variables. In: Proceedings of the Euro Noise Congress, Maastricht, Nederland, vol 31

Leonowicz A (2006) Two-variable choropleth maps as a useful tool for visualization of geographical relationship. Geografija 42:33–37

Miedema HME, Oudshoorn CGM (2001) Annoyance from transportation noise: relationships with exposure metrics DNL and DENL and their confidence intervals. Environ Health Perspect 109(4):409–416

Morel J (2012) Physical and perceptual characterization for indicators of annoyance due to urban road traffic noise in isolation and combined with industrial noise. PhD thesis. Lyon, University of Claude Bernard-Lyon 1, France, 311 p

Reinermann-Matako A (2013) Maps as decision support tool in political decision processes. Department of Cartography, University of Trier, Trier

Ricciardi P, Delaître P, Lavandier C, Torchia F, Aumond P (2015) Sound quality indicators for urban places in Paris cross-validated by Milan data. J Acoust Soc Am 138(4):2337–2348

Schiewe J, Weninger B (2013) Visual encoding of acoustic parameters – framework and application to noise mapping. Cartograph J 50(4):12 p – doi:10.1179/1743277412Y.0000000026

Weninger B (2013) Developing a color scale for traffic noise maps: design aspects for online mapping, Dresden, In: ICC 2013, pre-conference workshop on Map Design, 6p

WHO (2011) World Health Organization, Burden of disease from environmental noise http://www.who.int/quantifying_ehimpacts/publications/e94888/en/. Accessed 12 Apr 2016

Chapter 3
Evaluating the Current State of Geospatial Software as a Service Platforms: A Comparison Study

Benjamin G. Lewis, Weihe Wendy Guan, and Alenka Poplin

Abstract The goal of this chapter is to evaluate and compare Geospatial Software as a Service (GSaaS) platforms oriented toward providing basic mapping capabilities to non-GIS experts. These platforms allow users to organize spatial materials in layers, perform overlay and basic visual analysis, and share both final maps and the processes used to create them with remote collaborators. The authors gathered data on the characteristics of 15 platforms through an online survey, then summarized the results and created an Excel tool to enable users to sift through the data to identify platforms based on need. This study presents a snapshot of the current GSaaS landscape, summarizes current capabilities, points out weaknesses, and considers the potential of this class of application.

Keywords GIS • Web mapping • Collaborative mapping • Software as a service • Map service

3.1 Introduction

Scholars from a broad range of disciplines are interested in using geospatial information collaboratively in ways paralleling the joint editing of text documents and spreadsheets. Collaboration in this context includes the creation and organization of geospatial information, as well as joint curation, editing, and publishing online, which can result in the creation of new geospatial knowledge. In this context, the development of "Software as a Service" (SaaS) is crucial because of its potential

---

B.G. Lewis (✉) • W.W. Guan
Center for Geographic Analysis, Harvard University, 1737 Cambridge Street, Suite 350, Cambridge, MA, 02138, USA
e-mail: blewis@cga.harvard.edu; wguan@cga.harvard.edu

A. Poplin
Department of Community and Regional Planning, Iowa State University, Room 487, College of Design, Ames, IA, 50011, USA
e-mail: apoplin@iastate.edu

M. Leitner, J. Jokar Arsanjani (eds.), *Citizen Empowered Mapping*, Geotechnologies and the Environment 18, DOI 10.1007/978-3-319-51629-5_3

to lower barriers to entry, to provide researchers everywhere (assuming reasonable bandwidth exists) with the necessary infrastructure (network-connected hardware, software, and data hosting) to engage with each other's work productively. SaaS represents software hosted remotely and accessed through the Internet, most often via a web browser, and not needing to be downloaded and installed locally. Outside of the GIS arena, SaaS systems have been augmenting or replacing locally installed applications for many years. Examples include Hotmail, online calendars, and Google Docs.

Starting in the late 1990s, and accelerating since the release of Google Maps in 2005, the range of options available within the class of applications we will refer to as "Geospatial SaaS" (abbreviated as GSaaS), has increased along with adoption. A wide range of powerful GSaaS platforms have become available, bewildering in variety. Some systems are free and some are subscription-based. Many platforms are available at a range of price tiers, with each price level presenting a different set of capabilities. Some GSaaS systems provide rich sets of analytic functions, while others are simple. Some run on platforms that are themselves open source software packages, some do not. Some make it easy to share data and services between systems, some do not.

One could easily assume that the GSaaS field, like most other technology fields, is evolving under the driving force of user demand. However, it must be considered whether "user" refers only to geospatial technology professionals, or does it include those with no experience in geographic information technology? If one considers scholars of the humanities and social sciences as part of this user community, it is worth asking whether current GSaaS products meet their needs. In such a diverse field, how should any user, let alone one untrained in GIS, determine the most appropriate system for her needs? Are there capabilities that are missing from existing platforms? Could the identification of these capabilities be useful for guiding future development and, perhaps for expanding the community that benefits from such platforms?

The goal of this chapter is to review the development of GSaaS technology, analyze existing GSaaS platforms, and examine their capabilities and functions. Section 3.2 provides a brief overview of the online mapping applications that predate GSaaS and which serve a related but different user need. Section 3.3 introduces the emergence of GSaaS. Section 3.4 delineates the components essential for GSaaS development. Section 3.5 describes an empirical study on existing GSaaS, which is the core of our contribution. It focuses on a set of basic characteristics that, based on our experience, are essential for scholars who, while experts in their field, are not experts in GIS technology or software development. We use a standardized survey form to gather information on 15 GSaaS platforms. These systems allow scholars to organize spatial materials in layers, overlay layers with one another, and perform basic visual analysis, without needing specialized GIS skills. Subsequent analysis in this paper concentrates on the strengths and weaknesses of these platforms, and attempts to portray the current landscape of GSaaS, summarize its capabilities, and examine its strengths and weaknesses. Section 3.6 applies our findings to the creation of a practical tool for providing guidance to users without a background in geospatial technology. To address the problem that this study will quickly become dated, the authors will periodically publish updated survey results and new versions of the GSaaS Platform Selection Tool (GSaaS Platform Survey Support Materials 2016). Section 3.7 summarizes our comments and perspectives on the current and future GSaaS development.

3.2 Previous Work: Online Interactive Mapping

GSaaS platforms evolved from interactive mapping systems. Though early web-mapping applications did not support shared editing or spatial analysis, the systems did nonetheless serve mapping data to distributed users across the Internet. Users could not yet collaborate through them, but they could independently view the same geospatial information at the same time which was a start. These web-mapping applications laid the foundation for the development of today's GSaaS platforms.

3.2.1 Web-Mapping and Web 2.0

In the early era of the web, roughly 1995–2004, mapping was restricted to the delivery of static maps displayed in raster formats (GIF, JPEG, or PNG). Such maps were published online in a form of a picture, a graphical visualization of the selected area. Typically, maps were embedded within a web page, presenting one-way communication of content. Constraints that influenced the representation of maps on the web include transfer rate, color depth, screen resolution, and the capabilities of web clients and servers (Arleth 1999).

By mid-2000, the term Web 2.0 was coined to reflect important changes in the way software developers and end-users were starting to use web (O'Reilly 2005). Web mapping refers to web applications that have a special frame of reference (Gartner 2009) and the term is often associated with concepts such as the GeoWeb, Web-GIS, Internet GIS, Online GIS, and internet cartography. Web mapping applications began to enable users to view, search and/or browse spatial information represented in the form of online maps. This stage of online mapping application development resulted in functionality that started to go beyond representing static maps. Development of dynamic maps enabled users to interact with them, search for information, zoom-in and out, and use tools available in the application user interface. Such tools often supported simple spatial queries such as measuring distances, finding locations, or finding directions. Web-based GIS enhanced the use of GIS in the following directions: a) Spatial data access and distribution, b) spatial data exploration and visualization, and c) spatial data processing, analysis, and modeling (Dragićević 2004). Online maps not only became accessible (Elzakker 2001), interactive, highly dynamic, and visible to many users, but they also contributed to the democratization of mapping, and permitted new trends in the broader use of mapping techniques, tools, and applications (Kraak and Brown 2001).

Geographic applications built on Web 2.0 architectures offer a variety of capabilities without the need to install GIS software locally. Gartner (2009) listed the most popular application programming interfaces (API's) at the time as Google Maps, MS Virtual Earth, and Yahoo Maps. All of these systems enabled map exploration (Castelli et al. 2006) with the possibility of navigating an information space and interacting with this space in a new way. Users were able to access online maps and navigate in a flexible and location-independent way, find meaningful information about the surrounding physical world, and use online mapping services to access and manipulate geographic information. These online interactive maps were able to present mapping information that adapts to users' need. However, these early Web 2.0 online interactive maps also had some disadvantages, which included limited options for spatial analysis, inability to upload user's data, inability to combine user's data with data stored within the platform, and few options for collaborating online with maps.

3.2.2 Collaborative Online Mapping

Collaborative mapping, sometimes referred to as collaborative cartography, involves participation and collaboration by which users produce a product together. The product in this case is a map of user-generated geographic information. Goodchild in his earlier publications (Goodchild 2007a, b, c, d) focused on the phenomenon of citizens collecting and sharing geographic information via online platforms. Examples of such platforms include OpenStreetMap, Google Earth, and Wikimapia, platforms that have given rise to a global network of citizen mapmakers. Goodchild observed a "widespread engagement of large numbers of private citizens, often with little in the way of formal qualifications, in the creation of geographic information, a function that for centuries has been reserved to official agencies" (Goodchild 2007b). He termed this phenomenon "volunteered geographic information" (VGI) as a special case of the more general Web phenomenon of "user-generated content" exemplified by platforms such as Wikipedia. Goodchild (2007d) specified different levels of sophistication and distinguished among simpler projects, in which volunteers produce gazetteers and add descriptions and hyperlinks for locations (an example is www.wikimapia.com); projects in which volunteers contribute to the substantial technical content (such as openstreetmap.org); and those that allow contributors to make their own complex spatial information available to others (such as earth.google.com). Online mapping collaborations may happen in an asynchronous or synchronous mode (MacEachren 2001). A few years later, in addition to data generation, collaboration was occurring rapidly at levels of data integration, map services, spatial API's, spatial analytics, interface design, and even in the development of the hosting software itself.

3.3 Emergence of Geospatial Software as a Service

When Google Maps appeared in 2005, suddenly a web-based map viewer existed that was superior in many ways to existing desktop viewers. For the first time, except perhaps for obtaining driving directions on MapQuest, there was a clear advantage to exploring a map on web rather than the desktop. At the time of its release, no one (including Google) fully appreciated the level of excitement a fast and deep online map would generate. Just a few months after its release, Paul Radenmacher, developer at Dreamworks Animation, created the world's first mashup by hacking Google Maps and Craigslist to act as web services, and then feeding that content to his application housingmaps.com (DuVander 2010). Soon many other mashup efforts followed and companies began to start actually designing their sites to encourage reuse.

Google quickly released an official API to make it easy for a user, with a little coding, to have a Google Map show up in his or her own site (Google Maps 2015). Google's fast, high-resolution "slippy map" (Slippy Map 2015) was a free data-as-a-service (DaaS) application, which employed a new client-server technology called AJAX or "asynchronous JavaScript and XML". AJAX enabled web applications to respond more quickly and seamlessly, behave more like desktop applications, and it allowed the backend server hardware to handle more requests with fewer resources. Because of these innovations, mapping applications became better at scaling to handle larger datasets while also supporting more users simultaneously.

This new approach to deploying maps online pioneered by Google is arguably the single most important technological development in the creation of the niche we now call GSaaS. It is worth noting that Google did not invent the core technology used in Google Maps. Two brothers, Lars and Jens Rasmussen, developed the core software in an Australian startup called Where 2 Technologies, which Google acquired. Google took what was originally a desktop C++ application and turned it into a web application backed by Google's infrastructure (Google Maps 2015). The technology used in Google Maps became the technology behind all online mapping systems including those we are calling GSaaS today.

Neither Google Maps itself, nor the web based mapping platforms before it, are examples of the collaboration-enabling GSaaS platforms this chapter is focused on. However, Google Maps paved the way to make such GSaaS possible, namely by integrating the following four fundamental technologies, which are instructive to tease apart (Hutcheon 2015; Tiled Web Map 2015):

Maps are Broken into Manageable Chunks A hierarchical tiling scheme is used to represent a map using 256 $\times$ 256 pixel tiles viewable at a series of discrete scales. These pieces are easy to send across the web via HTTP.

Tiles Are Pre-generated A cache of tiles is created for a given layer at a series of discrete zoom levels, where each scale doubles the previous scale, going from level 0 with a scale of 1:2,236,330,260 (the earth in 1 tile) to theoretically any level of detail (Google Maps goes to level 22 (building level) with a scale of 1:533.

A New Projection Web Mercator is a variation of the Mercator projection but simpler and therefore faster to calculate than standard Mercator. While the formulas are fundamentally the same as for the Spherical Mercator projection, the difference is that before applying zoom, world coordinates are adjusted such that the upper left corner is (0,0) and the lower right corner is (256, 256). In Web Mercator at scale 0, the whole world fits into one 256 × 256 tile. Unlike most projections, which use units of meters or feet, the unit for Web Mercator is the pixel.

Asynchronous Communication AJAX is used in client/server communication. Among other things, this allows small image tiles for a given map view to be requested at slightly different times and returned for display at different times, all without needing to refresh the browser page. This, when coupled with #1 and #2 above, makes for a fast and smooth map viewing experience, namely the so-called "slippy map."

Within a few months of the release of Google Maps, open source software implementations appeared, such as the OpenLayers (openlayers.org) client and the TileCache (tilecache.org). The availability of these components, along with others such as the GDAL (gdal.org) spatial data libraries and the spatial relational database PostGIS (postgis.net), made it possible for any developer to start building collaborative online mapping systems.

Open source software has been critical to the evolution of GSaaS and for many other areas of web development. It allows for high quality software to be made available to any developer anywhere. Many, if not all, of the GSaaS platforms in our survey make significant use of open source software, even if organizations do not release their software as an open source product.

3.4 The Building Blocks of SaaS

In addition to open source web mapping software, Software as a Service often makes use of other types of cloud service building blocks, a common one being hardware or Infrastructure as a Service (IaaS), which enables hardware running in the cloud to be rented out to anyone at any time in virtually any conceivable configuration. An advantage of building an application on IaaS is the potential ability to scale dynamically as demand increases and decreases. Many SaaS application developers now take as a given the existence of IaaS infrastructures such as Amazon EC2, Microsoft Azure, and Google's Cloud Platform.

Another building block is Data as a Service or DaaS systems such as Google Maps and OpenStreetMap. They provide a critical content ingredient to GSaaS. In addition, Platform as a Service (PaaS) systems, another building block, provide developers with specialized environments hosted in the cloud on which to build new backend capabilities. PaaS takes away the burden for developers to individually set up and maintain development and testing environments (hardware, operating system, coding software, code repository, testing configuration, etc.), making

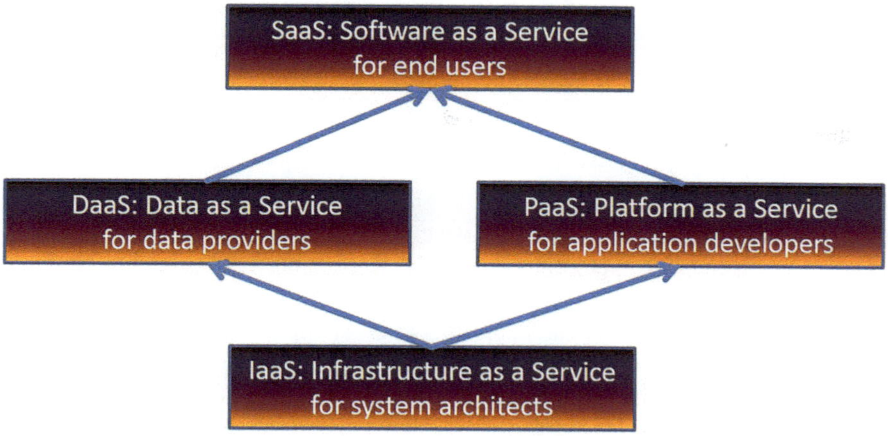

Fig. 3.1 Relationship between IaaS, PaaS, DaaS, and SaaS

collaborative software development more efficient. Software as a Service generally sits on top of one or more of these other pieces, which provide scalable, accessible, and redundant versions of traditional hardware, data, and backend applications to the user-facing SaaS product. Figure 3.1 shows the relationship between them.

Running in the cloud, SaaS applications have a number of characteristics that are driving their adoption over locally installed software. These include: (1) No additional hardware costs, (2) no initial setup costs, (3) pay for what you use, (4) usage is scalable, (5) updates are automated, (6) cross device compatibility, (7) accessible from any location, (8) customizability in terms of branding, (9) easier access to large datasets also in the cloud, and (10) scalable data processing and analysis capabilities.

A recent study by Goldman Sachs showed that spending on cloud computing infrastructure would grow at a compounded rate of 30% a year from 2013 through 2018. Compare this with a rate of 5% a year for traditional enterprise IT hardware and software (Columbus 2015; High 2014), and it is not hard to see where traditional installations are likely to be heading. While this trend is well-established in the broader IT world, it is just starting to catch on in the geospatial world.

There are other important technology trends, which increase the importance of the cloud for geospatial applications. These include:

- the emergence of inexpensive network connected mobile devices ($30 smart phones) that greatly increase the potential number of both content creators and consumers;
- the growth in popularity of open source software and its increasing importance in IT infrastructures of organizations, especially in the case of the U.S. Government.
- the explosion in using geo-located social media, which is a new form of computationally intensive spatial data to analyze;

- the growth in availability of big spatiotemporal datasets, such as remotely sensed imagery and LIDAR; and

3.5 Assessing the Current GSaaS Landscape

Given the rapid evolution of data, infrastructure, software, and platform as services (including DaaS, IaaS, SaaS, and PaaS), the Geospatial SaaS products available to end users are rapidly evolving. More products are becoming available, and more capabilities are added every year. People not familiar with the geospatial technology market find it difficult to keep up with these changes and even product developers must expend a significant effort keeping up with features rolled out by peers and competitors, identifying latest trends, and reassessing the strengths and weaknesses of the various players. The main goal of this section is to provide an overview of the Geospatial Software as a Service (GSaaS) platforms that are oriented toward providing basic mapping capabilities to non-GIS experts.

3.5.1 Qualifications for Inclusion in the GSaaS Survey

In order to portray the current GSaaS landscape objectively, the producers of GSaaS platforms are surveyed and a narrated summary of their responses is presented. The first step in creating the survey was to define what qualifies as a GSaaS. Based on experience supporting scholars across disciplines interested in applying geospatial technologies within their work, the following characteristics are essential:

Map Oriented User Interface The system should allow users to overlay multiple map layers, and be able to create, symbolize, and publish them. This eliminates from our survey some powerful DaaS systems such as OpenStreetMap and Wikimapia which are oriented toward organizing contributions to a single common dataset. This also eliminates in this case the very powerful MapBox platform, which focuses on enabling users to create cartographically refined base maps to use in other applications. MapBox could be used for creating layers to be used in GSaaS systems but it is not itself a general-purpose system.

Import and Export Vector Datasets The system should allow users to upload at least one type of vector data and export data in at least one vector format. This eliminates systems such as MapQuest.

Access Control The system should allow users to control access to their own layers. This eliminates the many read-only mapping sites.

No Programming Required The system should work out-of-the-box without needing to write JavaScript or SQL code. This eliminates some PaaS systems such as Google Maps API and Bing Maps API.

Hosted Platform The system should be available in a hosted configuration, such that a user does not have to install any software and can get all or most of what he/she needs via a web browser. This eliminates map server software such as GeoNode and ArcGIS Server.

Preservation Based on experience with libraries and archives there is often an assumption among scholars that digital materials, once published, will be maintained online and available long term. Permanence is a complex characteristic to measure, especially when it comes to hosted digital materials, so we have not included it directly in our survey, except to require that all systems have a way to export data in an open format to enable archiving elsewhere.

Based on the above criteria, we identified the following platforms to include in the survey:

ArcGIS Online (esri.com/software/arcgis/arcgisonline) was first released in 2012. It is a product of Environmental Systems Research Institute (Esri). The platform is comprised of applications and templates for creating and sharing interactive maps. It is an integral part of Esri's ArcGIS suite of applications. Two versions of the platform are available: A free public account for non-commercial use which provides basic functions for creating and sharing maps and which provides access to content shared by Esri and users. In addition, there is a subscription-based product, sold in tiers, differentiated by the number of user logins and the number of credits provided. Credits are the currency for paying for storage, analytics, and some types of data within the ArcGIS Online system. In addition to an end-user application, ArcGIS Online services can be built into browser-based and mobile applications via APIs.

Bing Maps (bing.com/maps) was first released in 2010. The system is a web mapping service with a public web site provided as a part of Microsoft's Bing suite of search engines and powered by the Bing Maps for Enterprise framework. In addition to consisting of an end-user application, Bing Maps services can be built into browser-based and mobile applications via APIs.

CartoDB (cartodb.com) is the product of the company CartoDB. It was first released in 2011. In addition to consisting of a hosted subscription-based service, CartoDB is open source software which can be installed on a server such that the underlying software can be built upon and extended, if a particular function is not available. The hosted system comes in several pricing tiers from free to Enterprise. These tiers vary primarily by data storage, access control to visualizations, and level of branding ability. In addition to consisting of an end-user application, CartoDB services can be built into browser-based and mobile applications via APIs.

The **eSpatial** (espatial.com) platform was released in 2010 by the company bearing the same name. It followed the release of the iSMART map server software platform in 2005. eSpatial is an online mapping and analytics tool that allows users to visualize and analyze a variety of data from spreadsheets and tables. eSpatial comes in a free version and three subscription tiers which vary mainly by storage space, number of records per dataset, and number of geocoded datasets allowed. A variety of data from eSpatial's library is available to be used within the system. The system is especially useful for sales and marketing applications.

GIS Cloud (giscloud.com) was first released in 2007 by the company bearing the same name. The platform supports visualization, analysis, and exploration of geographic information. The primary goals of GIS Cloud are to simplify the exchange of geographical information between users and to provide an easy way to analyze this information regardless of the location of its users. The system makes extensive use of HTML5. It offers a free and a paid version. In addition to consisting of an end-user application, GIS Cloud services can be built into browser-based and mobile applications via APIs.

Google Fusion Tables (google.com/fusiontables), which was first launched by Google in 2009, has a focus different from the other systems in this survey. It is a Platform as a Service (PaaS) system oriented toward data table management, but with enough mapping capabilities to be included in our survey. It offers novel capabilities for collaborating with and visualizing spatial data. For example, a user may join spatial (or non-spatial) tables with tables controlled by other groups to create new views of original tables, with all data being live and capable of being edited by those with permission.

Google MyMaps (google.com/mymaps/d/) was released by Google in 2007. It is a simple online mapping application, which enables the user to create and edit maps as well as to import data to create new maps. In addition to consisting of an end-user application, Google Maps API-based services can be built into browser-based and mobile applications in many ways via APIs.

Harvard WorldMap (worldmap.harvard.edu) was released in 2012 by the Center for Geographic Analysis (CGA) at Harvard University. This general purpose, open source platform may be installed on other servers and extended by other organizations. WorldMap evolved out of the CGA's experience building customizable web mapping environments for scholars desiring to publish their spatial materials online, even when their files are very large.

InstaGIS (instagis.com) was released by the company InstaGIS in 2013. The platform is oriented toward business and marketing applications with tools for exploring general patterns, setting up marketing campaigns, and performing site selection analysis.

iSpatial (t-sciences.com/product/ispatial) was released by Thermopylae in 2012. iSpatial is built around Google Earth and open standards with the aim of making Google Earth and Maps a powerful platform for commercial, military, humanitarian relief, and intelligence work.

MangoMap (mangomap.com) was released in 2010 by the company bearing the same name. The general purpose mapping system is offered in several tiers, which vary in price by number of maps allowed, storage space, and the ability to apply custom branding.

Map2Net (map2net.com) was released in 2012 by the company Almageo based in Morocco. Map2Net is a general purpose GSaaS mapping platform, which is sold in tiers based on the number of accounts purchased.

The authors are aware that the list above may not be complete. We hope readers will inform us using the survey form, if they know of other systems that should be included. We plan to periodically update the survey results and make them publicly available (GSaaS Platform Survey Support Materials 2016). Current survey results are listed in Appendices B and C.

3.5.2 Design of the GSaaS Survey

The objective of the survey is to capture major characteristics of the selected GSaaS platforms in order to evaluate platform service niche, its strengths, and weaknesses, and to generate practical guidance for platform selection. In designing the survey the authors had these goals: 1) yield information to address the most common needs of the non-technical user community; 2) be self-explanatory to technical professionals, such as GSaaS product managers; 3) be short and simple, i.e., it does not require much time to complete; and 4) provide results that are easy to compile for analysis.

The authors designed the questions and answer choices based on experience developing GSaaS platforms and providing geospatial consultation to non-technical users from a broad range of backgrounds. Google Forms were used to implement the survey, and producers of the selected GSaaS platforms were invited to describe their products by filling out the form for their product. The survey included 25 questions, most of which are multiple choice, with “Other” as a free text addition. See Appendix A for the list of survey questions.

3.5.3 Conducting the GSaaS Survey

The survey form was published online (GSaaS Platform Survey Support Materials 2016) in September of 2015. The authors sent an email invitation including the URL to contacts for all of the selected platforms. The invitation recipients were either product managers for the systems or, if contact information could not be found, sales representatives. In both cases, the invitation included the language “if you are not the most appropriate person to fill this out, please feel free to forward this to a more appropriate person within your organization.” Nine out of the eleven invited platform producers responded to the survey within 2 weeks.

The responders included the company CEOs, chief product managers, and program directors. Results from the survey were captured to a database from which a summary table was created. Two organizations (eSpatial and Google) did not fill out the survey, so the authors filled it out based on publicly available information. Many of the platforms surveyed have both free version and paid versions, which come in tiers based on capacity. We asked the representatives to fill out a form for the free version and a separate form for the highest level of the tiered version.

The survey-collected data are all public information. The authors could have compiled them through research without asking the platform producers but the advantage of the survey is to ensure the data collected are the most up-to-date and accurate. The disadvantage is survey responders may not have accurately understood the survey questions, thus resulting in answers not comparable between platforms. To verify the survey responses, the authors tested all systems and reviewed online statements from the respective companies about their products. No errors were detected in the responses. However, due to the complexity of having so many diverse systems, there is a chance of error caused by misunderstanding of the survey questions, or the authors misinterpreting the survey responses when making a summary.

3.5.4 Summary of Survey Results

Original survey responses are captured in a Google Sheet with the multiple-choice answers presented in Appendix B. Free text entries for the "Other" fields are presented in Appendix C.

Survey results revealed the availability of a series of functionalities for organizing, editing, analyzing, visualizing, and sharing geospatial data, provided via browser, without the need to set up server, install software, write code, or administer systems.

All 15 platform surveyed allow for some level of vector data uploading, but eleven of them have limitations in the number of vector features that can be uploaded per file or in total. Six of the surveyed GSaaS have limitations on the file size of the vector layers that can be uploaded and edited. Raster upload is not possible or is limited for the majority of the surveyed platforms. Only two of them, AGOL and GIS Cloud, enable an unlimited raster upload. Online data creation, in the form of geocoding address lists and geocoding tables with regions, is enabled in different forms by the majority of the surveyed platforms. Twelve of the GSaaS platforms allow digitizing and creation of vector features. Only one, Harvard WorldMap, offers the possibility to georeference raster images within the platform's capabilities. The majority, fourteen, enable HTML linking of the multimedia content online, and in eight platforms the users can store their multimedia content on the provider's server.

Most of the surveyed GSaaS platforms have at least two privacy control options, which include the option of viewing maps publically and viewing them by the author, only. A more strict limitation on who can view the maps is offered by nine platforms, and the ability to limit who can edit is offered by eight.

Common vector data analysis, such as setting attribute query filters on vector layers, or overlay analysis with vector layers is provided by many platforms. Network routing is included in seven platforms. Raster data analysis is not available for most, and only CartoDB supports raster algebra. None of the platforms surveyed claim to need additional software, though the authors are aware of commercial desktop software that enhances the user experience for at least one of the platforms.

Out of the 17 products from eleven producers included in the survey, nine are free. Among the free ones, three allow users to store more than 1GB of data. Four of them allow users to upload both raster and vector data. Four platforms allow users to create temporal animations and three of them are free. Four platforms enable buffer analysis and three of these have a free option. Three platforms allow users to work with their data in the field without an internet connection. Two of them have a free option.

According to survey results, five platforms have an open source software backend. This means these systems make some use of open source software. However, only two of the systems surveyed, as far as the authors can confirm, are themselves open source-licensed. The pricing schemes available fall into three categories, including free options offered by nine platforms, incremental cost usage implemented by five, and fixed cost pricing for limited use employed by five. Data storage capacity varies, and only six GaaSs offer unlimited data storage capacities. All offer less than 100 MB and the majority (14) offer between 100.1 MB and 1 GB of storage space. Unlimited storage is offered by AGOL, GISCloud, Harvard WorldMap, iSpatial, Mango Map, and Map2Net.

Most platforms are strong in vector data handling but are weak in, or missing, raster data handling. This is a critical unmet need for humanity scholars as well as environmental scientists, among others. The authors do not see unsurpassable technical barriers in providing raster analytical functions in GSaaS, but are keenly aware of the financial implications due to the usually large size of raster datasets. It is the authors' belief that raster-based functionality will become more available in the priced platforms, but will remain missing in the free tier for most.

3.6 Development of a User's Guide

The GSaaS survey describes the current landscape. However, for users without a geospatial technology background, it remains a challenge to navigate through this landscape. To enable a user to explore many possible combinations of criteria, the authors developed a "GSaaS Platform Selection Tool" tool in Excel, which allows a user to choose from any possible combination of criteria and generate a list of systems (GSaaS Platform Survey Support Materials 2016).

The authors considered the option of rephrasing technical terms in the survey, (which is for the professional product managers), to be more understandable in the platform selection tool, which is for non-technical end users. We decided to minimize "translation" to maintain a clear relationship between the survey and the selection tool, however it would be easy to adjust the terms once we release the selection tool for public use and receive user comments. It is the authors' intention to continue to update, maintain, and improve the survey and the platform selection tool over time (GSaaS Platform Survey Support Materials 2016).

3.6.1 Examples of Use Cases for the "GSaaS Platform Selection" Tool

Suppose a humanist is interested in finding a platform that allows him/her to upload scanned and georeferenced historical maps, add photos with place tags to these maps, control visual transparency for the different layers in the map, and invite colleagues in another country to improve it. The user has no funding to pay for a system. The person could enter the following choices as shown in Fig. 3.2, and learn there is one product available.

Categories	Criteria (Click to Select)	Platforms Meeting Selected Criteria
Backend source code	na	Harvard WorldMap (free)
Cost to users	Free	
Data storage capacity	na	
User uploaded vector	na	
User uploaded raster	Limited file size for raster layers	
Online data creation	na	
Link multimedia	HTML pointing to multi-media	
Privacy control	Data and maps can be edited by selected users	
Find & add online layers	na	
Data curation	na	
Spatial analysis	na	
Visualization	Change layer transparency	
Cartographic editing	Change layer transparency	
Publishing	Display a vector layer's attribute table	
Data export	Online editing of symbology Automatically create chloropleth maps	
Mobile integration	Label features based on attributes Generate heat map from vector features	
API availability	Render temporal features or layers using a time bar	
Additional software	Other	
Support available	na	

Fig. 3.2 Use Case A: Upload raster layer and share editing

Categories	Criteria (Click to Select)	Platforms Meeting Selected Criteria
Backend source code	na	Carto DB
Cost to users	Fixed cost for unlimited usage	Mango Map
Data storage capacity	na	
User uploaded vector	na	
User uploaded raster	Limited file size for raster layers	
Online data creation	na	
Link multimedia	HTML pointing to multi-media	
Privacy control	Data and maps can be edited by selected users	
Find & add online layers	na	
Data curation	na	
Spatial analysis	na	
Visualization	Change layer transparency	
Cartographic editing	Allow adding graphic elements	
Publishing	Save a map view as an image file	
Data export		
Mobile integration		
API availability		
Additional software		
Support available	na	

na
Save the state of a map view via permalink
Generate an embed snippet
Save a map view as an image file
Print a map view to a printer/plotter
Other

Fig. 3.3 Use Case B: Upload raster layer, share editing, and add graphic element to map for publishing

Let us assume the user also needs to add graphic elements such as a scale bar and north arrow to the map and create an image file to insert into a book the humanist is writing. The user would enter that choice and see that there is no product available, so the user could choose to purchase a product, as shown in Fig. 3.3 and finds that there are two systems available.

Suppose a social scientist is interested in overlaying his/her own shapefiles with some base maps, geocoding some data tables containing columns of canonical geospatial regions such as zip or FIPS codes, creating choropleth maps quickly, inviting others to comment on the data and map, and embed this map into a personal blog. The scientist also needs to access the map on his/her mobile phone in the field, and prefers not to pay for the software. He/she will be happy to find that four systems are available, as shown in Fig. 3.4.

If the same scientist also needs to store more than 10GB of data in this mapping system, he/she will realize that there is no free product to use, but will need to pay for it, as shown in Fig. 3.5.

These cases are hypothetical, however, the authors believe the Excel tool is useful for providing basic guidance on what systems exist and what they do. When the choice has been narrowed down to a few candidates, the user may then refer to Appendix B to see what other features this subset of platforms offers.

Categories	Criteria (Click to Select)	Platforms Meeting Selected Criteria
Backend source code	na	AGOL (free)
Cost to users	Free	CartoDB (free)
Data storage capacity	na	GIS Cloud (free)
User uploaded vector	Limited number of vector features	Google Fusion Tables (free)
User uploaded raster	na	
Online data creation	Geocode table with regions	
Link multimedia	na	
Privacy control	na	
Find & add online layers	na	
Data curation	Allow users to comment on data or map	
Spatial analysis	na	
Visualization	Automatically create chloropleth maps	
Cartographic editing	na	
Publishing	Generate an embed snippet	
Data export	na	
Mobile integration	Reformat map view to fit mobile devices	
API availability		
Additional software		
Support available		

na
Reformat map view to fit mobile devices
Add data layers to a map view from a mobile device
Create or upload data when disconnected & sync later
Other

Fig. 3.4 Use Case C: Geocode table, create choropleth map, and generate embed snippet

Categories	Criteria (Click to Select)	Platforms Meeting Selected Criteria
Backend source code	na	Carto DB
Cost to users	Fixed cost for unlimited usage	
Data storage capacity	10.1 - 100GB	
User uploaded vector	Limited number of vector features	
User uploaded raster	na	
Online data creation	Geocode table with regions	
Link multimedia	na	
Privacy control	na	
Find & add online layers	na	
Data curation	Allow users to comment on data or map	
Spatial analysis	na	
Visualization	Automatically create chloropleth maps	
Cartographic editing	na	
Publishing	Generate an embed snippet	
Data export	na	
Mobile integration	Reformat map view to fit mobile devices	
API availability		
Additional software		
Support available		

na
Reformat map view to fit mobile devices
Add data layers to a map view from a mobile device
Create or upload data when disconnected & sync later
Other

Fig. 3.5 Use Case D: Geocode table, create choropleth map, generate embed snippet, and store over 10GB of data online

3.6.2 Limitations and Long-Term Plan

The authors are aware that technology reviews like this one run the risk of becoming out of date quickly, possibly even before the book is published. We hope to mitigate this risk through the following strategy:

- Invite readers to send us corrections and suggestions.
- Continue to make the survey available to system owners who would like their system to be included.
- Make results of the study available online and if improvements are made, issue new versions of the results.

3.7 Conclusions and Discussions

This study focuses on assessing systems that enable people without a technical background, to start working with their research materials geospatially. Using these online systems, users can create, curate, organize, and publish mapped information using a web browser. We focus on GSaaS platforms, because we see them as offering potential advantages over desktop systems, including ease of use, availability of base maps, such as high-resolution satellite imagery with global extent, digital publishing options, such as embeddable maps, the ability to share live views with remote collaborators, shared editing of large datasets that are not public, and others. The selected systems support basic GIS operations, do not require software installation, often provide ready-to-use data layers, and are available anywhere and anytime through a browser as long as one has internet access. Of course, for those people around the world, who still do not have good access to the Internet, desktop mapping systems will generally remain preferable.

We attempted to provide a snapshot of this group of systems. We attempted to include the major capabilities scholars require, such as data import, visualization, analysis, shared editing, data export, publishing, storage space, etc. Although each system we examined supports basic data manipulation functionality, there is a wide range in the way these functions are implemented.

Though none of the GSaaS platforms included is equivalent in analytic capability to advanced desktop GIS software packages, GSaaS capabilities are expanding rapidly, with many basic analytical capabilities available for vector data. However, raster data analysis has not been emphasized in most systems. Among the systems surveyed, only two of them offer online georeferencing of scanned maps, and only one provides map algebra functions for raster analysis.

The authors are aware that this survey has some important limitations. For example, it does not measure user friendliness, quality of cartography, accuracy of analysis, file format for input/output, capabilities for long-term data storage and archiving, or scalability. Many of these qualitative properties should be measured by a user survey or experiments based on user experience rather than a producer survey. We see the need for a consumer report on GSaaS by a group of independent and objective evaluators. This work is beyond the scope of this study.

Within the systems we surveyed, one can clearly see the legacy of desktop GISs in their design. If all computing is destined for the cloud, then perhaps GSaaS is a transitional paradigm, with desktop GIS as its reference point. It remains to be seen whether GSaaS will move forward and become a complete replacement of desktop GIS.

Where may GSaaS be heading? Let us consider GSaaS within the context of the evolution of the web. From this perspective GSaaS is following the broader IT trends in information sharing led by platforms like Facebook (text and photos), GitHub (software code), SoundCloud (sonic media), and other online systems in which anyone can create a login and have access to a full featured, cloud-hosted application for creating and publishing a particular type of content. GSaaS also contributes to and consumes content from seminal mass collaboration platforms such as Wikipedia, OpenStreetMap, and Wikimapia. These are systems where mass brainpower, what Clay Shirky calls 'cognitive surplus' (Shirky 2010), collaboratively develops and maintains complex public goods. From this perspective, GSaaS systems are a kind of simplified onramp to the geospatial web for many people.

The systems surveyed in this study support a step beyond collaboration, allowing anyone to create specialized systems that enable others to voluntarily share and collaborate around geospatial information in targeted ways for specific audiences. Why is this important? As we are seeing with other online media creation platforms, new human potential is released when access is better distributed. We see GSaaS as a pivot point in the evolution of GIS, pointing to a future still being imagined, but which will almost certainly include more people engaging with the creation and sharing of geospatial information.

Acknowledgement This study is partially funded by the U.S. National Endowment for the Humanities Digital Humanities Implementation Grants, Award No: HK-50091-13.

Appendices

Appendix A – Survey Questions

(An asterisk indicates a required question.)

1. Name of the platform *:
2. URL of the platform *:
3. Contact person's name:
4. Contact person's email:
5. Is this the free version or the highest tier of the non-free version? *

   (If the mapping platform you represent is sold in a number of tiers as many products are, please fill out the form for the highest tier of the platform, and the free tier (if there is one) separately. You are also welcome to enter information about the intermediate tiers – each as a separate entry.)

   a) Free version
   b) Highest tier
   c) Other:

6. Cost to users *

   a) Free
   b) Incremental cost by usage
   c) Fixed cost for unlimited usage
   d) Other:

7. Data storage capacity per user account *

   a) < 100 MB
   b) 100.1 MB - 1GB
   c) 1.1 - 10GB
   d) 10.1 - 100GB
   e) 100.1GB - 1 TB
   f) Unlimited

8. User uploaded vector data capacity *

   a) No vector data upload enabled
   b) Limited number of vector features
   c) Limited file size of vector layers
   d) Unlimited

9. User uploaded raster data capacity *

   a) No raster data upload enabled
   b) Limited file size for raster layers
   c) Unlimited

10. Online data creation capabilities

    a) Digitize vector features
    b) Geocode address list
    c) Georeference raster image
    d) Geocode table with a column of canonical geospatial regions such as zip or FIPS codes
    e) Other:

11. Ability to link multimedia content to geographic features

    a) HTML pointing to multi-media contents (such as photos, video, maps) on web
    b) Multi-media documents stored on server
    c) Other:

12. Privacy control options *

    a) Data and maps are always viewed publicly
    b) Data and maps can be viewed by author only
    c) Data and maps can be viewed by selected users
    d) Data and maps can be edited by selected users
    e) Other:

13. Finding and adding online layers to a map view
    a) Layers within the same platform only
    b) WMS layers from anywhere
    c) Esri REST layers from anywhere
    d) WMTS layers from anywhere
    e) GeoRSS feeds from anywhere
    f) Other:
14. Data curation capabilities
    a) Create metadata online
    b) Upload metadata files
    c) Share metadata editing with selected users
    d) Allow users to rate data quality
    e) Allow users to comment on data layers or map views
    f) Other:
15. Spatial analysis capabilities
    a) Overlay analysis with vector layers
    b) Buffer analysis with vector layers
    c) Network routing
    d) Set attribute query filter on vector layers
    e) Map algebra on raster layers
    f) Other:
16. Visualization capabilities
    a) Change layer transparency
    b) Display a vector layer's attribute table
    c) Online editing of symbology
    d) Automatically create choropleth maps
    e) Label features based on attributes
    f) Generate heat map from vector features
    g) Render temporal features or layers using a time bar
    h) Other:
17. Cartographic editing options
    a) Print map view as it is, no layout editing
    b) Allow adding graphic elements such as legend, scale bar, north arrow, map title, etc.
    c) Advanced cartographic editing for professional publishing
18. Publishing capabilities
    a) Save the state of a map view via permalink
    b) Generate an embed snippet to bring a live iframe of map view into another application

c) Save a map view as a PDF file
d) Print a map view to a printer/plotter
e) Other:

19. Data export options

a) Allow export to file of vector data layers
b) Allow export to file of raster data layers
c) Allow a data layer to be used as a map service by another GIS system
d) Other:

20. Mobile integration options

a) Reformat map view to fit mobile devices
b) Add data layers to a map view from a mobile device
c) Create data or upload data to a map view from a mobile device without network connection and sync later
d) Other:

21. Is additional commercial software required to take full advantage of the system? *

(i.e. in order to import or export data or to perform analysis)

a) Yes
b) No

22. If additional commercial software is required to take full advantage of the system, please list any required applications:

23. API availability for developers *

a) Yes
b) No

24. Support available *

a) Commercial support from you, the SaaS provider
b) Commercial support from a 3rd party provider
c) Community support
d) Other:

25. Type of license (if any) for backend software package used to provide the service *

a) Open source
b) Commercial

Appendix B – Summary of Survey Responses

Categories	Name of the platform	AGOL (free)	AGOL	Bing Maps (free)	CartoDB (free)	Carto DB	eSpatial (free)	eSpatial	GIS Cloud (free)	GIS Cloud	Google Fusion Tables (free)	Google MyMaps (free)	Harvard WorldMap (free)	InstaGIS	iSpatial	Mango Map (free)	Mango Map	Map2Net
Backend source code	Commercial	x	x	x			x	x	x	x	x	x			x			
	Open source				x	x							x	x		x	x	x
Cost to users	Free	x		x	x		x		x		x	x	x			x		
	Incremental cost by usage		x					x		x				x	x			
	Fixed cost for < 10k requests/day																	
	Fixed cost for 10k–50k requests																	
	Fixed cost for 51k–150k requests																	
	Fixed cost for unlimited usage					x				x					x		x	x
Data storage capacity	< 100MB	x	x	x	x	x	x	x	x	x	x	x	x	x	x	x	x	x
	100.1MB–1GB	x	x			x		x	x	x	x	x	x	x	x	x	x	x
	1.1–10GB	x	x			x		x		x		x	x	x	x		x	x
	10.1–100GB		x			x				x		x	x		x		x	x
	100.1GB–1TB		x			x				x			x		x		x	x
	Unlimited		x							x			x		x		x	x

User uploaded vector	No vector data upload enabled																	
	Limited number of vector features	x		x	x	x	x	x	x		x	x		x			x	
	Limited file size of vector layers			x	x						x	x	x			x		
	Unlimited		x							x					x			x
User uploaded raster	No raster data upload enabled			x			x	x			x	x		x	x	x		x
	Limited file size for raster layers	x			x	x			x				x				x	
	Unlimited		x							x								
Online data creation	Digitize vector features	x	x	x	x	x			x	x		x	x	x	x		x	
	Geocode address list	x	x		x	x	x	x	x	x	x			x	x			
	Georeference raster image												x					
	Geocode table with regions	x	x		x	x			x	x	x			x				x
	Other	x							x	x				x				

(continued)

(continued)

Categories	Name of the platform	AGOL (free)	AGOL	Bing Maps (free)	CartoDB (free)	Carto DB	eSpatial (free)	eSpatial	GIS Cloud (free)	GIS Cloud	Google Fusion Tables (free)	Google MyMaps (free)	Harvard WorldMap (free)	InstaGIS	iSpatial	Mango Map (free)	Mango Map	Map2Net
Link multimedia	HTML pointing to multi-media	x	x	x	x	x			x	x	x	x	x	x	x	x	x	
	Multi-media documents stored on server	x	x						x	x				x	x	x	x	
	Other																	
Privacy control	Data and maps are always viewed publicly	x	x	x	x	x	x	x	x	x	x	x	x	x	x	x	x	x
	Data and maps can be viewed by author only	x	x	x		x	x	x		x	x	x	x	x	x	x	x	x
	Data and maps can be viewed by selected users					x		x		x	x	x	x	x	x		x	
	Data and maps can be edited by selected users					x		x		x	x	x	x	x			x	
	Other	x	x							x				x				x

Find & add online layers	Layers within the same platform only	x	x		x	x	x	x			x	x	x		x			x
	WMS layers from anywhere	x	x		x	x			x	x			x		x		x	
	Esri REST layers from anywhere	x	x										x					
	WMTS layers from anywhere				x	x			x	x					x			
	GeoRSS feeds from anywhere	x	x	x					x	x			x	x				
	Other	x	x							x				x				
Data curation	Create metadata online	x	x										x	x		x	x	
	Upload metadata files	x	x											x				
	Share metadata editing with selected users	x	x										x	x				
	Allow users to rate data quality	x	x										x					
	Allow users to comment on data or map	x	x		x	x			x	x	x	x	x					
	Other									x								

(continued)

(continued)

Categories	Name of the platform	AGOL (free)	AGOL	Bing Maps (free)	CartoDB (free)	Carto DB	eSpatial (free)	eSpatial	GIS Cloud (free)	GIS Cloud	Google Fusion Tables (free)	Google MyMaps (free)	Harvard WorldMap (free)	InstaGIS	iSpatial	Mango Map (free)	Mango Map	Map2Net
Spatial analysis	Overlay analysis with vector layers	x	x		x	x	x	x	x	x				x			x	
	Buffer analysis with vector layers	x	x		x	x			x	x							x	
	Network routing	x	x				x	x			x	x		x				
	Set attribute query filter on vector layers	x	x		x	x	x	x	x	x	x	x	x	x			x	x
	Map algebra on raster layers				x	x												
	Other	x																
Visualization	Change layer transparency	x	x	x	x	x	x	x	x	x	x	x	x	x	x	x	x	x
	Display a vector layer's attribute table	x	x		x	x	x	x	x	x	x	x		x		x	x	x
	Online editing of symbology	x	x	x	x	x	x	x	x	x	x	x	x		x	x	x	x
	Automatically create chloropleth maps	x	x		x	x	x	x	x	x	x	x	x	x		x	x	x

	Label features based on attributes	x	x		x	x	x	x	x	x		x	x	x		x	x	x
	Generate heat map from vector features	x	x		x	x	x	x	x	x	x			x	x			
	Render temporal features or layers using a time bar	x	x		x	x			x	x					x			
	Other	x	x							x				x				
Cartographic editing	Print map view as it is, no layout editing	x	x	x	x	x	x	x	x	x	x	x	x	x	x	x	x	x
	Allow adding graphic elements				x	x		x	x	x							x	x
	Advanced cartographic editing				x													
Publishing	Save the state of a map view via permalink		x	x	x	x	x	x	x	x	x	x	x	x		x	x	x
	Generate an embed snippet	x	x	x	x	x	x	x	x	x	x	x	x	x		x	x	
	Save a map view as an image file	x	x		x	x	x	x	x	x		x		x			x	

(continued)

(continued)

Categories	Name of the platform	AGOL (free)	AGOL	Bing Maps (free)	CartoDB (free)	Carto DB	eSpatial (free)	eSpatial	GIS Cloud (free)	GIS Cloud	Google Fusion Tables (free)	Google MyMaps (free)	Harvard WorldMap (free)	InstaGIS	iSpatial	Mango Map (free)	Mango Map	Map2Net
	Print a map view to a printer/plotter	x	x	x	x		x	x	x	x	x	x	x				x	x
	Other	x	x			x			x	x								
Data export	Allow export to file of vector data layers	x	x	x	x	x	x	x	x	x	x	x	x	x	x	x	x	
	Allow export to file of raster data layers	x	x						x	x			x				x	
	Allow a data layer to be used as a map service	x	x		x	x			x	x			x	x				
	Other	x	x							x								
Mobile integration	Reformat map view to fit mobile devices	x	x	x	x	x			x	x	x	x		x	x	x	x	
	Add data layers to a map view from a mobile device	x	x		x				x	x	x	x			x			

	Create or upload data when disconnected & sync later	x	x						x	x					x			
	Other																	
API availability	Yes	x	x	x	x	x			x	x	x	x	x	x	x			
	No						x	x								x	x	x
Additional software	Yes																	
	No	x	x	x	x	x	x	x	x	x	x	x	x	x	x	x	x	x
Support available	Commercial support from you, the SaaS provider	x	x		x	x		x	x	x				x	x	x	x	x
	Commercial support from a 3rd party provider		x		x	x			x	x	x	x						
	Community support	x	x	x	x	x	x	x	x	x	x	x	x					
	Other																	

Appendix C – Survey form "Other" Fields Results

ArcGIS Online (free)

- Online data creation capabilities (other): Feature layers, tile layers
- Privacy control options (other): Sharing and security are controlled by the owner for each map or layer and can be private, shared to a defined group, to the person's org or the public.
- Finding and adding online layers to a map view (other): Many other layers: http://doc.arcgis.com/en/arcgis-online/create-maps/add-layers.htm
- Spatial analysis capabilities (other): Many more robust capabilities for analysis: http://www.arcgis.com/features/features-analytics.html
- Visualization capabilities (other): Many other visualization capabilities including smart mapping: http://www.arcgis.com/features/visualization.html
- Publishing capabilities (other): App templates including story maps and many other purpose-built templates
- Data export options (other): Many other export capabilities: http://doc.arcgis.com/en/arcgis-online/use-maps/extract-data.htm

CartoDB

- Publishing capabilities (other): Access via API

GIS Cloud

- Online data creation capabilities (other): Mobile data collection with smartphones and tablets, import spreadsheet with coordinates
- Privacy control options (other): Separate permission roles for layers in the same map, data collection permission levels
- Finding and adding online layers to a map view (other): TMS, WFS
- Data curation capabilities (other): Google like labeling system in the works
- Visualization capabilities (other): The fastest map engine available today that allows you to render big maps with millions of features. Literally no one can match our speed and performance
- Publishing capabilities (other): Share with other users who can then access the map on their smartphone or tablet
- Data export options (other): Custom reports

GIS Cloud (free)

- Online data creation capabilities (other): Mobile data collection with smartphones and tablets, import spreadsheet with coordinates
- Publishing capabilities (other): Share with other users who can then access the map on their smartphone or tablet

Map2Net

- Privacy control options (other): The author can share his data with other users if he wants to

References

Arleth M (1999) Problems in screen map design. In: Proceedings of the 19th international cartographic conference, Ottawa, Canada, vol 1, pp 849–857

Castelli A, Rosi A, Mamei M, Zambonelli F (2006) Ubiquitous browsing of the world, Chapter 7 in the book titled The Geospatial Web. Springer, Verlag

Columbus L (2015) Roundup of cloud computing forecasts and market estimates. Forbes, http://www.forbes.com/sites/louiscolumbus/2015/01/24/roundup-of-cloud-computing-forecasts-and-market-estimates-2015

Dragićević S (2004) The potential of Web-based GIS. J Geogr Syst 6:79–81

DuVander A (2010) 5 years ago today the web mashup was born. http://www.programmableweb.com/news/5-years-ago-today-web-mashup-was-born/2010/04/08

Gartner G (2009) Applying Web Mapping 2.0 to Cartographic Heritage e-Perimetron, Vol. 4, No. 4, 2009 [234–239]

Goodchild MF (2007a) Citizens as voluntary sensors: spatial data infrastructure in the world of Web 2.0. Int J Spat Data Infrastruct Re 2:24–32

Goodchild MF (2007b) Citizens as sensors: the world of volunteered geography. GeoJournal 69(4):211–221. Reprinted in Dodge M, Kitchin R, Perkins C (eds) The map reader: theories of mapping practice and cartographic representation. Wiley, Hoboken, p 370–378. [441]

Goodchild MF (2007c) Citizens as sensors: Web 2.0 and the volunteering of geographic information. Geofocus 7:8–10. [439]

Goodchild MF (2007d) Citizens as voluntary sensors: spatial data infrastructure in the world of Web 2.0. Int J Spat Data Infrastruct Res 2:24–32. [437]

Google Maps (2015) In Wikipedia. Retrieved from 15 Sept 2015. https://en.wikipedia.org/wiki/Google_Maps

GSaaS Platform Survey Support Materials (2016) http://www.gis.harvard.edu/tools/software/gsaas-platform-survey-support-materials

High P (2014) Gartner: top 10 strategic IT trends for 2015. Forbes, http://www.forbes.com/sites/peterhigh/2014/10/07/gartner-top-10-strategic-it-trends-for-2015/

Hutcheon S (2015) The untold story about the founding of google maps. Medium.com. https://medium.com/@lewgus/the-untold-story-about-the-founding-of-google-maps-e4a5430aec92

Kraak M-J, Brown A (2001) Web cartography: developments and prospects. Taylor & Francis, London

MacEachren AM (2001) Cartography and GIS: extending collaborative tools to support virtual teams. Prog Hum Geogr 25(3):431–444

O'Reilly T (2005) What is Web 2.0 – Design patterns and business models for the next generation of software. O'Reilly Media, Inc., Sebastol

Shirky C (2010) Cognitive surplus: creativity and generosity in a connected age. GIA Reader 21(3) (Fall 2010) 216 pages, Penguin Press, New York/London

Slippy Map (2015) In OpenStreetMap Wiki. Retrieved from 15 Sept 2015. http://wiki.openstreetmap.org/wiki/Slippy_Map

Tiled Web Map (2015) In Wikipedia. Retrieved from 15 Sept 2015. https://en.wikipedia.org/wiki/Tiled_web_map

van Elzakker C (2001) Users of maps on the web. In: Kraak MJ, Brown A (eds) Web cartography. Taylor and Francis, London, pp 37–52

Chapter 4
Big Geo-Data Handling Based on Parallel and Distributed System's Strategies

E. Stylianidis, I. Kapouranis, and E. Valari

Abstract Nowadays, information handling is among the biggest scientific challenges. Information is freely offered in great amounts almost everywhere, through internet and other easily accessible sources, constituting an enormous pool of available data. The process of exploiting available data is not an easy task, since it involves not only finding the proper means to do so, but it also needs to be done in reasonable time. With the continuously increasing number of satellite images, geo-data is progressively increasing both in quantity and in resolution, thus, there is a need of faster and more robust techniques of processing them. Exploiting geo-information is of paramount importance, as it comprises the source of many useful applications such as mapping, property bordering, area discovering, and land use, as well as other geospatial applications. While the presented techniques can be implemented for a variety of big data problems, our study apply them in map update, proposing a methodology that successfully tackles the big data problem by taking advantage of the processing power of many units simultaneously. The main goal of this work is to initiate distributed and parallel techniques in mapping systems by implementing efficient map updating algorithms based on road network extraction methodology on satellite/aerial images. In particular, the proposed algorithms are used to update digital maps, exploiting the distributed and parallel systems' capabilities while applying various single-machine techniques to greatly increase the overall performance. The performance results confirm our initial assumption that a parallel and distributed system can significantly reduce the processing time of big data images, without any loss of output quality, having the necessary robustness of a market-ready product. More specifically, our study shows that even though the base algorithm (without any optimizations) is used, by adding more processing nodes the computation time can even be reduced to one third of the single machine processing time.

E. Stylianidis (✉) • E. Valari
Faculty of Engineering, School of Spatial Planning & Development, Aristotle University of Thessaloniki, Thessaloniki, Greece
e-mail: sstyl@auth.gr; evalarig@auth.gr

I. Kapouranis
Department of Informatics, Aristotle University of Thessaloniki, Thessaloniki, Greece
e-mail: kapouranis.ilias@gmail.com

M. Leitner, J. Jokar Arsanjani (eds.), *Citizen Empowered Mapping*,
Geotechnologies and the Environment 18, DOI 10.1007/978-3-319-51629-5_4

Keywords Big geo-data • Satellite images • Map updating • Image processing • Parallel and distributed systems

4.1 Introduction

Satellite images are one of the main sources for Geographical Information Systems (GIS) and despite the fact that technological advancements provide high resolution, the issue of handling them is still open for research. The processing of satellite images can provide valuable geospatial information. According to the published report of Euroconsults, "Satellite-Based Earth Observation: Market Prospects to 2023", 353 Earth Observation (EO) satellites are expected to be launched over the next decade, more than double of those launched from 2004 and 2013 (http://www.euroconsult-ec.com/). As the production of satellite images grows, the need of compact and robust algorithms that are able to effectively process them in reasonable duration also increases. In this study, we introduce a methodology for handling big geo-data in a fast and robust way with the aid of parallel and distributed systems. While the methods and techniques described in the following refer to geo-data, they can also be applied to a variety of application areas with heavy computational tasks, such as in agriculture, geology, forestry, biodiversity conservation, regional planning, certain areas in medical imaging, etc.

In the last two centuries, we have witnessed drastic population changes from rural to urban areas. With the systemic growth of urban areas, the continuous changes of the road network and urban sprawl, navigation and mapping systems need continuous update. One of the big geo-data handling applications is the map update, which can be performed in different ways. With this book chapter, we propose a method for fast map update having satellite, or other source (e.g. aerial) images as the only input. In general, this procedure needs to use expensive equipment. However, the advances in computer science and the availability of low cost equipment for digitizing and digital processing of aerial images as well as satellite images, open a wide range of applications for the production of geo-information. Geo-data is handled via image processing techniques. Image processing is a very complicated research field, since images can be of various sizes, resolutions, and content, while they may also contain noise of any kind that hinders processing. Pattern recognition and classification are just two of the many important issues that are addressed through image processing (Wilhelm and Burge 2008; Bernd 2002; Acharya and Ray 2005). Despite the fact that any image processing technique is applicable to satellite images, their vast size makes them very hard to be processed, as a simple "run" of an algorithm can last for hours, or even days. This is where distributed systems step in.

A distributed system is a software, the components of which are in the same network and communicate with each other through message passaging, to achieve the same goal (Coulouris et al. 2011). Until now, the term "distributed" in conjunction with GIS has only been used for "distributed storage". Distributed storage, also known as distributed databases or distributed warehousing, is a network

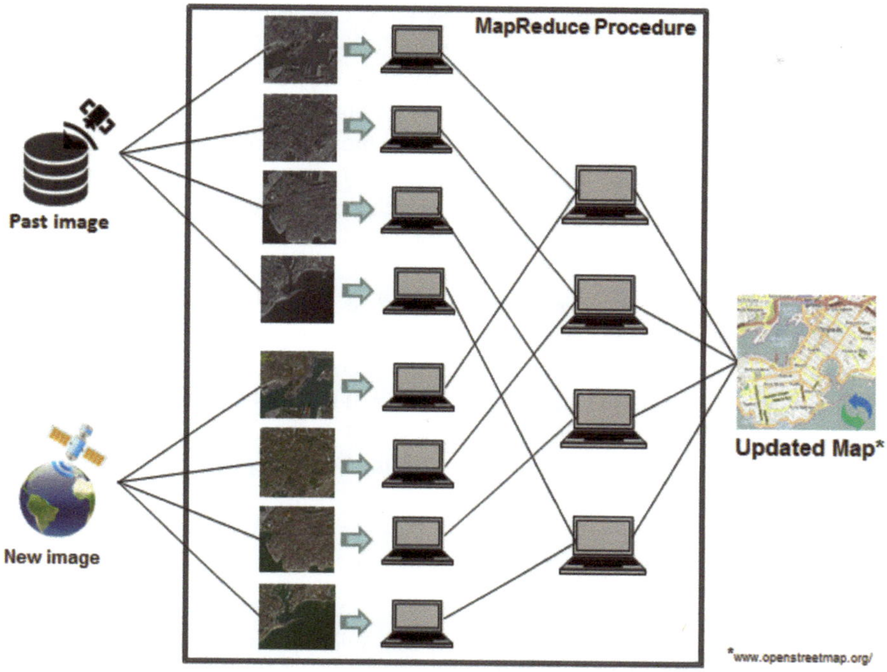

Fig. 4.1 Basic architecture of the MapReduce procedure

of computers which shares parts of a huge database that a single computer is not sufficient to handle. The structure offers transparency, giving the feeling that a single machine database is used.

However, distributed databases are not designed for heavy processing, but only for data retrieval and query support. In addition, algorithms related to geo-data are currently applied on single machines and only a minority of them is taking advantage of distributed computing (Hawick et al. 2003).

The proposed approach, as illustrated in Fig. 4.1, exploits the advantages of distributed systems by using a large number of computers, each with multiple processors. The core idea is based on the "divide and conquer" logic and the framework developed by Google, MapReduce (Dean and Ghemawat 2008). The initial input is divided into pieces and each piece is sent over the network to a computer that is responsible for its processing. After every machine has finished processing, the output is aggregated to form the final result. This method can be effectively applied to image processing, having every high resolution image divided into parts and every part of it is separately processed by different computers/processors. For each piece, a feature vector containing the information about road network, buildings, or any other class of data is extracted. After all feature vectors have been extracted, they are unified into one image containing all the information parsed.

The distributed methods implemented accelerate the processing of satellite images, leading to easier map update of existing digital maps, enriching navigation systems, and keeping the urban space view up to date. More specifically, in order to achieve a map update, the current image of the landscape is employed, as well as its (past) counterpart which had been used to create the existing digital map. After ensuring that these images correspond to exactly the same area, a feature vector is extracted for each one and those vectors are compared with certain similarity measures to identify landscape changes. According to landscape alterations detected, a threshold criterion aids the algorithm to decide whether the map area needs to be updated or reconstructed from the beginning, so that the map update procedure is optimized.

The rest of this book chapter is organized as follows: In Sect. 4.2, a brief description of the distributed systems is presented. In Sect. 4.3, there is an elaborated research on related works that were taken into account for this study. In Sect. 4.4, the structure of the proposed technique is explained in detail and the results from its implementation in a map update case are presented. In Sect. 4.5, conclusions of the current work and a future work plan is closing the book chapter.

4.2 Distributed Systems

Distributed computing is a field of computer science which studies distributed systems. As mentioned before the components of a distributed system are in the same network and communicate with each other through message passaging (O'Brien and Marakas 2008). Three main characteristics of distributed systems are: (i) The synchronization of the components, (ii) the lack of a global clock, and (iii) the independent failure of components. The last characteristic means that an error or failure on a specific machine does not affect the others.

A computer program that runs on a distributed system is called a distributed program and distributed programming is the process of writing these programs (Andrews 2000). There are many alternatives for the message exchange mechanism, such as remote procedure call (RPC) connections and message queues. As for the message queues, each computer saves in a local queue all messages that it receives from the rest. Every time a new message comes, or per interval, the queue is checked and depending on the message type the appropriate action is taken.

A significant goal and challenge of distributed systems is location transparency. Components of a distributed system are connected in the same network in order to communicate. The concept of the network is not limited in a local house network or a database warehouse with thousands of computers. A distributed environment can exist with computers in different cities or continents, provided that they can communicate. By location transparency, the programmer or the end user has no knowledge about the exact location of the computer used, only that it is in the network and available to be used.

The term distributed computing also refers to the use of distributed systems to solve computational problems. In distributed computing, a problem is divided into several parts, each of which is solved by one or more computers that communicate with each other by exchanging messages.

4.2.1 Distributed Environment Architectures

Several hardware and software architectures are used for distributed computing. The system architecture is divided into two basic levels, namely the low level, which is the hardware level and the high level, which contains the programming interfaces used. At the low level, machines used need to be connected via a network so that communication is established and at a higher level, processes running on these machines are required to interconnect with a communication system. Distributed programming structure can vary based on many architectures or categories:

- **Client-server**: Code running on the computer of the client requests data from the server and displays them appropriately to the user. The data input to the client affects the elements found on the server, only if it is a permanent change.
- **3-tier architecture**: The 3-tier architecture introduces a new level, which is the database server. The communication is as follows: The client only communicates with the server that manages user requests and the client then communicates with the database server for any data access.
- **Peer-to-peer**: An architecture where there is no special machine or machines that provide a service or manage the network resources. Instead, all responsibilities are evenly distributed among all machines, known as peers. Peers can serve both as clients and servers.
- **Tight coupling**: This will normally be a cluster of machines which work closely together to run a common procedure in parallel. The work is divided into sections which are run individually by a machine and then joined together again to generate the final result.

The method of communication between concurrent processes is another key aspect of the distributed environment architecture. Usually, through messaging protocols, processes communicate directly with each other in a master/slave relationship. The master process is managing the cluster and dividing the work among the slaves. Alternatively, using a common database can enable distributed systems to communicate without server interconnection (Lind and Alm 2006).

4.2.2 Hadoop Programming Environment

Hadoop is one of the most widely used open source programming frameworks in a distributed environment. It is used to process huge data volumes, which could not be handled by a single machine. The idea came from a published work of Google and the open source implementation started by Cafarella and Cutting (2004) – project Nutch at the time (https://en.wikipedia.org/wiki/Apache_Hadoop#History). Hadoop is using the master/slave architecture and differs from the other programming environments in:

- **Simplicity**: Hadoop offers an application programming interface (API) which is a set of classes and functions already implemented for convenience during programming. The development of a program is easy and fast because programmers deal with the infrastructure itself. In addition, providing a ready API greatly reduces errors that can arise because the supplied code is tested and guaranteed to work properly with standards set.

- **Robustness**: It is designed to run on machines that can be used by everyone. Despite the diversity and the low reliability offered by low-end machines, it has been implemented in such a way as to address problems that could be caused by machines and continue normal operation without being perceived (mostly) by the user. Some problems that can arise include: a machine shuts down due to a problem in its system, data congestion, and memory overflows. Thus, the machine does not respond, network congestion and proper communication cannot be established, etc.
- **Scaling**: The data processing in Hadoop results in almost linear reduction of processing time compared to the time needed on a single machine. We use the expression "almost linear reduction" and not just "linear", because there is always a cost synchronization and communication (overhead) which increases the total time. However, the additional overhead time is insignificant compared with the time reduction. Moreover, new nodes can be added to the cluster without having to change the data representation, i.e. how data is loaded and the execution manner of processes or the code of the programmer.
- **Flexibility**: Hadoop has no standard for the data it receives, thus, it can process any type of data, structured or not, from any source. Data from multiple sources can be combined with many methods allowing better and more in-depth analysis.

Hadoop is divided into two main parts, each of which performs separate functions. These are the Hadoop Distributed File System (HDFS) and MapReduce.

- **HDFS**: It is used for storing data in a cluster. It undertakes the registration of files, how many copies exist and where each copy resides. It is responsible to provide every requested file without any problem.
- **MapReduce**: MapReduce is the environment that manages processes, on which machines they will run, and their synchronization features.

4.2.3 How MapReduce Works

The MapReduce methodology is based on divide-and-conquer methods mostly used by parallel and distributed architectures. As the term indicates, the model relies on two important phases, the Map and the Reduce and the mapper and reducer tasks, which take care of the processing at each phase, respectively. The data in MapReduce are represented as key-value pairs <*K*, *V*>. Mappers accept as input and emit as output <*K*, *V*> pairs. Similarly, reducers accept as input the same structure and output, also key value pairs.

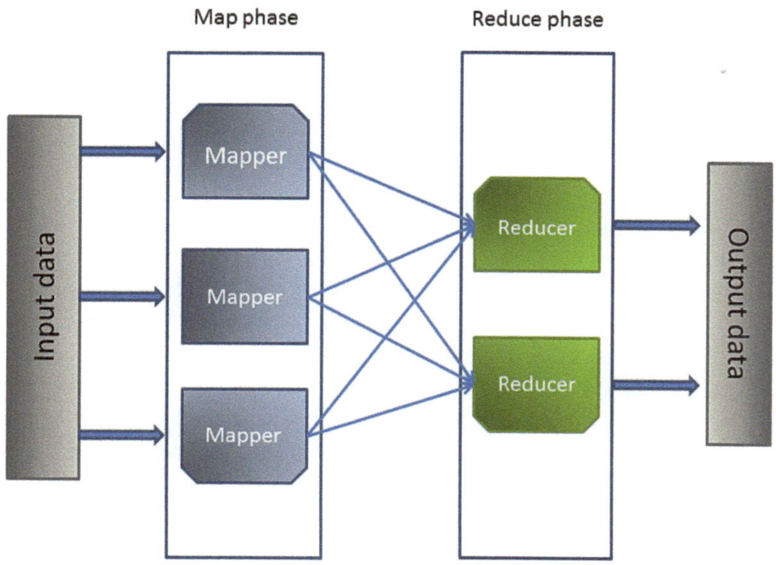

Fig. 4.2 Map and reduce phase

Before the Map phase, the input is split, based on user preferences or the system default, to the mappers in order to initialize the procedure. The main function of each mapper is called to process the input $<K, V>$ pairs and produce the output that will be sent to reducers, for example $<K_1, V_1>$, $<K_2,V_1>$, $<K_1, V_2>$, ..., $<K_n, V_m>$.

After every mapper task has finished processing and emitted its output, the $<K, V>$ pairs are sent to reducers for final computations. Every pair that has the same key value K is sent to the same reducer. This way, if there are data chunks of the same entity in different mappers, they can be aggregated during the reduce phase to produce the final result. For example, reducer #1 will take as input the pairs $<K_1, V_1>$ and $<K_1, V_2>$, reducer #2 will take as input $<K_2, V_1>$, etc. When a reducer finishes its processing, the output is written into the HDFS until all reducers have finished and the MapReduce process is complete. Figure 4.2 presents the described process in a graphical way.

4.3 Related Work

The aim of this work is to introduce distributed computing in topographical systems by implementing a simple map update pipeline in road network extraction from aerial and satellite imagery. Map updating and road network extraction is a challenge for Information and Communications Technology and the geospatial domain, based on the traditional classification algorithms, mainly due to the huge size of aerial and satellite imagery.

4.3.1 Map Update

The detection and definition of the road network from aerial or satellite images for map update is always an intriguing issue. Digital road information is the prerequisite for a wide variety of applications (Li et al. 2003), especially in cases where GIS data exists for illustrating the past road network condition and new data are necessary for making maps up to date. In general, map update methods can be summarized in two main categories, the automatic and the semi-automatic.

The research on automated feature extraction from aerial and satellite images in the last few years has been performed due to the need for data acquisition, updating, and information correction for various GIS applications. Grote and Heipke (2008) dealt with road extraction in suburban areas from high resolution aerial images for updating a road database. They suggested a region-based approach as a road extraction algorithm. The cartographic data extraction from aerial or satellite imagery is a very important application in geospatial sciences. In fact, the automatic extraction of objects, such as road networks, from digital images is both scientifically challenging and of high significance for data acquisition and GIS databases update. The scientific community is well aware that this is a big issue, which has mostly remained unsolved.

Over the last decades, various attempts both for automatic and semi-automatic road network extraction have taken place, having as main input either aerial or satellite images. Gruen and Li (1995, 1997) introduced dynamic programming and least-squares B-splines snakes for linear features extraction, such as road networks. Hu et al. (2000) proposed a semi-automatic road extraction scheme that is based on template matching and optimization using a Hopfield neural network. Bong et al. (2009) suggested a hybrid simple color space segmentation and edge detection (Hybrid SCSS-EDGE) algorithm to extract roads automatically from satellite imagery. Bacher and Mayer (2005) developed an approach for automatic road extraction from high resolution multispectral imagery, such as IKONOS or QUICKBIRD, in rural areas. Gecen and Sarp (2008) proposed an automated road extraction technique that was applied to four different satellite images (SPOT, IKONOS, QUICKBIRD, ASTER) with different resolutions. Zhang and Couloinger (2006) suggested a new approach for road network extraction from multispectral imagery by using a spectral clustering algorithm. Results from all these studies show that the attempt to detect the road network from high resolution images has not always been successful, so far. In cases where the study area contains structures, like trees and vehicles, the road extraction is a bit more problematic. In fact, tree line bounding and shadows of high buildings are the main obstacles regarding road network extraction.

4.3.2 Road Network Extraction

There are several proposed techniques facing the road extraction challenge. This study proved that by combining spatial and spectral information, the amount of overlap between classes can be decreased, providing higher classification efficiency and more accurate urban land cover maps. In another interesting approach, Mnih and Hinton (2010) studied the effect of the resolution on the automatically extracted roads individually on each satellite image. The accuracy of the generated results was tested with GIS data layers that represent reality.

There are many studies based on Artificial Neural Networks (ANNs). Hongbin et al. (2008) used pre-treatment which was done by color hue, saturation, and value (HSV) transformation followed by a Support Vector Machine (SVM) to extract the road skeleton and used the seed growth algorithm to extract the median road line that has certain length and direction. In addition, mathematical morphology is used in order to assist the seed growing method for the detailed road information extraction. Reis et al. (2014) proposed road detection using a neural network with millions of trainable weights that looks at a much larger context than previous attempts at learning the task. The neural network is trained on massive amounts of data using a consumer graphics processing unit (GPU). This method demonstrates that predictive performance can be substantially improved by initializing feature detectors using recently developed unsupervised learning methods, as well as by taking advantage of the local spatial coherence of output labels.

Regarding the map update, there are a lot of new ideas and new approaches promoted in this area. There are three main approaches: (i) Road maps could be updated by ground surveying, either by using traditional methods (e.g., total station or GPS), or by using a more automatic method (e.g., mobile mapping system). This approach is unrelated to our work. The other two approaches are: (ii) Road network extraction and changes detection based on new remotely-sensed images. This technology has been widely researched for many years. Although there are few successful fully automated techniques, there are many partially automated feature extraction techniques available to detect road network changes (Hinz and Baumgartner 2003). The last approach (iii) requires the use of a more recent map to update an old road map. By feature matching, the unchanged and changed roads can be determined during the mapping time interval. This is an ad-hoc technology to maintain a road network database. In an assessment paper, Zarrinpanjeh et al. (2013), proposed a new framework for road map updating from remotely-sensed data. Three main computational entities of an ant-agent, seed extractor, and algorithm library are designed and road map updating is performed through three main stages of verification of the old map, extraction of possible roads, and results' grouping of both stages.

4.3.3 Distributed Systems

There are some papers discussing spatial problems using distributed techniques. Di (2004) presents an approach and a system for automated information extraction and knowledge discovery under the interoperable distributed web service framework. Zhang et al. (2007) present the design and implementation of a Distributed Virtual

Geographic Environment (DVGE) system. The DVGE system is an Internet based virtual 2D and 3D environment that provides users with a shared space and a collaborative platform for publishing multidimensional geo-data, and simulating and analyzing complex geo-phenomena. Furthermore, this work analyzes grid services, Open Geo-data Interoperability Specifications (OpenGIS), and Geography Markup Languages (GML). Hofmann (1999) introduces a Multi-Tier Framework for accessing distributed, heterogeneous spatial data in a federation based Environmental Information System (EIS). In particular, this work deals with the EIS redesign in order to handle heterogeneous and distributed data in various geospatial formats. Components of that framework provide a system and data abstraction that uses concepts of the upcoming OpenGIS standardization efforts. Finally, an important goal of this framework is that all GIS components can be easily integrated into network-based information systems architectures.

Tait (2005) describes the implementation of geoportals for distributed GIS. Geoportals is a key application of distributed GIS which provide a gateway to discover and access geographic web services. In this work, four geo-portal projects are presented that help to define distributed GIS and illustrate the challenges to be encountered in order to achieve the goal of wider GIS usage. Jhummarwala et al. (2014) give an overview of the parallel and distributed GI systems. In particular, this work explains that the focus on the development should be shifted from traditional GIS to parallel and distributed GIS as the traditional GI systems have become quite mature and saturated, while technologies such as MPI (Message Passing Interface) and GPGPUs (General Purpose Graphics Processing Units) can be readily utilized for faster geo-data processing.

Finally, Eldaw and Mokbel (2015) introduce SpatialHadoop which is a full-edged MapReduce framework with native support for spatial data. SpatialHadoop is a comprehensive extension to Hadoop that injects spatial data awareness in each Hadoop layer, specifically, the language, storage, MapReduce, and operations layers. This work offers an extensive experimental evaluation of real datasets which shows that SpatialHadoop achieves much better performance than Hadoop for spatial data processing. In our case, the use of Hadoop is more suitable than SpatialHadoop. The main reason is that we handle satellite-aerial images as simple images without taking into consideration the geo-localization which carries each one of them.

4.4 A New Approach: Distributed and Parallel Map Update Techniques

In this section, we present some fundamental concepts related to map update techniques in a distributed and parallel environment.

Table 4.1 Symbols used in this book chapter

Symbol	Interpretation
$S_{img(t)}$	The set of aerial images
t	Time period of an image
DiffList	Set of road network differences between a pair of images
$imgPartName_t$	The filename of the image part in time period t
imgPixels	The RBG values of every pixel in imgPartName_t
imgPixelFeature	The output of the feature extraction algorithm for every pixel
HS	Hashset which holds the pixels which are road
$FE(p_{i,j})$	Feature extraction applied to a pixel
$N(p_{i,j})$	The neighborhood of the $p_{i,j}$
L_{cr}	List of candidate road pixels of the image
Ep	Counter of the examined pixels
P_{end}	Termination percentage
k-hop	Value to an integer k

Problem Definition

Given a pair of aerial map images S_{img} in different time intervals t_s, monitor all differences between the pair of map images in terms of the road network, in a distributed and parallel way.

To facilitate efficient processing, every image is split into parts to achieve proper division of labor leading to increased efficiency. In particular, in our approach two main ways of image segmentation are considered: (i) every image is partitioned in as many pieces as the existing map processes in the cluster. On the one hand, this technique ensures the labor division, on the other hand, however, the I/O operations are increased and (ii) every image is partitioned into as many pieces as the number of computers in the cloud. In this way the labor division is not assured but the I/O operations are kept at a low level. Table 4.1 summarizes symbols used in the book chapter.

In this study, our main focus is on the image comparison algorithm rather than on the method used for feature extraction techniques. As a result, our approach makes use of efficient and tested feature extraction techniques described by (Kirthika and Mookambiga 2011) for the road extraction in aerial images. Neural networks are applied on high resolution aerial images for road detection. In particular, ANNs are found to be superior to several previous techniques due to their ability to incorporate both spectral and textural information. The road detection procedure is performed based on neural network classifiers. Different combinations of texture and spectral parameters are used in the road extraction process and the functionality of the neural network is evaluated comparing the road map and the manually detected road pixels that constitute our ground truth.

4.5 Algorithmic Techniques

In this section, we present the algorithmic techniques developed for the map update in a parallel and distributed environment. Initially, we provide a baseline approach to solve the problem, followed by a more efficient algorithm that can reduce the comparison time between the two maps. In addition to the presentation of algorithms, we present the performance of similar algorithms for comparisons. We propose two approaches. The first one is a more sophisticated approach and aims to reduce the number of image feature computations by using a specific k-hop examination pixel step. The final approximate approach uses a sample of pixels in order to apply the feature extraction computation followed by comparisons. The approximate algorithm significantly reduces the execution time, while the efficiency is highly related to certain parameters and the type of image data.

4.5.1 The Map Update Baseline Algorithm (MBASE)

The simplest algorithm developed to solve the map update problem in a distributed and parallel way directly derives from the Hadoop - MapReduce philosophy. In particular, in this baseline algorithm the input are aerial images from the same area in different time periods. The output of this procedure are differences of the road network between a pair of images. The method consists of two main parts: (i) The Map phase in which the road extraction algorithm is applied to images. The output of this phase for every image is a list which stores the computed values of every pixel, and (ii) the Reduce phase in which the comparison procedure takes place. The input of this phase is the output of the Map phase. The key of the Map phase input ($imgPartName_t$) is the id of an image part including its time signature t. In order to correlate the corresponding image parts from different time periods, the time signature is removed leaving only the id of the image part which is the output of the Map phase (imgPartName). Each pair of images that covers the same area has the same id. The output of the Reduce phase is a list which stores pixel values with differences between corresponding pixels of the pair of images.

The outline of MBASE is given in Algorithm 1 (compare Appendix at the end of the chapter). The road extraction operation is performed at line 5 and the comparison between the images, in order to find differences, is executed at line 12.

Although the use of the MapReduce framework reduces the execution time, more improvements can be applied in order to have additional computation time reduction of the method.

4.5.2 The Map Update Advanced Algorithm (MAD)

The main drawback of the **MBASE** algorithm is the large amount of data produced from the Map phase and transferred to the Reduce phase, where the final comparisons take place. The MAD algorithm is a method based on the key idea to reduce the amount of data emitted per image from the Map phase to the Reduce phase. More specifically, in this algorithm only a *Map <idPartImage, list<pixelRoad>>* is emitted from the Map phase. For every image part, a *list<pixelRoad>* which contains only pixels that are recognized as "road" is generated. Thus, the only work needed in the Reduce phase is the crosscheck of every pixel emitted, between the pair of *Map <idPartImage, list<pixelRoad>>*. If a pixel exists only in one of the lists, it is stored as a difference in the output of the Reduce phase. The comparison between Map images is very fast due to the map data structure advantages (Mehlhorn and Sanders 2008). Algorithm 2, whose outline is given in Appendix, shows the main steps of the MAD algorithm. In line 5, the calculation of pixel values is performed and if the pixel represents a road, then it is stored to the map structure. The map structure emitted to the Reduce phase is shown in line 10. Finally, the comparison between maps is made at line 12.

4.5.3 The Map Update K-Hop Algorithm (KH-MAD)

Although MAD is more efficient than **MBASE**, it is designed to calculate features for all image pixels, to check if every pixel represents a road part or not. However, our goal is to minimize the number of feature calculations per image. As such, we developed yet another technique, the **KH-MAD**, the key point of which is to use two parameters in order to skip some pixels. There is a parameter *k* (*k - hop*) which defines the number of pixels that are skipped at every hop. The other parameter is the pixel's neighborhood $N(p_{i,j})$, with size *NxN*, which will be marked as candidate for further examination. In our approach the neighborhood parameter is fixed to 3*x*3 square centered at the current examined pixel.

The Map phase starts with a list $L_{cr}(p_i)$, which holds candidate pixels for examination. This list is initialized with one element which is the pixel at the upper left corner of the image. The initialization pixel was randomly picked to be the upper left. Through the Map phase, iteration takes place for elements of the $L_{cr}(p_i)$ list as long as there is at least one element. In every iteration, feature extraction is performed for an element-pixel. If the current pixel is a road, then the algorithm takes the 3*x*3 neighborhood around the pixel and stores it in the $L_{cr}(p_i)$. If the current pixel does not represent a road, the algorithm continues to other pixels of the image using a specific step, based on the parameter *k* (*k - hop*) which is located at the $(i + k, j)$, $(I, j + k)$ position. The output of the Map phase is the same as in the MAD algorithm, a map structure for every part of an image. Regarding the Reduce phase of the **KH-MAD** approach, it is the same as in the **MAD approach**. In Algorithm 3 the main steps of the **KH-MAD** are given. Lines 5 to 16 provide the methods, which are followed in order to prune the feature extraction operation by using the k-hop technique.

4.5.4 *The Map Update Sample Hop Algorithm (SH-MAD)*

The last proposed algorithm is a slight variation of the **KH-MAD** approach. More specifically, **SH-MAD** uses the same pruning technique for feature extractions, however, one more parameter S_p is added, which determines the percentage of the pixels to be examined. The termination condition of the iteration process is changed and it is now the size $L_{cr}(p_i) < S_p$. The **SH-MAD** algorithm can be used in conjunction with several different sampling algorithms in order to ensure that the total computations do not surpass the desired level. In our approach, the **SH-MAD** algorithm uses as sampling method the **KH-MAD** algorithm.

Algorithm 4 presents the main steps of **SH-MAD**. In line 9, the algorithm performs the termination condition of the iteration process based on the E_p and P_{end} values, which are initialized in lines 5 and 6 respectively. For clarification reasons, E_p is the number of pixels that have been examined during each iteration of the algorithm. P_{end} is the maximum percentage of image pixels that can be examined before termination. In every iteration the formula at line 9 is used to check whether pixels examined are more than the specified percentage to continue or terminate the iteration.

Each algorithm proposed in this study provides examples on how a framework for distributed and parallel processing, like Hadoop, can be useful in geospatial-based problems. The *k-hop* image traversing algorithm is also a simple paradigm of how to travel across pixels of an image. One drawback, due to its simplicity, is that if the starting pixel is the first at the top left corner and the termination percentage of the **SH-MAD** is too small, then there is a possibility of leaving the pixels at the bottom right corner of the image unchecked. Any sampling or image traversing algorithm can be chosen to fill that gap and results may vary depending on the landscape of captured images.

4.6 Performance Evaluations

This section illustrates performance results and showing at the same time the efficiency and scalability of the proposed approaches. All algorithms have been implemented in Java and the experiments have been conducted on a ten node cluster of 80 Gigabytes of total RAM and 40 processing cores, running CentOS 6. We study the performance of algorithms in terms of their runtime and pruning capabilities, by varying the most important parameters such as nodes used, to determine the impact of using Hadoop, as well as the hop length (k - hop), and the maximum number of calculations before the termination (P_{end}). The computational cost of algorithms is given by the number of times the feature extraction algorithm was applied on the images as well as the number of the total comparisons required to determine differences. These values set the processing capabilities of the algorithms, by being able to reduce the number of the most time-consuming action of the process, the feature extraction, as well as not parsing the whole image for comparisons in the reduce phase. The default values for parameters, if not otherwise specified, are: $k\text{-}hop = 5$, $P_{end} = 0.8$, and $N(p_{i,j}) = 3x3$.

The proposed theoretical study is validated by a thorough experimental evaluation, based on real-world, as well as, synthetically generated images due to insufficient real-world data.

4.6.1 Data Description

Images used in this research date back to 2008 and they were captured with a Microsoft Vexcel UltraCam camera. According to the calibration report of the camera, the calibrated focal length was 100.500 mm (both for panchromatic and multispectral camera), the pixel size 7.2 m for the panchromatic camera and 21.6 m for the multispectral camera, while the image format was 9420pixels x 14430pixels for the panchromatic camera and 3140 pixels x 4810pixels for the multispectral camera. For the dataset used in this research, the Ground Sampling Distance (GSD) was 20 cm, while the (average) total size per image is 500 MB, leading to 225GB of total dataset size. The average altitude above WGS84 ellipsoid for the overall dataset was between 2800 m and 3900 m, depending on the area of image acquisition. A snapshot of the images used, is illustrated in Fig. 4.3.

4.6.2 Experimental Results for Real-Life and Synthetic Data

The first experiment includes the number of machines which are available in order to run map update algorithms. In particular, Fig. 4.4 illustrates the performance of the baseline (**MBASE**) map update algorithm by varying the number of the running

Fig. 4.3 Snapshot of the images used

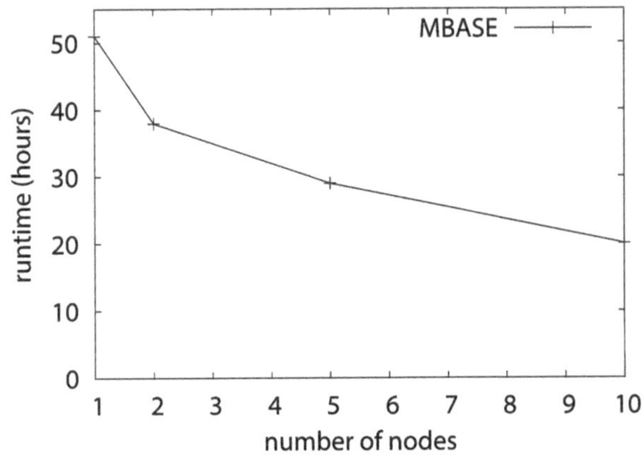

Fig. 4.4 Running time vs. number of nodes

Table 4.2 Number of calculations & comparisons

		No. calculations		No. comparisons
Algorithm	Image size (pixels)	Before	After	Total
MBASE	271,748,160	271,748,160	271,748,160	271,748,160
MAD	271,748,160	271,748,160	271,748,160	97,013,873
KH-MAD - 5	271,748,160	217,398,528	220,116,009	95,691,256
KH-MAD - 20	271,748,160	203,811,120	209,246,083	95,167,239
SH-MAD – 0.9	271,748,160	225,550,972	233,703,417	96,476,291

machines (nodes). The number of the used nodes is 1 to 10. Every machine node has 4 cores with 8 GB RAM. We run this experiment only with the **MBASE** algorithm in order to demonstrate the difference of the Hadoop usage even without using any pruning technique and the real data with a total size of 50 GB.

It is obvious that when the number of nodes increases, the runtime decreases. However, the runtime reduction is not proportional to the number of nodes because of two main reasons. The first reason is the amount of information sent between nodes. Even if 100 nodes are available, the time needed for the execution would still depend on the network traffic. In the **MBASE** algorithm, every pixel is checked if it represents a road or not and that value is sent to the Reducers making the information exchange almost the same as the input size. This is the purpose of the **MAD** algorithm, to decrease the size of the information being sent over the network. The second reason has to do with the size of the input and the overhead of using the MapReduce framework. MapReduce is designed to process big data such as Terabytes of data, thus the communication and coordination of its processes add a computational overhead. With huge datasets this overhead is insignificant, but becomes noticed at smaller input size.

Table 4.2 shows the number of calculations for a pair of images as well as the total number of comparisons executed by **MBASE**, **MAD**, **KH-MAD,** and **SH-MAD** with specific parameter values.

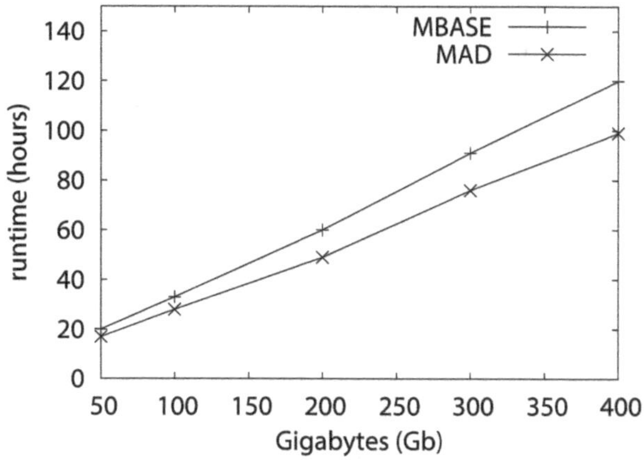

Fig. 4.5 Running time MBASE vs. MAD

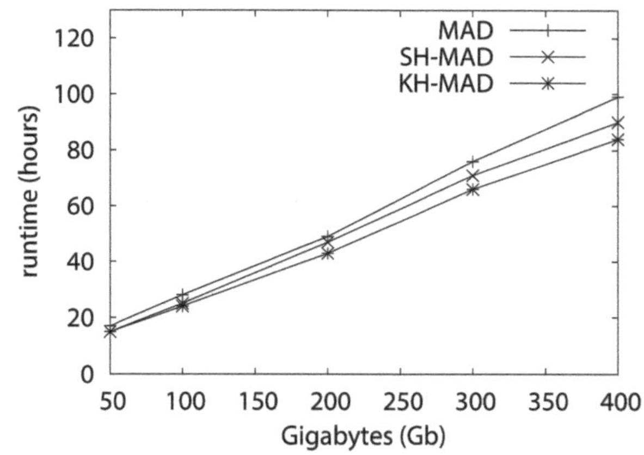

Fig. 4.6 Running time vs. input set size

To counter the huge data transfer problem through the network, we implemented the aforementioned **MAD** algorithm. As illustrated in Fig. 4.5, the **MAD** algorithm runtime is better as the input increases and thus we use **MAD** as the base algorithm for all the following comparisons.

Moreover, Fig. 4.6 illustrates the impact of the image data amount on the performance of the algorithms. In this set of experiments, we use the maximum number of available nodes from the existing infrastructure, i.e. ten nodes. Every algorithm was launched and tracked multiple times with the same input size. Then the average running time was picked for every algorithm, to show the overall performance with various input size.

It is noticed that with a small input the difference between runtime is not that distinct due to the overhead added by the framework. As the input increases, the gap between algorithms increases too, as the effects of the different approaches take place. In our experiment, **KH-MAD** performed better than **SH-MAD** without undermining **SH-MAD** algorithm's logic.

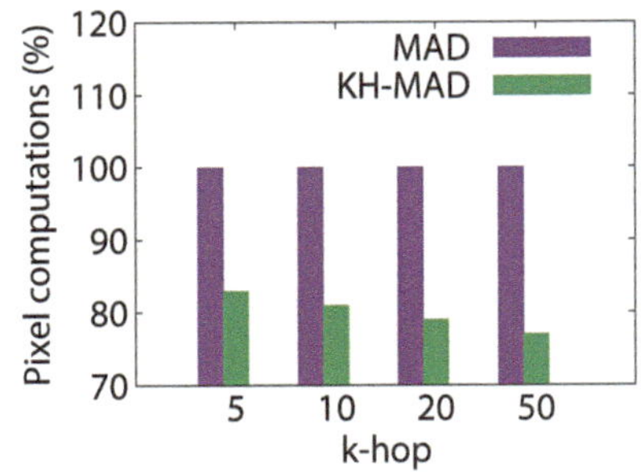

Fig. 4.7 Pixel computations based on k-hop

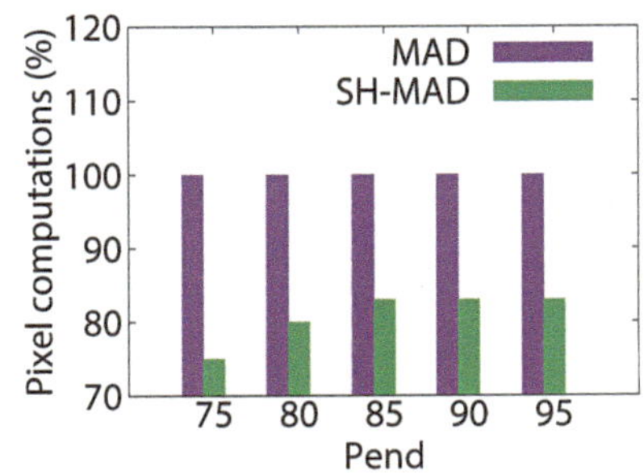

Fig. 4.8 Pixel computations based on P_{end}

Figure 4.7 presents the number of total calculations that each algorithm executed and the effectiveness of the pruning sequence proposed. The changeable variable is *k-hop* and results are completely representative of the **KH-MAD** theory. As the *k-hop* variable increases, so does the number of pruned calculations with only a small effect on the algorithm"s accuracy. Even with a small *k-hop* value, the pruned calculations are almost at 20%, showing **KH-MAD**"s effectiveness. The pruned calculations for the **SH-MAD** algorithm compared to the standard **MAD** algorithm are given in Fig. 4.8. In this case the *k-hop* value is fixed at five for every P_{end} value. When the P_{end} termination percentage is low, the algorithm terminates based on the fixed percentage as noticed at values 75 and 80. This can happen when the number of total computations needed is greater than the defined threshold. At the last three values (85, 90, 95) the pixel computations are the same due to the termination of the **KH-MAD** before it reaches the termination percentage P_{end}. The **SH-MAD** algorithm is used in conjunction with another sampling algorithm to ensure that the total computations do not exceed the desired level.

In order to ensure that algorithms are working as intended and pruned computations are not affecting the output accuracy, Fig. 4.9 illustrates the accuracy of the **KH-MAD** with a varied *k-hop* value compared to the **MAD** algorithm. The accuracy for values 5 and 10 remain at 99%, considering that there is a decrease in computations of about 20% each. For a *k-hop* value of 20 the accuracy drops to 98% and for a *k-hop* value of 50 the accuracy drops to 97%. Comparing the different results, we can observe that for each *k-hop* value the accuracy drop is only 1% but

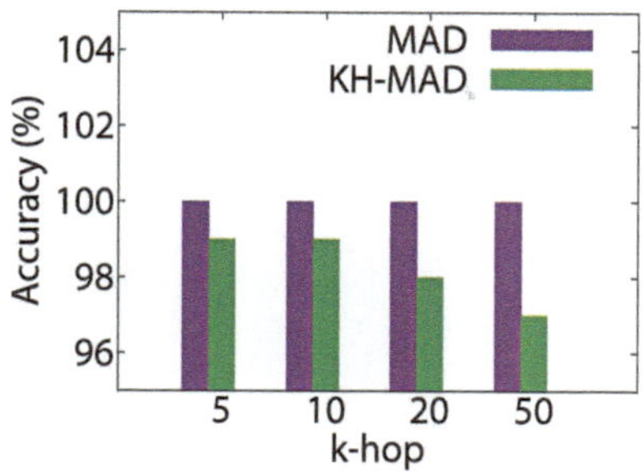

Fig. 4.9 Accuracy based on k-hop

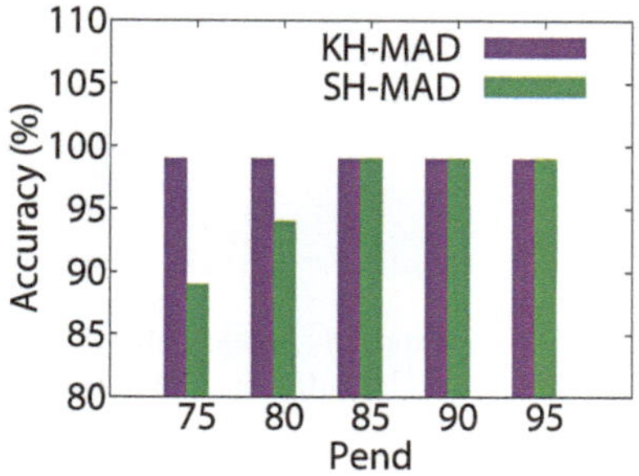

Fig. 4.10 Accuracy based on P_{end}

the pruned calculations percentage increases by about 3%. We can assume that this is a safe and efficient trade-off depending on the input. If the input consists of rural area images, then pruned calculations will be higher as the number of road pixels in rural areas is far less than the number of road pixels in urban regions.

Finally, the accuracy of the **KH-MAD** with a *k -hop* value set to 5 and the **SH-MAD** is given in Fig. 4.10.

Once more, the variable that changes is the P_{end} percentage of the **SH-MAD**. At 75% the accuracy falls by a large amount as there are "road pixels" that were left unchecked. The same goes for the 80% value as it increases the accuracy but the result may still not seem satisfying. For the last three P_{end} values (85, 90, 95), the accuracy remains at around 99%. Even though 15% less calculations may seem a small number, when dealing with the large sizes of today's real world datasets, then the decrease is rather significant.

4.6.3 Examples of Map Updates

In this section we present the results of two examples, as extracted from the implementation of the described algorithms. Figure 4.11 shows part of the aerial image at two different time periods, namely in the left and in the center of the figure. The algorithms' application took place in order to determine the differences between the images. The right image illustrates with dots the pixels which were assigned by the algorithm as part of the road-change. The proposed approach can determine

Fig. 4.11 Example I: Image before (*left*); image after (*centre*); image with algorithm's results (*right*; it should be noted that the image is magnified inside the box)

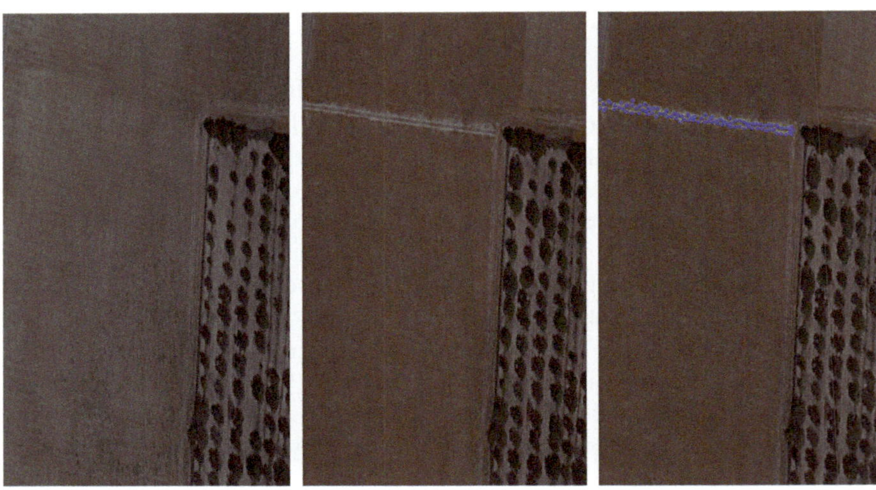

Fig. 4.12 Example II: Image before (*left*); image after (*centre*); image with algorithm's results (*right*)

correctly which pixels belong to the road in order to properly report the changes between images.

One more representative example is shown in Fig. 4.12. In this figure, the left and the center images present two parts of aerial images from the same area but, again, from different time periods. The differences which were found by applying our algorithms are presented in Fig. 4.12 (right). As it can be seen, the proposed methodology successfully determined the new road addition.

4.7 Conclusions and Future Work

In the present work we described several techniques to get a grasp of how distributed computing can be used to solve map update problems that, due to the size of the input, cannot be dealt with using existing GIS software. Those techniques are based on parallel and distributed strategies that split the workload to several computers whose processing units work simultaneously for the same cause, saving a lot of time. Our research focuses on one of the most interesting geospatial applications, namely the automatic road network update from aerial or satellite imagery.

In our research we used different algorithms for identifying differences between two datasets of the same area but from a different acquisition time. In practice, we tested four algorithms, namely **MBASE**, **MAD**, **KH-MAD**, and **SH-MAD**. Our research showed that the **KH-MAD** approach outperformed the others, while the **MAD** algorithm reduces the running time by a considerable amount. **SH-MAD** seems to have better performance than **KH-MAD** with the same accuracy depending on the data type. Finally, **MBASE** is not considered in this comparison, since it is just a baseline approach as the relevant figures have shown in the previous sections.

Feature extraction algorithms were not part of this research and this is the reason we are not evaluating the road network extraction results. For the implementation of this study we chose a feature extraction algorithm based on ANN, which is considered as the most appropriate and efficient approach. The proposed framework is open to use any feature extraction technique.

In this research we used the typical (image) partitioning as provided by the Hadoop framework. Even though Hadoop offers many other possibilities to customize the partitioning algorithm in order to better handle the image files, we have not used them as it was not part of this research. It is our aim to test various partitioning approaches with the same datasets, in order to assess the system performance with respect to these parameters, as well.

One of the main areas that present great interest in parallel and distributed systems is the medical image processing. In particular, there is a lot of research going on lately in Intravascular Ultrasound (IVUS) image processing that is considered to be big data, as each human vessel is constituted by hundreds of high resolution images. Similar to road detection in satellite maps, it is important to be able to identify different components found in the vessel wall in the stenosis in short time. There are four major tissue components, therefore the tissue characterization discriminates four classes rather than two (road – not road) in our first attempt to test big-data handling techniques presented in this book chapter. Taking into account the fact that doctors need to know soon, if there is any artery prone to break and cause a heart attack, fast tissue characterization in vessel walls is very important and the authors intend to expand the presented methodology into the even more difficult tissue characterization field.

Appendix

Algorithm 1

Algorithm 1 MBASE

Input: $S_{img(t)}$: the set of aerial images, t: time period of an image
Output: Set of road network differences between a pair of images $DiffList$

Splitting images and feeding them to the $Mappers$;
Map Phase: input $\langle imgPartName_t, imgPixels\rangle$, output$\langle imgPartName, imgPixelFeature\rangle$;
$imgPixelFeature \leftarrow \{\}$;
for each pixel $p_{i,j} \in imgPixels$
  Apply feature extraction to $p_{i,j}$ // $FE(p_{i,j})$
  $imgPixelFeature.add(FE(p_{i,j}))$
end for
Emit output:$\langle imgPartName, imgPixelFeature\rangle$
Reduce Phase: input: $\langle imgPartName, imgPixelFeature\rangle$, output$\langle imgPartName, DiffList\rangle$;
$DiffList \leftarrow \{\}$;
for each $i < imgPixelFeature_{height}$ && $j < imgPixelFeature_{width}$
  if $imgPixelFeature_{t_0}(i,j)$!= $imgPixelFeature_{t_1}(i,j)$
    $DiffList.add(Pair(i,j))$;
  end if
end for
Emit output: $\langle imgPartName, DiffList\rangle$
Merge output for all image parts;

Algorithm 2

Algorithm 2 MAD

Input: $S_{img(t)}$: the set of aerial images, t: time period of an image
Output: Set of road network differences between a pair of images $DiffList$

Splitting images and feeding them to the $Mappers$;
Map Phase: input $\langle imgPartName_t, imgPixels\rangle$, output$\langle imgPartName, HS\rangle$;
$HS \leftarrow \{\}$;
for each pixel $p_{i,j} \in imgPixels$
  Apply feature extraction to $p_{i,j}$ // $FE(p_{i,j})$
  if $FE(p_{i,j})$ *is road*
    $HS.put(FE(p_{i,j}))$;
  end if
end for
Emit output:$\langle imgPartName, HS\rangle$
Reduce Phase: input: $\langle imgPartName, HS\rangle$, output$\langle imgPartName, DiffList\rangle$;
$DiffList \leftarrow HS_{t0} \cup HS_{t1} - HS_{t0} \cap HS_{t1}$;
Emit output: $\langle imgPartName, DiffList\rangle$
Merge output for all image parts;

Algorithm 3

Algorithm 3 KH-MAD

Input: $S_{img(t)}$: the set of aerial images, t: time period of an image
Output: Set of road network differences between a pair of images $DiffList$

Splitting images and feeding them to the $Mappers$;
Map Phase: input $\langle imgPartName_t, imgPixels \rangle$, output$\langle imgPartName, HS \rangle$;
$HS \leftarrow \{\}$;
$L_{cr} \leftarrow p_{0,0}$; // initialize the list with the first pixel of the image
Initialize the $k - hop$ value to an integer k;
Set $N(p_{i,j})$ as a $3x3$ square;
while $L_{cr} \neq \emptyset$
    Apply feature extraction to $p_{i,j}$ // $FE(p_{i,j})$
    if $FE(p_{i,j})$ *is road*
        $HS.put(FE(p_{i,j}))$;
        $L_{cr}.add(N(p_{i,j}))$;
    else
        $L_{cr}.add(N(p_{i+h,j}))$;
        $L_{cr}.add(N(p_{i,j+h}))$;
    end if
end while
Emit output:$\langle imgPartName, HS \rangle$
Reduce Phase: input: $\langle imgPartName, HS \rangle$, output$\langle imgPartName, DiffList \rangle$;
$DiffList \leftarrow HS_{t0} \cup HS_{t1} - HS_{t0} \cap HS_{t1}$;
Emit output: $\langle imgPartName, DiffList \rangle$
Merge output for all image parts;

Algorithm 4

Algorithm 4 SH-MAD

Input: $S_{img(t)}$: the set of aerial images, t: time period of an image
Output: Set of road network differences between a pair of images $DiffList$

Splitting images and feeding them to the $Mappers$;
Map Phase: input $\langle imgPartName_t, imgPixels \rangle$, output$\langle imgPartName, HS \rangle$;
$HS \leftarrow \{\}$;
$L_{cr} \leftarrow p_{0,0}$; // initialize the list with the first pixel of the image
$Ep \leftarrow 0$; // initialize the percentage of the examined pixels
Initialize the termination percentage P_{end};
Initialize the $k-hop$ value to an integer k;
Set $N(p_{i,j})$ as a $3x3$ square;
while $Ep < P_{end}$
    Apply feature extraction to $p_{i,j}$ // $FE(p_{i,j})$
    if $FE(p_{i,j})$ *is road*
        $HS.put(FE(p_{i,j}))$;
        $L_{cr}.add(N(p_{i,j}))$;
    else
        $L_{cr}.add(N(p_{i+h,j}))$;
        $L_{cr}.add(N(p_{i,j+h}))$;
    end if
end while
Emit output:$\langle imgPartName, HS \rangle$
Reduce Phase: input: $\langle imgPartName, HS \rangle$, output$\langle imgPartName, DiffList \rangle$;
$DiffList \leftarrow HS_{t0} \cup HS_{t1} - HS_{t0} \cap HS_{t1}$;
Emit output: $\langle imgPartName, DiffList \rangle$
Merge output for all image parts;

References

Acharya T, Ray AK (2005) Image processing: principles and applications. Wiley, Hoboken

Andrews GR (2000) Foundations of multithreaded, parallel, and distributed programming. Addison-Wesley, Reading, 0-201-25752-6

Bacher U, Mayer H (2005) Automatic road extraction from multispectral high resolution satellite images. Int Arch Photogramm Remote Sens XXXVI, Part 3/W24, pp 29–34

Bernd J (2002) Digital image processing. Springer, Berlin

Bong DBL, Lai KC, Joseph A (2009) Automatic road network recognition and extraction for urban planning. Int J Comput Electr Autom Control Inf Eng 3(5)

Coulouris G, Dollimore J, Kindberg T, Blair G (2011) Distributed systems: concepts and design, 5th edn. Addison-Wesley, Boston

Dean J, Ghemawat S (2008) MapReduce: simplified data processing on large clusters. Commun ACM 51(1):107–113

Di L (2004) Distributed geospatial information services architectures, standards, and research issues.

Eldaw A, Mokbel MF (2015) SpatialHadoop: a MapReduce framework for spatial data. In: IEEE 31st international conference on data engineering (ICDE), pp 1352–1363

Gecen R, Sarp G (2008) Road detection from high and low resolution satellite images. Int Arch Photogramm Remote Sens Spat Inf Sci XXXVII. Part B4, pp 355–358

Grote A, Heipke C (2008) Road extraction for the update of road databases in suburban areas. Int Arch Photogramm Remote Sens Spat Inf Sci XXXVII. Part B3b

Gruen A, Li HH (1995) Road extraction from aerial and satellite images by dynamic programming. J Photogramm Remote Sens 30:11–20

Gruen A, Li HH (1997) Semi-automatic linear feature extraction by dynamic programming and LSB-snakes. Photogramm Eng Remote Sens 63(8):985–995

Hawick KA, Coddington PD, James HA (2003) Distributed frameworks and parallel algorithms for processing large-scale geographic data. Parallel Comput 29(10):1297–1333

Hinz S, Baumgartner A (2003) Automatic extraction of urban road networks from multi-view aerial imagery. ISPRS J Photogramm Remote Sens 58:83–98

Hofmann C (1999) A Multi-tier Framework for Accessing Distributed, Heterogeneous Spatial Data in a Federation Based EIS. In: Proceedings of the 7th international symposium on advances in geographic information (ACM-GIS), pp 140–145

Hongbin M, Yahong Z, Qun H (2008) Road extraction from high resolution remote sensing image based on mathematics morphology and seed growth. Int Arch Photogramm Remote Sens Spat Inf Sci. XXXVII. Part B3b, Beijing

http://www.euroconsult-ec.com/. Accessed 21 Sept 2015

https://en.wikipedia.org/wiki/Apache_Hadoop#History

Hu X, Zhang Z, Zhang J (2000) An approach of semi-automated road extraction from aerial image based on template matching and neural network. Int Arch of Photogramm Remote Sens XXXIII, Part B3, pp 994–999

Jhummarwala A, Potdar MB, Prashant C (2014) Article: parallel and distributed GIS for processing geo-data: an overview. Int J Comput Appl 106:9–16

Kirthika A, Mookambiga A (2011) Automated road network extraction using artificial neural network. In: International conference on recent trends in information technology (ICRTIT), pp 1061–1065

Li X, Qiao Y, Yi W, Guo Z (2003) The research of road extraction for high resolution satellite image. IEEE Geosci Remote Sens Symp 6:3949–3951

Lind P, Alm M (2006) A database-centric virtual chemistry system. J Chem Inf Model 46:1034–1039

Mehlhorn K, Sanders P (2008) Hash tables and associative arrays. algorithms and data structures: the basic toolbox (PDF), Springer, pp 81–98

Mnih V, Hinton GE, (2010) Learning to Detect Roads in High-resolution Aerial Images. In: Proceedings of the 11th European conference on computer vision: part VI (ECCV)

O'Brien J, Marakas GM (2008) Management information systems. McGraw-Hill Irwin, New York, pp 185–189

Reis JC, Pruski C, Reynaud-Delaître C (2014) State-of-the-art on mapping maintenance and challenges towards a fully automatic approach. Expert Syst Appl:1465–1478

Tait MG (2005) Implementing geoportals: applications of distributed GIS. Comput Environ Urban Syst 29:33–47

Wilhelm B, Burge M (2008) Digital image processing: an algorithmic approach using java. Springer, Dordrecht

Zhang Q, Couloinger I (2006) Automated road network extraction from high resolution multi-spectral imagery. ASPRS 2006 Annual Conference, Reno, Nevada, May 1–5, 2006

Zarrinpanjeh N, Samadzadegan F, Schenk T (2013) A new ant based distributed framework for urban road map updating from high resolution satellite imagery. Comput Geosci 58:337–350

Zhang J, Gong J, Lin H, Wang G, Huang J, Zhu J, Xu B, Teng J (2007) Design and development of distributed virtual geographic environment system based on web services. Inf Sci 177:3968–3980

Chapter 5
Productive Networks and Indirect Locations

André Sabino and Armanda Rodrigues

Abstract Discovering interesting locations to users is a challenge for social and productive networks. The evidence of the content produced by users must be considered in this task, which may be simplified by the use of the metadata associated with the content, i.e., the categorization supported by the network, namely – descriptive keywords and geographic coordinates. In this book chapter we present a productive network representation model, designed to discover indirect keywords and locations. The spatial dimension of the model enables indirect location discovery methods through the interpretation of the network as a graph, solely relying on keywords and locations that categorize or describe productive items. The model and indirect location discovery methodology presented in this chapter avoid content analysis, and are a new step towards a generic approach to the identification of relevant information, otherwise hidden from the users. The evaluation of the model and methods is accomplished by an experiment that performs a classification analysis over the Twitter network.

Keywords Data mining • Location recommendation • Social networks • Productive networks

5.1 Introduction

There are several online services for content sharing. In fact, the number of services is increasing, with some focusing on different contexts of the social life, and others on the nature of the content media. This process of specialization is an indicator of the diverse nature of both content and sharing context.

These services are usually referred to as *social network services*. By definition, a social network "*refers to the ways in which people are connected to one another and how these connections create and define human society on all levels: the individual, the group, and the institutional*" (Eisenberg and Houser, 2007). In actuality, the

A. Sabino (✉) • A. Rodrigues
NOVA LINCS, Faculdade de Ciências e Tecnologia, Departamento de Informática, Universidade NOVA de Lisboa, Lisbon, Portugal
e-mail: amgs@campus.fct.unl.pt; a.rodrigues@fct.unl.pt

M. Leitner, J. Jokar Arsanjani (eds.), *Citizen Empowered Mapping*, Geotechnologies and the Environment 18, DOI 10.1007/978-3-319-51629-5_5

focus of social network services varies from the support of human relationships to the sharing of content. While supporting a set similar features, these services do differ. We identified three types of services: *Social Networks*, *Content Sharing Networks*, and *Content Indexing Networks*.

In this book chapter, we refer to the set of Social Networks, Content Sharing Networks, and Content Indexing Networks as *productive networks*.

Productive networks enable users to annotate content using keywords, which produces classification and categorization systems with many potential applications, such as profiling users according to interests, enabling advertising and e-commerce customization, identifying popular topics, or even *locations of interest*.

Networks have a natural representation as a graph. In the context of productive networks, each node is a network element, i.e., user, item, keyword, or location. Edges of this graph may represent relationships between users, between users and content items, between items and annotation keywords, and, particularly, between locations and users. We are interested in the structure and visibility of these later relationships.

The broad hypothesis of our work is that *relationships between users and items, and annotation processes present in productive networks constitute evidence of human behavior*. This evidence refers to structure and organization of human relationships, content production, and content geo-reference, which is present in productive networks independently of the content media type.

Our focus is on *indirect relationships*, which we define as relationships that are not represented as a single edge or path of edges between the same type of elements in the network graph. These relationships are actually represented by a path involving all types of elements: From user to items; between items through annotation keywords and locations; and back to users.

We systematically evaluate productive networks to define a theoretical model, which enables the construction of tools for the identification of potential indirect relationships between different network elements. We present a methodology to discover indirect locations, which are potentially relevant to users.

One of the applications of our model and methodology is in the emergency management domain, for population density variation estimation. Monitoring productive networks to identify indirect relationships between users and locations may help improve risk assessment in disaster forecasts. Currently, this assessment relies on census (static) data for risk estimation (Fortes et al. 2014a), which could be improved with population density variation estimations for the same forecast time window.

5.2 Related Work

Laere et al. (2010), use learning methods to automatically assign geographic coordinates to Flickr photos. The authors also use a clustering approach to obtain regions of interest, and present a method that successfully predicts the location of a previously unseen photo. The authors also provide a discussion on the effects of spatial granularity on the meaning of the location recommendation for a particular photo.

Peregrino et al. (2013) present a method to infer location from Twitter posts. It is based on text analysis, and cross referencing with Wikipedia[1] entries. Our model and this work can be integrated in a solution that first infers the geographic location of a Twitter post, and then discovers related indirect locations for recommendation.

Ozdikis et al. (2013) use evidential reasoning techniques over Twitter data to estimate locations, with the ultimate goal of event detection. Using the Dempster-Shafer Theory, the authors use the Twitter post location, text, and user profile declared location to construct belief intervals for sets of locations where certain events might have happened. This approach enables the discovery of locations that may be relevant to a user interested in a particular event. The evaluation is presented using belief percentage as effectiveness metric, and cannot be compared with our results.

Ho et al. (2012), Hu and Ester (2013), and Son et al. (2013), present methods for mobile user profiling and event prediction, based on the cross reference between the user production and location. These authors are mainly focused on predicting the next location where the user will be, or an event will occur, from the evidence of past production. In contrast, we are interested in suggesting locations that user did not yet consider.

Ye and Yin (2010), Ference et al. (2013), and Wang et al. (2013), present methods for location recommendation in location-based social networks. The methods are usually dependent on the type of post available on these networks, but provide a good insight over the location recommendation problem.

5.3 Productive Networks

We propose a three terms to categorize network services, namely *Social Networks*; *Content Sharing Networks*; and *Content Indexing Network*. Individual definitions of these three network services are:

Social Network (SN): is a system which aims to replicate the network behavior through which human beings relate with each other. Examples are Facebook[2] and Google Plus[3];

Content Sharing Network (CSN): is a system where the main goal is to host and make available the content posted by users. We include, in this category,

[1]http://www.wikipedia.com

[2]https://www.facebook.com

[3]https://plus.google.com

content sharing networks through which communities emerged, and in which some aspects of a social network behavior happen. Examples are Flickr[4] and Instagram[5];

Content Indexing Network (CIN): is a system that is mainly focused on enabling the search of content which may be hosted somewhere else, or which is not freely available. The goal is to create awareness that such content exists. Examples are the ACM Portal[6] or the IEEE Explore.[7]

To guide our study, we propose two dimensions to classify networks, which are the *discovery focus* and *upload policy*. The discovery focus is determined by the main information type that is being delivered to the user. Regardless of the type, the main discovery focus of these networks is either users or content. Most networks actually enable both discovery focuses, but are tailored to promote one over the other. The upload policy determines if users are able to freely upload content or if this is regulated by a curation process. Figure 5.1 illustrates the classification of the three types of network according to these dimensions.

We propose to name the set defined by these three types of network as *productive networks*. We coined the term *productive network* to shift the focus of our definition from the social aspect of user relationships and organization, particularly focused by some networks, to the evidence of human production, found in all networks of our study.

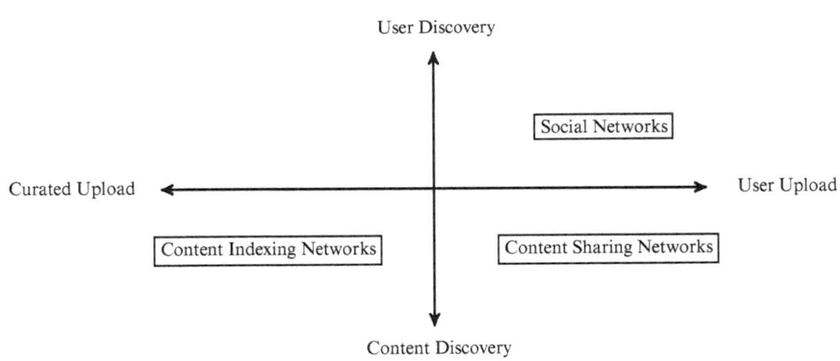

Fig. 5.1 Types of productive networks according to upload policy (*horizontal axis*) and main discovery focus (*vertical axis*)

[4]https://www.flickr.com

[5]https://www.instagram.com

[6]http://portal.acm.org

[7]http://ieeexplore.ieee.org

5.3.1 Productive Network Survey

The aim of the study is to provide evidence that supports a model for productive networks. The initial set of networks (the *seed*) was composed of the top sixteen networks listed at Wikipedia's *Social Networking Websites List*,[8] on January 1, 2013. The list was ordered by the Global Alexa Page Ranking,[9] and by the declared registered number of users.

All networks which were considered candidates to be included in the study were preliminarily evaluated to determine whether they enable searching for items or users, and provide free registration. Networks enable annotation through different user interfaces, which differ considerably between services. These differences may be motivated by the content media type, network scope, or particular design guidelines, which may include hiding the annotation system on free textual descriptions of the content.

The study systematically describes and categorizes several productive networks according to the following dimensions: supported user relationships, content media type, context and policy for annotation keywords, context and policy for spatial annotation of content, and focus of the search tool. These dimensions can be more specifically explained as follows:

Supported user relationships: Users may relate with individual users or with groups of users.

Content media type: Media types considered may be *text*, *image*, *video*, or *url* (a particular case of text).

Annotation context and policy: Annotation keywords may be used to describe or classify content. Keywords may be part of the content, or separate from it. The network may provide a specific taxonomy for keywords, and users may or may not be able to create new keywords.

Spatial annotation context and policy: The content may be annotated with specific locations, which may be represented by coordinates or with the added semantic of geographical places. Locations may refer to points or areas, and may be part of the content or separate from it.

Focus of the search tool: The network may provide search tools for users, items, keywords and/or locations.

5.3.2 Survey Results

To the initial set of sixteen networks, we added 25 related networks, for a total of 41 productive networks. Each network is categorized according to its subtype, which may include Content Indexing Network (CIN), Content Sharing Networks (CSN), and Social Network (SN).

---

[8] https://en.wikipedia.org/wiki/List_of_social_networking_websites

[9] http://www.alexa.com/topsites

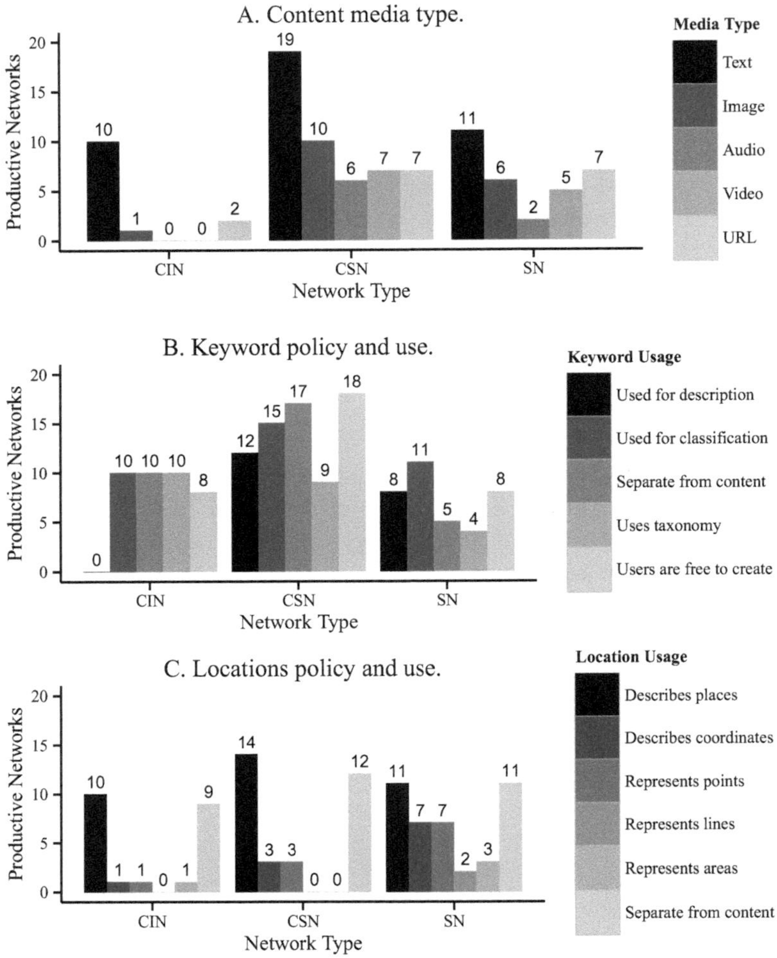

Fig. 5.2 Productive network survey result summary, clustered by network type. There are 41 networks in the survey, with 10 CIN, 20 CSN, and 11 SN. (**a**) Each network may support several media types. (**b**) Keywords may be used to *describe* or *classify* content. They may be *separate from the content*, be regulated by a *taxonomy*, and/or the users may be *free to create* them. (**c**) Locations may refer to a *place* or a only set or *coordinates*. Each location may describe a *point*, *polyline*, or *area*. Locations may also be *separate from the content*

Figure 5.2a presents the content media type for all types of network, showing a wide range of media types. Given our focus on meta-data analysis, these results suggest a variety of contexts where our indirect relationship identification methods may be implemented.

The keyword policy and use, presented in Fig. 5.2b, show that keywords are used for the purposes of content description and classification, simultaneously.

Table 5.1 Summary of the survey results describing the data set counts, and the differences between the seed and final set of networks. Results about keyword policy and use refer to the final set of networks. All results are clustered by network type

	Count		Keywords		
	Seed	Survey	Free to create	Searchable	With taxonomy
SN	7 (44%)	11 (27%)	8 (72%)	11 (100%)	4 (36%)
CSN	8 (50%)	20 (49%)	18 (90%)	20 (100%)	9 (45%)
CIN	1 (6%)	10 (24%)	8 (80%)	10 (100%)	10 (100%)
TOTAL	16	41	34 (85%)	41 (100%)	23 (56%)

Figure 5.2c presents the location policy and use, showing that most networks enable the annotation of content with place references.

The evidence produced by the survey enables the definition of a set of statements that summarize our findings. The first statement is that every *productive network has a representation of user, content item, and annotation keyword.*

There are two contexts where a person's identity, or alias, may appear in a network: as an author of content; or as the user who shared that content. For the purposes of representing users and personal interests, we consider that *users have an ownership relationship with content items, which they may or may not have authored themselves.*

Table 5.1 presents results on keyword policy for the three types of network. While fixed taxonomies are associated with curated upload policies and content discovery networks, the ability to freely create keywords is available in the majority of the networks. All networks, either focused on content or user discovery, enable a search by keyword.

From the results in Table 5.1, we conclude that *keywords are available for annotation, and establish relationships between content items.* These relationships imply that *users become associated with the keywords they use to annotate content.*

All networks enable search by keywords, which produces listings of content items and/or users. Relationships between users and content items, and between content items and keywords imply that *users that annotate content with the same keywords are related through those keywords.*

A subset of networks in the survey enable support for the annotation of content items with locations. Table 5.2 presents results related to location policy and use.

Table 5.2 implies that *some productive networks have a representation for locations which are used to annotate content items.*

All networks that support locations enable the annotation of content items with places, which provides semantic to the annotation. In fact, locations represent a specific case of annotation keyword, and enable the same type of relationships as keywords. Therefore *users that annotate content with the same locations are related through those locations.*

Table 5.2 Summary of the survey results about location keyword policy and use, presenting the differences between seed and final set of networks on location support. All results are clustered by network type

	Supports locations		Describes places	Describes coordinates
	Seed (n = 16)	Survey (n = 41)		
SN	7/7 (100%)	11/11 (100%)	11 (100%)	7 (64%)
CSN	7/8 (88%)	14/20 (70%)	14 (70%)	3 (15%)
CIN	1/1 (100%)	10/10 (100%)	10 (100%)	1 (10%)
TOTAL	15/16 (94%)	35/41 (85%)	35/35 (100%)	11/35 (31%)

Table 5.3 Evidence statements inferred from the productive network survey

Evidence statement	Description
E1	Productive networks have a representation of user, content item, and annotation keyword
E2	Users have an ownership relationship with content items, which they may have or not authored
E3	Keywords are available for annotation, and establish relationships between content items
E4	Users are associated with the keywords they use to annotate content
E5	Users that annotate content with the same keywords are related through those keywords
E6	Some productive networks have a representation for locations, which are used to annotate content items
E7	Users that annotate content with the same locations are related through those locations

5.3.3 Discussion and Evidence Statements

Survey results enable the identification of several statements which guide the definition of a productive network model. Table 5.3 summarizes the evidence statements resulting from the survey.

Statements in Table 5.3 represent a list of requirements for both model construction and validation.

5.4 Productive Network Model

Basic elements of the model are users, U, items, I and keywords, K. Items are owned by users, and annotated with keywords.

Let us define U, I and K such as:

$$U = \{U_1, \cdots, U_n\} \textit{ is a finite set of users, } n \geq 1$$

$$I = \{I_1, \cdots, I_m\} \textit{ is a finite set of items, } m \geq 1$$

$$K = \{K_1, \cdots, K_o\} \textit{ is a finite set of keywords, } o \geq 1$$
$$L = \{L_1, \cdots, L_u\} \textit{ is a finite set of locations, } u \geq 1$$

Note: Subscripts used in definitions serve to distinguish between elements of the same set. We use i, j, k for users, p, q, r for keywords, t, u, v for items and m, n, o, u for set dimensions.

The ownership of an item, $I_t \in I$, by one user, $U_i \in U$, is defined by:

$$O(U_i) = \{I_u \mid I_u \in I \wedge I_u \text{ is owned by } U_i\}$$
$$Own(U_i, I_t) \Rightarrow I_t \in O(U_i)$$

The annotation of an item, $I_t \in I$, by a keyword, $K_p \in K$, is defined by:

$$A(I_t) = \{K_q \mid K_q \in K \wedge K_q \text{ annotates } I_t\}$$
$$Annotate(K_p, I_t) \Rightarrow K_p \in A(I_t)$$

We refer to keywords that are used in annotations as the user's *direct keywords*. The set of all direct keywords, $U\!K$, of an user, $U_i \in U$, is defined by:

$$U\!K(U_i) = \{K_p \mid K_p \in K \wedge \exists I_t \in O(U_i) : (Annotate(K_p, I_t))\}$$

We now define relationships that items and keywords enable between users. We begin with the definition of a direct relationship, $D\!R$, between two users, $\underset{i \neq j}{U_i, U_j} \in U$, which is defined by:

$$D\!R(U_i, U_j) = \{K_p \mid K_p \in K \wedge \underset{t \neq u}{\exists I_t, I_u} \in I : (\begin{Bmatrix} Own(U_i, I_t), Own(U_j, I_u) \\ Annotate(K_p, I_t), Annotate(K_p, I_u) \end{Bmatrix})\}$$

Based on the previous expression, we now define indirect relationships, $I\!R$, between two users, U_i and U_j, is defined as follows:

if

$$D\!R(U_i, U_j) = \{\emptyset\},$$
$$\underset{k \neq i \neq j}{\exists U_k \in U} : \begin{Bmatrix} D\!R(U_k, U_i) \neq \{\emptyset\} \\ D\!R(U_k, U_j) \neq \{\emptyset\} \end{Bmatrix}$$

then

$$I\!R(U_i, U_j) = \{K_p \mid K_p \in K \wedge K_p \in D\!R(U_k, U_j)\}$$
$$I\!R(U_j, U_i) = \{K_p \mid K_q \in K \wedge K_q \in D\!R(U_k, U_i)\}$$

Table 5.4 All graphs that may be defined using the concepts of the model, each with an unique combination of node ($\mathscr{V}$) and edge ($\mathscr{E}$) sets

$\mathscr{G}_{id} = \langle \mathscr{V}, \mathscr{E} \rangle$
$\mathscr{G}_1$: Users connected through their items $\mathscr{V} = \{U_i \mid U_i \in U \wedge \underset{j \neq i}{\exists U_j} \in U : (\mathit{DR}(U_i, U_j) \neq \{\emptyset\})\}$ $\mathscr{E} = \{I_t \mid I_t \in I \wedge \underset{i \neq k}{\exists U_i, U_k} \in \mathscr{V} : (Own(U_i, I_t) \wedge Own(U_k, I_t))\}$
$\mathscr{G}_2$: Users connected through keywords, which annotate their items $\mathscr{V} = \{U_i \mid U_i \in U \wedge \underset{j \neq i}{\exists U_j} \in U : (\mathit{DR}(U_i, U_j) \neq \{\emptyset\})\}$ $\mathscr{E} = \{K_p \mid K_p \in K \wedge \left\{ \begin{array}{l} \exists U_i \in \mathscr{V} : Own(U_i, I_t) \wedge Annotate(K_p, I_t) \\ \underset{j \neq k}{\exists U_j, U_k} \in U : K_p \in \mathit{DR}(U_j, U_k) \end{array} \right\}\}$
$\mathscr{G}_3$: Items connected through their common users $\mathscr{V} = \{I_t \mid I_t \in I \wedge \left\{ \begin{array}{l} \underset{i \neq j}{\exists U_i, U_j} \in U : Own(U_i, I_t) \wedge Own(U_j, I_t) \\ \exists K_p \in K : Annotate(K_p, I_t) \end{array} \right\}\}$ $\mathscr{E} = \{U_i \mid U_i \in U \wedge \underset{t \neq u}{\exists I_t, I_u} \in \mathscr{V} : (Own(U_i, I_t) \wedge Own(U_i, I_u))\}$
$\mathscr{G}_4$: Items connected through their common keywords $\mathscr{V} = \{I_t \mid I_t \in I \wedge \exists K_p \in K, \underset{u \neq t}{\exists I_u} \in I : (Annotate(K_p, I_t) \wedge Annotate(K_p, I_u))\}$ $\mathscr{E} = \{K_p \mid K_p \in K \wedge \underset{t \neq u}{\exists I_t, I_u} \in \mathscr{V} : (Annotate(K_p, I_t) \wedge Annotate(K_p, I_u))\}$
$\mathscr{G}_5$: Keywords connected through users which use them to annotate items $\mathscr{V} = \{K_p \mid K_p \in K \wedge \exists I_t \in I : (Annotate(K_p, I_t))\}$ $\mathscr{E} = \{U_i \mid U_i \in U \wedge \underset{p \neq q}{\exists K_p, K_q} \in \mathscr{V} : (K_p \in \mathit{UK}(U_i) \wedge K_q \in \mathit{UK}(U_i))\}$
$\mathscr{G}_6$: Keywords connected through common items $\mathscr{V} = \{K_p \mid K_p \in K \wedge \exists I_t \in I : (Annotate(K_p, I_t))\}$ $\mathscr{E} = \{I_t \mid I_t \in I \wedge \underset{p \neq q}{\exists K_p, K_q} \in \mathscr{V} : (Annotate(K_p, I_t) \wedge Annotate(K_q, I_t))\}$

It is now possible to describe graphs implicitly defined by the network. Table 5.4 presents all possible graphs.

Locations enable similar definitions, from direct to indirect relationships. For a user, $U_i \in U$, the set of all direct locations, UL, of all of the user's items is defined by:

$$\mathit{UL}(U_i) = \{L_l \mid L_l \in L \wedge \exists I_t \in O(U_i) : (GeoRef(L_l, I_t))\}$$

We are now able to define direct relationships based on locations, DRL, between two users, U_i and U_j, such as:

$$\mathit{DRL}(U_i, U_j) = \{L_l \mid L_l \in L \wedge \underset{t \neq u}{\exists I_t, I_u} \in I : \left\{ \begin{array}{l} Own(U_i, I_t), Own(U_j, I_u), \\ GeoRef(L_l, I_t), GeoRef(L_l, I_u) \end{array} \right\}\}$$

Table 5.5 Extension of the graphs presented in Table 5.4, using the location concept. Each graph represents an unique combination of node and edge

$\mathcal{G}_{id} = \langle \mathcal{V}, \mathcal{E} \rangle$
$\mathcal{G}_7$: Locations connected through users
$\mathcal{V} = \{L_l \mid L_l \in L \wedge \exists I_t \in I : (GeoRef(L_l, I_t))\}$
$\mathcal{E} = \{U_i \mid U_i \in U \wedge \exists L_l \in \mathcal{V}, \exists I_t \in I : (Own(U_i, I_t) \wedge GeoRef(L_l, I_t))\}$
$\mathcal{G}_8$: Locations connected through items
$\mathcal{V} = \{L_l \mid L_l \in L \wedge \exists I_t \in I : (GeoRef(L_l, I_t))\}$
$\mathcal{E} = \{I_t \mid I_t \in I \wedge \exists_{l \neq m} L_l, L_m \in \mathcal{V} : (GeoRef(L_l, I_t) \wedge GeoRef(L_m, I_t))\}$
$\mathcal{G}_9$: Locations connected through keywords
$\mathcal{V} = \{L_l \mid L_l \in L \wedge \exists I_t \in I : (GeoRef(L_l, I_t))\}$
$\mathcal{E} = \{K_p \mid K_p \in K \wedge \exists_{t \neq u} I_t, I_u \in I : (\left\{\begin{matrix} Annotate(K_p, I_t) \wedge Annotate(K_p, I_u) \\ \exists_{l \neq m} L_l, L_m \in \mathcal{V} : \left\{\begin{matrix} GeoRef(L_l, I_t) \\ GeoRef(L_m, I_u) \end{matrix}\right\} \end{matrix}\right\})\}$
$\mathcal{G}_{10}$: Users connected through locations
$\mathcal{V} = \{U_i \mid U_i \in U \wedge \exists U_j \in U : (U_i \neq U_j \wedge \mathbb{DRL}(U_i, U_j) \neq \{\emptyset\})\}$
$\mathcal{E} = \{L_l \mid L_l \in L \wedge \exists U_i \in \mathcal{V} : \left\{\begin{matrix} \exists I_t \in I : Own(U_i, I_t) \wedge GeoRef(L_l, I_t) \\ \exists U_j \in U : L_l \in \mathbb{DRL}(U_i, U_j) \end{matrix}\right\}\}$
$\mathcal{G}_{11}$: Items connected through locations
$\mathcal{V} = \{I_t \mid I_t \in I \wedge \left\{\begin{matrix} \exists U_i, U_j \in U : U_i \neq U_j \wedge Own(I_i, I_t), Own(U_j, I_t) \\ \exists K_p \in K : Annotate(K_p, I_t) \end{matrix}\right\}\}$
$\mathcal{E} = \{L_l \mid L_l \in L \wedge \left\{\begin{matrix} \exists I_u \in \mathcal{V} : GeoRef(L_l, I_u) \\ \exists U_j \in U : L_l \in \mathbb{DRL}(U_i, U_j) \end{matrix}\right\}\}$
$\mathcal{G}_{12}$: Keywords connected through locations
$\mathcal{V} = \{K_p \mid K_p \in K \wedge \exists I_t \in I : (Annotate(K_p, I_t))\}$
$\mathcal{E} = \{L_l \mid L_l \in L \wedge \exists I_t \in I, \exists K_q \in \mathcal{V} : (GeoRef(L_l, I_t) \wedge Annotate(K_q, I_t))\}$

Table 5.5 presents all possible graphs with these new concepts.

Finally, an indirect relationship, $\mathbb{RL}$, between two users, U_i and U_j, based on locations, is defined by:

if

$$\mathbb{DRL}(U_i, U_j) = \{\emptyset\},$$

$$\exists U_k \in U : \left\{\begin{matrix} \mathbb{DRL}(U_k, U_i) \neq \{\emptyset\} \\ \mathbb{DRL}(U_k, U_j) \neq \{\emptyset\} \end{matrix}\right\}$$

then

$$\mathbb{R}(U_i, U_j) = \{\forall L_l \in L \mid L_l \in \mathbb{DRL}(U_k, U_j)\}$$

$$\mathbb{R}(U_j, U_i) = \{\forall L_m \in L \mid L_m \in \mathbb{DRL}(U_k, U_i)\}$$

5.4.1 *Trivial Operations*

Operation definitions presented by the model focus on relationships between users and other network concepts, with the objective of enabling the discovery of indirect relationships. There are several operations involved in the development of methods for indirect relationship discovery which we consider trivial, such as:

Obtain all items of a keyword: For a keyword, K_p, produce a list of all content items annotated with K_p.

Obtain all users of a keyword: For a keyword, K_p, produce a list of all users who annotate content items with K_p.

Obtain all items of a location: For a location, L_l, produce a list of all content items annotated with L_l.

Obtain all users of a location: For a location, L_l, produce a list of all users who annotate content items with L_l.

Obtain a rank ordered list of elements: Sort a list with rank values by total order, or reverse total order.

5.4.2 *Model Validation*

Model concepts and operators should address the complete set of requirements summarized by the evidence statements. Table 5.6 presents a cross reference between model definition and those statements. Definitions of indirect relationships are not included in the table, because they are a consequence of evidence based operators and are not explicitly based on evidence statements.

Table 5.6 Cross reference between evidence statements inferred from the productive network survey and the model definitions

Evidence statement	Model concept and operators
E1	Definitions of the sets U, I, and K
E2	The operator *Own*
E3	The operator *Annotate*
E4	The user keywords set, $\mathbb{UK}$
E5	The direct relationships set, $\mathbb{DR}$
E6	Definition of set L. The operator *GeoRef*
E7	The direct relationships set, $\mathbb{DRL}$

5.5 Indirect Relationship Discovery

In the context of graphs in Tables 5.4 and 5.5, indirect relationships refer to shortest paths of size two (Fig. 5.3). Considering graph $\mathcal{G}_2$, which relates users through keywords, these paths are defined by one keyword used by the user – a *direct keyword* –, and one that is not – an *indirect keyword.*

The definition of indirect relationships enables the definition of a set of indirect keywords, $\mathbb{K}$, of a user, U_i, such that:

$$\mathbb{K}(U_i) = \{K_p \in K \mid \forall U_j \in U, U_j \neq U_i, \mathbb{R}(U_i, U_j) \neq \{\emptyset\} : K_p \in \mathbb{R}(U_i, U_j)\}$$

We may formulate a similar description for indirect locations, $\mathbb{L}$, of a user, U_i:

$$\mathbb{L}(U_i) = \{L_l \in L \mid \forall U_j \in U, U_j \neq U_i, \mathbb{RL}(U_i, U_j) \neq \{\emptyset\} : L_l \in \mathbb{RL}(U_i, U_j)\}$$

Based on a user's (U_i) set of indirect keywords or locations, building an ordered list of users with which U_i has indirect relationships is a trivial operation.

Both definitions of indirect elements (keywords and locations) represent the same principle for indirect relationships discovery, namely *to build a list of indirect relationships for a user, U_i, we should discover indirect keywords or locations for user U_i*. This is the theoretical framework for our set of experiments.

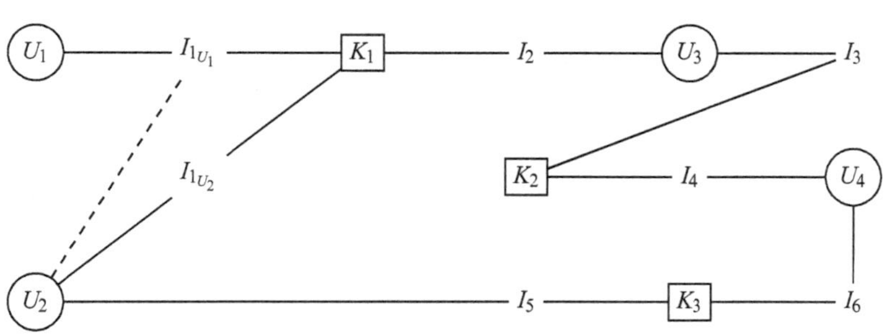

Fig. 5.3 Partial graph illustrating the representation of a shared item. It represents users in *circles*, keywords in *squares* and items without decoration. Although $I_{1_{U_1}}$ and $I_{1_{U_2}}$ both represent the same item in the network, they are actually distinct in the graph. $I_{1_{U_2}}$ defines the missing edge between U_2 and $I_{1_{U_1}}$ – represented by a *dashed line*

5.6 Discovering Indirect Locations

We focus our experiments on indirect location discovery. For a user, U_i, the list of indirect locations with rank values, $\mathbb{L}_r$, is defined by:

$$R(L_l) = | \{I_t \mid GeoRef(L_l, I_t)\} |$$
$$\mathbb{L}_r = \{\langle L_l, |R(L_l)|\rangle \mid L_l \in \mathbb{L}(U_i)\}$$

Our method is to train a classifier able to decide if a keyword or location is relevant to a user. The classifier model used is the Support Vector Machine (SVM). A SVM is a supervised learning model which assigns data items to one of two categories. The model is built around a training set of examples of both categories.

The SVM was fit to elements we want to identify as indirect, i.e., keywords or locations. Training is performed with same size sets of elements which are either related or not related to the user.

The SVM model determines if a particular element belongs to the user. The success of the classifier is determined by the training conditions, i.e., the set of features used to infer data patterns and the training set. The challenge is in determining if the training set of elements accurately represents the user's interests and in selecting a robust set of element features.

Features are represented by the pair, $\mathscr{F}$, determined by cardinalities of feature sets $\mathscr{A}$ and $\mathscr{B}$, such that:

$$\mathscr{F} = \langle |\mathscr{A}|, |\mathscr{B}| \rangle$$

For a location, L_l, of user's (U_i) direct locations, we propose one pair of feature sets, $\mathscr{F}_a$, defined by:

$\mathscr{F}_a$ Each location is represented by its absolute number of items ($\mathscr{A}$) and its absolute number of users ($\mathscr{B}$).

$$\mathscr{A} = \{I_t \mid \forall I_t \in I : L_l \in G(I_t)\}$$
$$\mathscr{B} = \{U_j \mid \forall U_j \in U : L_l \in \mathbb{UL}(U_j)\}$$

To sort the list of indirect locations we propose a method, R_{L_l}, which for every location, L_l, with a positive match, is defined by:

$$R_{L_l} = \sum_{L_g \in \mathbb{UL}(U_i)} |\{L_g \mid \exists I_t \in I : L_g \in G(I_t), L_l \in G(I_t)\}|$$

R_{L_l} calculates the sum of the number of co-occurrences between L_l and the user's locations.

Following the same approach as the indirect keyword ranking strategy, we refined the method to compute, R'_{L_l}, which is the normalization of R_{L_l} by the frequency of L_l, F_{L_l}, and is defined by:

$$F_{L_l} = \frac{|\{I_t \mid L_l \in G(I_t)\}|}{|I|}$$

$$R'_{L_l} = \frac{R_{L_l}}{F_{L_l}}$$

5.6.1 Network Sampling

To evaluate our methodology we created network graph samples. We are interested in building graph samples which replicate the network structure at a particular moment in time, which is different from samples which represent the same network only with fewer nodes and edges. The main difference between these goals is that the former preserves the network growth properties, which is ideal to evaluate our methodology with networks at different growth stages.

We followed the approach of Leskovec et al. (2005), where authors discuss a set of characteristics present in social networks' graphs, which led to the definition of a graph generation model, the Forest Fire model. This model later inspired a network sampling algorithm, proposed by Leskovec and Faloutsos (2006), which produces a graph preserving the same growth properties of the original graph, and which are:

Densification power law: The number of edges increases over time at a rate described by a power law of the number of nodes, i.e., $e(t) \propto n(t)^\alpha$, where $\alpha > 1$, with $e(t)$ edges at time t, and $n(t)$ nodes at time t.

Shrinking diameter: The graph diameter reduces over time.

We adapted the sampling method in Leskovec and Faloutsos (2006) to deal with the structure of the information present in networks used in the evaluation, yielding graph samples which verify both properties which we are interested in preserving. Algorithm 1 presents the sampling method.

Algorithm 1

```
Randomly select a seed node to visit.
while a node to visit exists do
    Visit one node.
    Decide, with α probability, if links to the node should be visited.
    if links should be visited then
        All outgoing nodes are selected to visit.
    end if
end while
```

Algorithm 2 Experiment outline. Presents the removal of the association between user and location, which the *experiment* is designed to recover. The *rank* function returns the position of a location in a list

Require: $O(U_i) \neq \{\emptyset\}$
$G'(I_t) = list()$
for all $I_t \in O(U_i)$ **do**
  if $G(I_t) \neq \{\emptyset\}$ **then**
    for all $L_l \in G(I_t)$ **do**
      for all $L_m \in G(I_t), L_m \neq L_l$ **do**
        $G'(I_t).append(L_m)$
      end for
      $\mathbb{L}_r = experiment(G'(I_t))$
      print $L_l \in \mathbb{L}_r$?
      print $rank(L_l, \mathbb{L}_r)$
    end for
  end if
end for

5.6.2 Experimental Protocol

We propose that if a user annotated items with a location, this location would be a valid suggestion in a scenario where the network was modified in such a way that this location is indirect to the user. Therefore, the experiment strategy is to remove one location from the set of locations associated to the user, and creating a new set of locations that annotate the user's items. The experiment outputs the ranked list of indirect locations, in which the removed location should appear. We refer to this procedure as *Location Recovery*.

Algorithm 2 presents the outline of experiments for indirect keyword and location discovery, for a particular user, U_i.

Given that the user explicitly annotated items with the location, we are sure that it is relevant to the user. Therefore, as presented in Algorithm 2, for each location whose relationship with the user is removed, the experiment goal is twofold:

1. Use the *experiment*() method to identify the location as indirect.
2. Attribute a high rank value to that location in the list of indirect locations.

We thus propose to instantiate the *experiment()* method with a machine learning approach, using support vector classifiers.

5.6.3 Metrics

To evaluate classification analysis results, we adopted standard metrics. Our focus is also on the evaluation of ranked lists of results, which are evaluated with specific metrics. The classification analysis identifies *relevant* locations, which are part of a whole set of *retrieved* elements. All metrics are defined in terms of relevant and retrieved elements.

These metrics measure the effectiveness with which information retrieval systems answer queries.

Precision (P) is the proportion of retrieved elements that are relevant. It is determined by

$$P = \frac{|relevant \cap retrieved|}{|retrieved|}$$

Precision at a rank K ($P@K$) reports the proportion of the top K retrieved elements that are relevant. Precision at a specific rank is relevant because only the top results are ultimately returned to the user. Given that our goal is to present a list of recommendations, which cannot be too long to be effectively delivered by most user interfaces, we show the precision at rank 1, 5, 10, and 20.

Recall (R) is the proportion of relevant keywords that are retrieved. It is determined by

$$R = \frac{|relevant \cap retrieved|}{|relevant|}$$

The main focus of our methods is not to retrieve all elements that are relevant to the user, but instead to ensure that the ones that are retrieved are indeed relevant. However, to access the quality of the approach we include the recall computation, which provides a measure of missed relevant results.

Mean reciprocal rank (MRR) informs where the first relevant element occurs in the ranking. It is determined by

$$MRR = \frac{1}{N} \sum_{1=1}^{N} \frac{1}{rank_i}$$

5.7 Results

In this section we present the results of a set of experiments, focused on indirect location discovery.

5.7.1 Dataset Description

The evaluation of the model and procedure uses six datasets built with Twitter data. Table 5.7 summarizes the datasets.

Table 5.7 Twitter datasets available for evaluation. Each dataset was obtained by collecting the live feed resulting from filtering the Twitter stream with the given queries

ID	Event description	Query
E1	Rock in Rio Lisboa music festival	#rirlx
E2	Lisbon summer holidays ("Santos Populares")	#santospopulares
E3	Lisbon Mega Picnic	#megapicnic
E4	Paredes de Coura music festival	#rirlx
E5	Paredes de Coura music festival V2	#paredesdecoura #vodafoneparedesdecoura
E6	Sol da Caparica music festival	#soldacaparica

Table 5.8 Description of the datasets. The number of *places* is indicated in the locations' column, in parenthesis

ID	Items	Users	Keywords	Locations
E1	47114	26750	1820	743 (287)
E2	558	356	546	79 (14)
E3	16	14	5	2 (2)
E4	375	177	203	0 (7)
E5	908	325	365	0 (13)
E6	303	188	168	0 (6)

We designed a live Twitter feed capture tool that collects and organizes the information according to our information model. All datasets originated from a particular event that we were able to monitor (live music summer events in Portugal).

The information collected contains users, items, keywords, locations, and places, where places are locations augmented with semantic. However, for datasets available, the number of places is relatively low. We consistently obtained a very low percentage of geo-referenced information. The low (or absent) number of located tweets has a significant impact in indirect location recommendation. To provide a measure of that, we include the description of datasets which do not yield results. This is a relevant constraint when designing applications which rely on indirect location discovery.

Table 5.8 describes datasets, with counts of several dimensions being available.

All datasets contain the complete set of tweets associated with events, starting 72 h before the event begins, and ending 72 h after it closes. However, only the first two, E1, and E2, contain enough spatial data to enable indirect location identification.

5.7.2 Location Clusters

Although datasets do contain geo-referenced items, these correspond to a low percentage of the total number of items (as expected), and the relationship pattern between item, keyword and location proved to be insufficient to enable our evaluation. The method requires a set of keywords to be associated with locations, through items. In most cases, however, each location corresponds to only one item. Such is caused by the granularity operated by the GPS sensor on mobile devices used to create the item. The same user, posting twice from the same location, a few minutes apart, is likely to produce two different coordinate pairs.

The solution to the problem is clustering locations. Instead of running the evaluation directly on locations, we compute a set of location clusters, using the DBSCAN algorithm, by Ester et al. (1996).

Algorithm 3 presents the clustering procedure outline, including the information cross referencing computation between single locations and respective clusters.

The choice of parameter values for DBSCAN was not focused on optimal behavior in terms of clustering. The problem lies with the high amount of locations, most with only one associated keyword, so we are mainly looking to significantly reduce the number of locations. The informal heuristic we followed was to obtain a number of clusters equal to around 10% of the number of locations in the datasets, merging items, users, and keyword sets of cluster members, thus producing clusters with more than 1 keyword on average. Figure 5.4 shows the clustering results (and parameters used) on the E1 dataset, reducing the number of location clusters from 743 to 79.

Table 5.9 shows the results of the experimental procedure on datasets with and without clustering.

Given the low percentage of keyword-location association, only the first (E1) dataset allows results without the clustering approach. However, clustering significantly improves the Mean Reciprocal Rank results, and allows indirect location discovery on smaller datasets.

Algorithm 3 Procedure used to build location clusters and associate the information needed for the classification analysis with clusters (instead of single locations)

Require: $\mathit{UL}(U_i) \neq \emptyset$
$clusters = \mathit{DBSCAN}(\mathit{UL}(U_i), eps, minpts)$
{*clusters* is a collection of location clusters.}
{Each *cluster* contains a set of locations.}
for all *cluster c* **in** *clusters* **do**
    $c.items = \{I_t \mid \forall L_l \in c.locations, G(L_l, I_t)\}$
    $c.users = \{U_i \mid \forall I_t \in c.items, O(I_t, U_i)\}$
    $c.keywords = \{K_p \mid \forall I_t \in c.items, T(K_p, I_t)\}$
end for

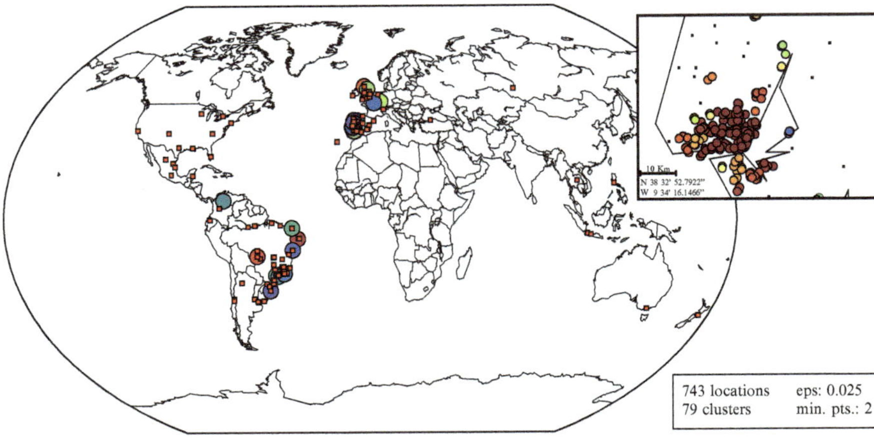

Fig. 5.4 Clustering results for dataset E1. DBSCAN parameters are set to produce around 10% of the initial amount of locations. *Circles* represent locations in clusters, *red squares* represent noise. The world map (Kavrayskiy VII projection) shows all 79 clusters and noise, and the *top right map* (orthographic projection) shows results around Lisbon, Portugal (coordinates displayed for the *lower left corner*)

Table 5.9 Results of the indirect locations classification analysis. E1 and E2 represent datasets with clustering. E1* is the original dataset, without clustering

ID	MRR	P@1	R@1
E1	0.5390	0.6415	0.4351
E1*	0.3785	0.6259	0.4118
E2	0.1365	0.7371	0.5804

5.8 Conclusions

We presented a productive network survey, which enabled an evidence supported productive network model. The productive network model provides a theoretical background with many potential applications. It enables a systematic approach for the identification of indirect relationships.

Among its potential applications, the model supports a methodology for indirect location discovery, which was presented, and evaluated in this book chapter, yielding good results. This methodology enables the construction of lists of locations which are potentially relevant to users, without any previous explicit relationships between the user and those locations. Potential applications of this result are varied, from advertisement, to public space and urban planning, or emergency management.

Indirect location discovery is also a potential step towards population density variation estimation and forecast, from productive network data. Inverting the focus of the recommendation, users could be suggested to locations, i.e., focusing on locations, which users are potentially relevant? Framing these recommendations in a time window could enable the detection of users who are temporarily and indirectly related with locations, and enable population density variations forecasting. This approach is particularly useful for the emergency management domain, and future work will focus on population density variation forecasting in coastal areas to improve risk assessment on wave runnup and overtopping scenarios (Poseiro et al. 2014a,b; Fortes et al. 2014b, 2015).

References

Eisenberg AF, Houser J (2007) Social network theory. In: Ritzer G (ed) Encyclopedia of sociology. Blackwell Pub., Malden, pp 4492:4

Ester M, Kriegel H, Sander J, Xu X (1996) A density-based algorithm for discovering clusters in large spatial databases with noise. In: Proceedings of 2nd International Conference on Knowledge Discovery and Data Mining

Ference G, Ye M, Lee W-c (2013) Location recommendation for out-of-town users in location-based social networks. In: Proceedings of the 22nd ACM International Conference on Information & Knowledge Management, San Francisco, pp 721–726

Fortes C, Reis M, Poseiro P, Capitão R, Santos J, Pinheiro L, Craveiro J, Rodrigues A, Sabino A, Silva SF, Ferreira J, Raposeiro P, Silva C, Rodrigues M, Simões A, Azevedo E, Reis F (2014a) HIDRALERTA Project – a flood forecast and alert system in coastal and port areas. In: Proceedings of the IWA World Water Congress and Exhibition, Lisbon

Fortes C, Reis M, Poseiro P, Capitão R, Santos J, Pinheiro L, Rodrigues A, Sabino A, Rodrigues M, Raposeiro P, Ferreira J, Silva C, Simões A, Azevedo E (2014b) O Projeto HIDRALERTA – Sistema de previsão e alerta de inundações em zonas costeiras e portuárias. In: Proceedings of the 8th Jornadas Portuguesas de Engenharia Costeira e Portuária, Lisbon

Fortes CJ, Reis MT, Poseiro P, Santos JA, Garcia T, Capitão R, Pinheiro L, Reis R, Craveiro J, Lourenço I, Lopes P, Rodrigues A, Sabino A, Ferreira JC, Silva S, Raposeiro P, Simões A, Azevedo EB, Vieira F, Rodrigues MDC, Silva CP (2015) Ferramenta de apoio à gestão costeira e portuária: o sistema hidralerta. In: Proceedings of the VIII Congresso sobre Planeamento e Gestão das Zonas Costeiras dos Países de Expressão Portuguesa, pp 1–18

Ho S, Lieberman M, Wang P, Samet H (2012) Mining future spatiotemporal events and their sentiment from online news articles for location-aware recommendation system. In: Proceedings of the First ACM SIGSPATIAL International Workshop on Mobile Geographic Information Systems, Redondo Beach, pp 25–32

Hu B, Ester M (2013) Spatial topic modeling in online social media for location recommendation. In: Proceedings of the 7th ACM Conference on Recommender Systems, Hong Kong, pp 25–32

Laere OV, Schockaert S, Dhoedt B (2010) Towards automated georeferencing of Flickr photos. In: Proceedings of the 6th Workshop on Geographic Information Retrieval, Zurich

Leskovec J, Faloutsos C (2006) Sampling from large graphs. In: Proceedings of the 12th ACM SIGKDD International Conference on Knowledge Discovery and Data Mining, Philadelphia

Leskovec J, Kleinberg J, Faloutsos C, Management HD, Applications D (2005) Graphs over time: densification laws, shrinking diameters and possible explanations. In: Proceeding of the 11th ACM SIGKDD International Conference on Knowledge Discovery and Data Mining. ACM Press, New York, pp 177–187

Ozdikis O, Oguztuzun H, Karagoz P (2013) Evidential location estimation for events detected in Twitter. In: Proceedings of the 7th Workshop on Geographic Information Retrieval, Orlando, pp 9–16

Peregrino F, Tomás D, Llopis F (2013) Every move you make I'll be watching you: geographical focus detection on Twitter. In: Proceedings of the 7th Workshop on Geographic Information Retrieval, Orlando, pp 1–8

Poseiro P, Reis M, Fortes C, Sabino A, Rodrigues A (2014a) Aplicação do sistema HIDRALERTA de previsões e alerta de inundações: caso de estudo da Costa da Caparica. In: Proceedings of the 3rd Jornadas de Engenharia Hidrográfica, Lisbon

Poseiro P, Sabino A, Fortes CJ, Reis MT, Rodrigues A (2014b) Aplicação do sistema HIDRALERTA de previsão e alerta de inundações: Caso de estudo da Praia da Vitória. In: Proceedings of the 12th Congresso da Água, Number 1

Son J, Kim A, Park S (2013) A location-based news article recommendation with explicit localized semantic analysis. In: Proceedings of the 36th International ACM SIGIR Conference on Research and Development in Information Retrieval, Dublin, pp 293–302

Wang H, Terrovitis M, Mamoulis N (2013) Location recommendation in location-based social networks using user check-in data. In: Proceedings of the 21st ACM SIGSPATIAL International Conference on Advances in Geographic Information Systems, Orlando

Ye M, Yin P (2010) Location recommendation for location-based social networks. In: Proceedings of the 18th SIGSPATIAL International Conference on Advances in Geographic Information Systems, San Jose, Number c, pp 458–461

Part II
Data Quality and Reliability

Chapter 6
Assessment of Volunteered Geographic Information Data Quality in The National Map Corps Project of the U.S. Geological Survey (USGS)

Erin Korris, Lily Niknami, and Elizabeth McCartney

Abstract In an effort to provide accurate and authoritative spatial data for The National Map of the U.S. Geological Survey (USGS) National Geospatial Program, the USGS began implementation of a new crowdsourcing project in 2010. The National Map Corps (TNMCorps) enlists volunteers to update structures data across all 50 states as well as Puerto Rico and the U.S. Virgin Islands. Volunteers collect and improve structures data by adding new features, removing obsolete points, and correcting existing data using a web-based mapping interface. Newly collected and modified point features become part of the USGS National Structures Dataset (NSD), a part of The National Map, which supplies data to US Topo maps, USGS cached base maps, and other derived products and services. Concern over the ability of volunteers to deliver high-quality data instigated a data quality study in 2012 during the Colorado pilot project, and a second nationwide quality study in July 2014. These data quality studies explore the quality of volunteered geographic information (VGI) within TNMCorps by assessing horizontal positional errors, attribute errors, and errors of commission. Results of the studies conclude that the quality of volunteered geographic data is significantly higher than baseline data, and the hierarchical editing approach improves the data at each stage. The results of the quality studies validate the overall data collection model of the project, and affirm that volunteers can provide high-quality geographic data with limited USGS monitoring.

Keywords Volunteered geographic information • Citizen science • Data quality • Crowdsourcing • Public participation GIS

E. Korris (✉) • L. Niknami
U.S. Geological Survey, P.O. Box 25046, MS 510, Denver, 80225-0046, CO, USA
e-mail: ekorris@usgs.gov

E. McCartney
U.S. Geological Survey, 1400 Independence Road, MS 554, Rolla, 65401, Missouri
e-mail: emccartney@usgs.gov

M. Leitner, J. Jokar Arsanjani (eds.), *Citizen Empowered Mapping*,
Geotechnologies and the Environment 18, DOI 10.1007/978-3-319-51629-5_6

6.1 Related Work

Throughout the progression of the geospatial industry, volunteered geographic information (VGI) has become ever more prevalent. Interest in participatory geospatial data collection is diverse, both in subject matter and demographics. The current discourse on VGI originates from various disciplines within the field of geography. Predominant bodies of literature can be categorized into three main groups. The first group is concerned with the presence of VGI in media. The second group is concerned with public participation in citizen science. The third group is concerned with methodologies, concerns, and approaches in addressing quality of VGI.

Ubiquitous uses of handheld global positioning system (GPS) devices have enabled the widespread collection of georeferenced feature data by citizens. Global positioning technologies have enabled the transition from citizens to "citizens as sensors" (Goodchild 2007). Increased collection of VGI stems from multiple needs, including transparency in data collection and dissemination, data accessibility, and crisis mapping (Neis and Zielstra 2014). Brovelli et al. (2016) examine developments in VGI and its presence in participatory data collection in mobile environments. Easy accessibility to mobile devices paired with growing interest in citizen science may increase the potential for participatory data collection. Mobile applications have tremendous potential to increase collection and dissemination of VGI. The sustainability of participation in citizen science projects is being explored in the context of volunteer motivation (Craglia and Shanley 2015). Gamification techniques are being deployed to increase user engagement with citizen science projects. Assessing the motivation of volunteers is difficult in certain projects due to privacy restrictions. However, some previously identified motivators include skill development, personal interest, and desire for social engagement (Craglia and Shanley 2015).

At the forefront of the research agenda in VGI and citizen science is quality assessment and monitoring of volunteer contributed data. As discussed by Hunter et al. (2012), criticism of collection and validation methods has increased the need for research in VGI data quality. Specific problems in data quality have been identified by previous authors, including static data review processes and lack of communication with volunteers regarding data quality concerns (Hunter et al. 2012). (Goodchild and Li 2012) suggest three different approaches to quality assurance, namely crowdsourcing, social, and geographic approaches. A crowdsourcing approach implies improving data quality through group validation and correction of data. A social approach implies "moderators" or "hierarchical" data review processes (Goodchild and Li 2012). A geographic approach implies the comparison of "geographic fact" with user contributed data (Goodchild and Li 2012). Various methodologies have been used to assess the accuracy of volunteer contributed data. Common errors in contributed data include positional, attribute, and formatting errors (Hunter et al. 2012). The monitoring of such errors is somewhat fluid and many projects implement automated scripts that identify erroneous values within the contributed datasets.

Previous literature on VGI notes various ways of approaching quality management. Bordogna et al. (2014) identify four approaches that are relevant to our exploration: "Ex ante, Ex post, cross-referencing, and the wiki approach." "Ex ante" is a preventative approach to quality management, whereas the "ex post" approach is applied after data collection (Bordogna et al. 2014). The "wiki approach" involves a collective effort to monitor and improve data quality (Bordogna et al. 2014). The National Map Corps (TNMCorps) project uses similar techniques in data quality management. The "Ex ante" approach is applied through the use of Standard and Peer Review editing guides that users read before data collection. "Ex post" is applied through daily quality monitoring and a U.S. Geological Survey (USGS) review of all feature data. The "wiki approach" is applied through a hierarchical editing process through which users peer review and update data. Authoritative datasets are often used to gain a quantitative measure of data quality in citizen science projects. TNMCorps data is not cross-referenced with an authoritative dataset because at the time of these studies (2012–2014) an authoritative dataset did not exist. The unique nature of the TNMCorps project has motivated our interest in further assessment of the quality of our volunteer contributed data.

A review of literature spanning from 2000 to 2015 identified the need for continuing research on the topic of VGI data quality, including the TNMCorps project. The hierarchical editing structure and quality monitoring procedures of the TNMCorps project makes it unique among other citizen science initiatives. Assessment of data contributions and quality over time will help identify strengths and weaknesses in the data collection process. Results of this study will be used to validate and improve future quality monitoring endeavors within the TNMCorps project, as well as contribute to the data quality dialogue within the field of VGI.

6.2 Overview of The National Map Corps

The USGS National Geospatial Program has a long history of citizen science. The progression of projects has closely been tied to available resources, changing technology, and, in many cases, required substantial USGS resources to use the data submitted. After the 7.5-min topographic map base series was completed in 1992, traditional topographic maps continued to be revised on a limited basis. In 1994, the Earth Science Corps was established to assist with the revision process. Over 3000 volunteers "adopted" 7.5-min topographic maps. Volunteers physically verified all features on the map, made annotations on the actual map where corrections were needed, and, within a one-year timeframe, mailed the map back to the USGS. Between 1994 and 2001, volunteers annotated between 100 and 300 maps each year, but the project had limited success due to the long map revision cycle.

In the mid-1990s the focus shifted from creating maps to building national geographic information systems (GIS) databases and later The National Map.

During this time, the process for acquiring data also changed from using primary data that USGS scientists collected or created to acquiring secondary data from other sources.

In the late 1990s, handheld GPS units became affordable for the average person. Embracing this available technology and focusing on building national GIS databases, TNMCorps was launched in 1998. Volunteers concentrated on collecting 36 different structure types (including courthouses, town halls, museums, libraries, schools, hospitals, and cemeteries) using handheld GPS units. Between 1998 and 2003, approximately 1000 volunteers collected over 22,000 points. Challenges included the following: A variety of positional accuracies due in part to differences between devices and expertise in using the devices; data integration challenges due to submission of data in a variety of formats including email, spreadsheets, and handwritten notes; and, finally, points were usually collected from the street or sidewalk, which resulted in the need for USGS technicians to move almost every point submitted to the middle of the structure to accurately represent its location. In 2006, hoping to mitigate challenges facing the GPS project, a web-based platform was introduced with some success, but as USGS priorities began to change, all citizen science efforts in the National Geospatial Program were suspended in 2008.

Operating in the sphere of "doing more with less," the US Topo project was launched in 2009 (Moore 2011). Working with national GIS databases, the USGS began mass producing US Topo maps on a 3-year cycle with a third of the conterminous United States in production during any 1 year (https://nationalmap.gov/ustopo/index.html). Due in part to the success of OpenStreetMap (OSM), Wikipedia, and other emerging crowdsourced web-based efforts in conjunction with the need for current, consistent data to support The National Map and the US Topo programs, the USGS began to explore crowdsourcing once again. In 2010, the USGS conducted a workshop on VGI to explore the feasibility of using VGI to support The National Map. As a direct result, a series of pilot projects were launched to answer vital questions including the following: Was technology indeed at a point where a VGI project could be successful using minimum resources, and, more importantly, would the quality of the data be good enough to ingest with minimal manipulation by the USGS?

The first pilot, OpenStreetMap Collaborative Prototype (OSMCP), Phase One (Wolf et al. 2011), demonstrated the ability to deploy a web application based on the OSM's web editor, Potlatch, where multiple editors could work simultaneously in a distributed environment. Phase One editors included experienced GIS professionals from the Kansas Data Access and Support Center and the USGS. The data focus was transportation. Building on lessons learned, a second pilot, OSM Collaborative Prototype, Phase Two was deployed in 2011 (Poore et al. 2012). Phase Two focused on the collection of 30 structure types over four USGS quadrangles in the Denver metropolitan area primarily using volunteer editors from surrounding universities. A data quality study analyzing horizontal positional accuracy and attribute accuracy from Phase Two determined that volunteers improved positional accuracy, and to lesser extent attribute accuracy, likely due to stringent and complex requirements during the pilot project. Following the results of the data quality study and building on knowledge gained during the pilot, the requirements and user materials were greatly simplified for the next pilot project. In 2012, the USGS streamlined the process to include only ten structure types and rebranded the effort as The National Map Corps (TNMCorps), while simultaneously expanding the efforts to the entire state of Colorado and welcoming anyone to participate. In addition, a quality study of the data collected for the State of Colorado suggested that volunteers were exceeding expectations for data quality standards, and it was demonstrated that TNMCorps project was also scalable when the project was expanded nationwide in 2013. In fiscal year 2014 (FY14), a national data quality study was conducted to reinforce the previous study and determine if volunteers were continuing to consistently meet or exceed quality goals.

As of 2015, TNMCorps volunteers are able to edit ten structure types in all 50 states, as well as Puerto Rico and the U.S. Virgin Islands. The results of the quality studies and the success of volunteer recruitment have allowed the program to continue and grow. As of July 2015, TNMCorps has over 2900 registered users who have contributed more than 150,000 points (Fig. 6.1).

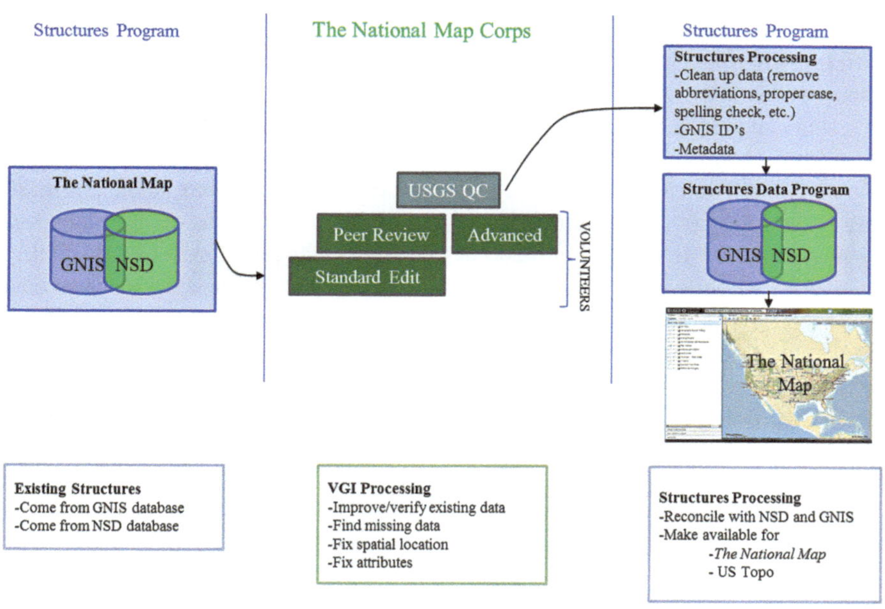

Fig. 6.1 The National Map Corps data workflow

6.2.1 Quality Assurance and Quality Control Procedures

As discussed in "Related Work," limited research has been done on the quality of crowdsourced and VGI data. Because of this concern about quality, we have implemented several quality assurance and quality control procedures.

6.2.2 Tiered-Editing Approach (Quality Assurance)

The first quality assurance measure put into place is a tiered-editing approach using three editing levels: Standard Editor, Peer Reviewer, and Advanced Editor. Tiered editing is a collaborative method that is most similar to the "wiki approach" to data quality assessment cited by Bordogna et al. (2014). When someone first registers as a TNMCorps volunteer, he/she is automatically assigned the Standard Editor role. The next level is Peer Review. After a volunteer has contributed 25 or more points through the Standard Editor, they have the option to peer review contributions from other Standard Editors. Having a more experienced volunteer review all points and multiple volunteers check each point increases the quality of the data as described in this book chapter. In addition to checking the validity of edited points, and catching any errors, Peer Reviewers are also asked to further improve data by making sure the attribution conforms to standards for inclusion in the National Structures Dataset (NSD). According to an internal standards document (USGS 2012), attribution standards include accurate point placement, accurate feature type, feature name, city, state, and zip code. An Advanced Editor role was introduced in the summer of 2014. Points edited by an Advanced Editor are not required to be peer reviewed. This role was initially created to facilitate faster processing of data contributed by trusted users (such as other government agencies) to the NSD, but was soon expanded and made available to volunteers with 200 or more edits who have passed a data quality assessment with roughly 90% or greater accuracy. To determine accuracy, a sample of the editor's points is taken, and a USGS TNMCorps team member carefully checks each point in the sample. Points that pass data quality assessment have accurate placement, feature type, feature name, city, state, and zip code attributes, as checked against authoritative sources. Thus, the points adhere to the inclusion standards of the NSD. The Advanced Editor role allows data from trusted editors to be added to The National Map more quickly and also improves data quality as shown in the quality study results. As of fiscal year 2015, all points must be checked by either a Peer Reviewer or an Advanced Editor before going into the NSD.

6.2.3 User Guides and Online Materials (Quality Assurance)

Another important quality assurance measure taken is the availability of detailed user guides and supplementary materials online. Implementation of user guides and supplementary materials is a preventative method; most similar to the "Ex ante" approach to data quality assessment cited by Bordogna et al. (2014). TNMCorps is open to anyone, and although no formal training is given to volunteers, those who sign up are encouraged to read through TNMCorps user guides that include information about the editing process, what to collect, what not to collect, and how to format the data.

6.2.4 Daily Quality Monitoring (Quality Control)

Daily contributions are monitored with automated scripts and semi-manual quality-control methods. Data from the editor is downloaded daily to an Esri® file geodatabase. Included in the daily downloads are a file of all changes from the past 24 h, a current copy of the data, a feature class of all deleted points, a view of the data edited through each of the user roles, and a vandalism table. Changes from the previous 24 h are run through a custom-built Safe Software FME® tool, which outputs a geodatabase of summary tables and feature classes for review when any of the following criteria are met: A user edits for the first time, a user edits 15 or more points in a 24-h period, or a user deletes five or more points in a 24-h period. A USGS TNMCorps team member then reviews this output for any potential problems. Volunteers are given immediate feedback addressing any issues that have been identified. Care is taken to return the information to participants in a positive way, and volunteers are typically appreciative of the guidance as well as the recognition that their contributions are valued. User engagement and feedback helps to increase transparency in data collection in the project, which potentially increases data quality over time. Additionally, contact with volunteers creates a dialogue between frequent contributors and staff members that further engages the user and allows them to provide feedback. As discussed by Craglia and Shanley (2015), sustaining user engagement is a crucial part in the sustainability of citizen science projects.

Any points that are found to have errors through daily quality monitoring are corrected by USGS staff at that time. This process improves the data in two ways: Volunteers' editing improves through feedback and correction, and data being reviewed daily by staff is corrected before being processed into the NSD.

As part of the daily download of data, a vandalism check is performed using a custom Python script that searches all fields against a dictionary table of inappropriate words. Matches are output to a table for review. Since many acceptable words may contain a portion of an inappropriate word, false positives are occasionally output to the table for review. When a false positive is triggered, that word is added to the dictionary as an exception to not be flagged in the future. At the time of this publication, no inappropriate contributions have yet occurred.

These measures were initially put in place for fear of potential data vandalism or malicious editing. In the several years since the beginning of the pilot projects, no malicious editing has been confirmed. In one instance, in the summer of 2015, a volunteer saved a large number of points without making any changes. This was detected through the quality monitoring methods described above, the user's account was suspended, and the user was contacted. The points that had been saved were rolled back to the previous version. This incident demonstrated that the quality safeguards put in place are working and that ultimately no harm was done to the data.

6.2.5 Structures Processing (Quality Control)

The USGS conducts a limited review of volunteer edited points that are ready to be processed into the NSD. As part of processing, all points are run through simple Visual Basic or Python scripts within Esri® ArcMap to locate and review spelling errors, punctuation, and apply standardization of naming format. A manual review may be triggered by things such as a large spatial disparity, an inconsistency noted between the type of feature and visual cues in an image background, missing information, or an attribute difference requiring clarification. All deleted points are reviewed along with the relevant NSD features against an imagery background. A few points are validated to ensure the volunteer is correctly applying delete actions. Obviously correct deletions are applied without further review. This comparison and validation is performed because deleted points are not peer reviewed. Additional internal validation is usually not applied due to the overall positive results of the quality studies, and because, currently, only edits that have been checked by a Peer Reviewer or Advanced Editor are processed into the NSD. Edited features are also reviewed for relevance and association with the USGS Geographic Names Information System (GNIS).

6.3 Quality Study

Since the beginning of the structures VGI pilot projects in 2011, three quality studies have been conducted. The results of the first, which was conducted following the second pilot project, are discussed briefly in the "Overview of The National Map Corps" section, and are explored further in Poore et al. (2012). The results of the two more recent studies are discussed below. The methodology for the two studies was similar, but with some differences that will be discussed in this section.

In 2012, after the expansion of structures collection to the State of Colorado, a data quality study was conducted during that pilot. During the fiscal year 2012 (FY12) Colorado pilot project, there were three "data status stages" in the TNMCorps project model. In this section, the following will be referred to as "stages":

1. Unedited: Baseline data from the USGS database commonly referred to as the GNIS, loaded to the TNMCorps editor, and not yet touched by volunteers.
2. Standard Edited: Features verified, edited, added, or deleted by volunteers.

3. Peer Reviewed: Features peer reviewed by another volunteer. Known during this pilot as Adopt-a-Quad (AAQ). Volunteers who had edited at least 20 points could sign up as Peer Reviewers. Peer Reviewers would check all points within a selected ("adopted") USGS quadrangle to assess positional accuracy and attribute accuracy (see https://navigator.er.usgs.gov/help/WebHelp/tnmcorps_attribute_formatting.pdf for attribute review guidelines). Peer Review volunteers were responsible for editorial formatting beyond the expectations for other volunteers.

Following the results of the FY12 Colorado pilot project data quality assessment, TNMCorps began a phased expansion to collect structures features in all 50 states. An ongoing goal of the project is to understand the overall quality of volunteered data collected, and so another quality study was conducted during the summer of 2014. The "stages" described in the FY12 Colorado pilot project have changed slightly, as follows:

Peer Reviewed: Features peer reviewed by another volunteer. Volunteers who have edited at least 25 points are automatically given the option to Peer Review. Peer reviewers check points that have gone through standard editing. Peer Review is equivalent to the role of AAQ during the FY12 Colorado pilot project, but the component of adopting a specific geographic area has been eliminated.

Advanced Edited: Features verified, edited, added, or deleted by an Advanced Editor. Advanced Editors have been offered this role based on results of a quality assessment of their edits as a Standard Editor or Peer Reviewer. Points checked by Advanced Editors do not require peer review. Because the Advanced Editor role had been created just prior to this quality study, only a small number of the total sampled points were advanced edits.

As part of the FY12 Colorado pilot project, results of this model were studied, and two hypotheses were tested:

1. Hypothesis 1: As data move from the Unedited to the Peer Reviewed stage, the overall quality and completeness improves with each step.
2. Hypothesis 2: The USGS can monitor and measure quality increases at acceptable internal cost.

The hypotheses tested for the FY14 national data quality study were the same, with the addition of an analysis of data contributed during the Advanced Edited stage, and the removal of the measure of completeness. This decision to not measure completeness in the FY14 study is explained in the Results section.

The domain for both the FY12 Colorado pilot project and the nationwide project was ten structure types (see https://navigator.er.usgs.gov/help/WebHelp/structure_def_table.pdf for a list of structure types). A random sample was selected from the population of all ten structure types for both studies. In the FY12 Colorado pilot project study, 100 points out of a total population of 4159 edited and unedited points were sampled for detailed USGS inspection. In the FY14 national data quality study, the dataset, which consisted of 377,595 points, was first separated by whether the data were edited or unedited. The resulting unedited dataset contained 325,211 points and the edited dataset contained 50,777 points. Points that had previously gone through USGS review in the editor were removed before the samples were taken. A total of 96 points each were sampled from the unedited and edited datasets. For each sampled point, in both studies, USGS project staff evaluated the validity of the identification, and the position and attribute accuracy. For the FY12 Colorado pilot project study, this allowed the overall accuracy of unedited data to be compared to standard-edited data and standard-edited data to be compared to peer-reviewed data. For the nationwide study, this allowed the overall accuracy of unedited data to be compared to edited (Standard, Peer Reviewed, Advanced) data.

For both studies, Safe Software FME® 2012 was used to track each structure point that was contributed to the map and record user login information, changes to location, attributes, and the time the change was made. Changes are stored in an OSM Extensible Markup Language (XML) standard planet file that the USGS downloads nightly. The file from the previous day is not overwritten, so a complete audit history of every point is available for a detailed analysis of changes to each structure as it was edited and quality checked.

Esri ArcGIS Data Reviewer® was used to create random samples. Using a confidence level of 95% and a 10% margin of error, points were randomly selected. Esri Data Reviewer® uses this standard formula

$$\textit{sample size} = \frac{Z^2 \times p \times (1 - p)}{m}$$

where

Z is the Z value (e.g., 1.96 for a 95% confidence interval)
p is the estimated incidence of the characteristic of interest in the population, expressed as a decimal percent (e.g., 0.3 = 30%). 0.5 maximizes the value $p(1-p)$ and is commonly used as a default.
m is the acceptable margin of error, expressed as a decimal (e.g., 0.10 = 10%)

Three types of errors were evaluated in both studies and one (errors of omission) was evaluated for only the FY12 Colorado pilot project study, as follows:

1. Horizontal positional errors: Points are positioned relative to The National Map orthoimagery (see https://nationalmap.gov/ortho.html for specifications) in the TNMCorps editor. For the FY12 Colorado pilot project study, a point passes the horizontal accuracy test if it was placed on the correct building using aerial imagery. The results of this evaluation are in columns five and six of Table 6.1. For the FY14 national data quality study, a point passes the horizontal accuracy test if it falls within the visual footprint of the correct building at a scale of 1:18,056. This scale was used to evaluate positional accuracy because one of the primary derivative products of The National Map data, the US Topo map, is created at a scale of 1:24,000. The closest, larger scale in the set zoom levels in the TNMCorps editor is 1:18,056. This methodology for measuring horizontal positional errors is slightly different than what was used during the 2012 study.

Table 6.1 Attribute and positional accuracy by stage, fiscal year 2012 Colorado pilot project

		Passed points			
		Attribute (Name)		Position	
Stage	Number of points sampled	Number	Percent	Number	Percent
Unedited	54	40	74	41	76
Standard edited	32	24	75	26	81
Peer reviewed	14	14	100	14	100

Table 6.2 Errors of Commission, fiscal year 2012 Colorado pilot project

		Errors of commission	
Stage	Number of points sampled	Number	Percent
Unedited	54	7	13
Standard edited	32	4	13
Peer reviewed	14	0	0

The 2012 study did not take into account the zoom level of the editor when the point was checked, and points were considered to have failed if they were not on the building regardless of the zoom level. The results of this evaluation are in columns five and six of Table 6.1.

2. Attribute errors: Name is a required attribute and street address is optional. For a point to pass the attribute accuracy test, the name must agree with the name found in an independent authoritative source (such as the official website for the facility). Because street address is an optional attribute, it was not included in these evaluations. The results of this evaluation are in columns three and four of Table 6.1.
3. Errors of commission: Points are identified falsely where no actual feature exists. It is usually possible, though often expensive, to determine whether or not a given feature exists in the real world. Independent confirmation of feature existence was attempted for every point sampled in these studies. The removal of a false point by a volunteer is a data improvement, and counts as a "pass" during evaluation. The results of this evaluation are in Table 6.2. Any points that are found to not exist in the real world are counted as a "fail" for this evaluation. Subsequently, the number of points deemed "errors of commission" are removed from the total sample size for the positional and attribute accuracy calculations. The results of this evaluation are in Table 6.2.
4. Errors of omission (FY12 Colorado pilot project study only): Not identifying a point where a relevant real-world feature does exist. These errors are difficult to measure because it is generally not possible to establish an absolute baseline for completeness of structures data. However, the ten structures within the domain of this project include one feature type for which a complete baseline could be closely approximated during the FY12 Colorado pilot project study. The U.S. Postal Service (USPS) publishes an online directory of post office locations, so for any zip code area it is possible (though time consuming) to find the

Table 6.3 Errors of omission, fiscal year 2012 Colorado pilot project

Stage	Identified post offices	Errors of omission	Completeness, in percent
Unedited	29	20	59
Standard edited	45	4	92
Peer reviewed	49	0	100

locations of all post offices. This was done for selected quads after they had been through Peer Review to evaluate the effects of the VGI process on feature class completeness. The study of errors of omission was, therefore, done over a different population than the other three types of errors. The population of interest in this case is the set of all real-world post offices as identified by the USPS, over the area of three 7.5-min quadrangles that had completed the Peer Review AAQ process (17 post offices, as shown in Table 6.3). These points were not sampled and every post office location was inspected for a corresponding USGS data point. For the data point to pass, the post office feature had to be present, positioned accurately, and correctly named. The results of this evaluation are in Table 6.3. Because a national authoritative dataset to compare against did not exist at the time of the FY14 national data quality study, this measure was not evaluated.

6.3.1 Results

Table 6.1 shows attribute and positional accuracy at each of the data stages during the FY12 Colorado pilot project quality study. Each of these two types of accuracy was improved by each successive stage, reaching 100% for points that had been checked by Peer Review volunteers.

Errors of commission (Table 6.2) are the number of points that needed to be removed during USGS inspection. The same trend as in Table 6.1 is apparent, which again validates Hypothesis 1. The data in Tables 6.1 and 6.2 also confirm Hypothesis 2 that an additional stage of internal USGS quality control can monitor the data quality at each of the other stages. (Note that Table 6.1 measures accuracy, whereas Table 6.2 measures errors; in Table 6.2, low numbers are good.)

It is impossible to obtain a 100% complete dataset for most feature types. Post offices were an exception (or nearly so) within the domain of the FY12 Colorado pilot project because the USPS publishes post office locations. The official USPS website has a locator tool (https://tools.usps.com/go/POLocatorAction_input) that was used to determine a complete dataset for quads that had been through peer review. For these quads, the set of USPS locations was compared to the USGS set of post office points. This population was not sampled, but was completely inspected. Table 6.3 shows the results of this evaluation. The true complete dataset for this area, established by internal quality assurance for the sample area, includes 49 post offices. Again, the accuracy of the data improved with each successive stage.

Table 6.4 Attribute and positional accuracy by stage, fiscal year 2014 national data quality study

		Passed points			
		Attribute (Name)		Position	
Data Sample	Number of points sampled	Number	Percent	Number	Percent
Unedited	**88**	**75**	**85**	**70**	**80**
Source:					
GNIS	41	34	83	32	78
NSD	47	41	87	39	83
Edited	**89**	**85**	**96**	**86**	**97**
Stage:					
Standard edited	69	65	94	66	96
Peer reviewed	11	11	100	11	100
Advanced edited	9	9	100	9	100

Table 6.5 Errors of Commission, fiscal year 2014 national quality study

		Errors of commission	
Data sample	Number of points sampled	Number	Percent
Unedited	**96**	**8**	**8**
Source:			
GNIS	42	1	2
NSD	54	7	13
Edited	**96**	**7**	**7**
Stage:			
Standard edited	73	4	5
Peer reviewed	14	3	21
Advanced edited	9	0	0

Table 6.4 shows attribute and positional accuracy for unedited data versus edited data during the FY14 national data quality study. The data sampled is further broken down by the source and the edit stage. Each of the two types of accuracy was improved by volunteers, showing great improvement just from Standard Editing, and reaching 100% for points that had been checked by Peer Reviewers or Advanced Editors.

The same trend is apparent, again validating Hypothesis 1. Overall, errors of commission were lower in the edited dataset with a slight increase in the number of errors of commission in peer review. In the advanced edit dataset, no errors of commission were found, meaning that Advanced Editors found and removed all points that should not have been on the map. (Note that Table 6.4 measures accuracy, whereas Table 6.5 measures errors; in Table 6.5, low numbers are good.)

For the FY14 national data quality study, errors of omission were not calculated. Because of the difficult nature of measuring completeness, and the lack of a truly complete dataset to measure against, it was decided not to repeat this portion of the study.

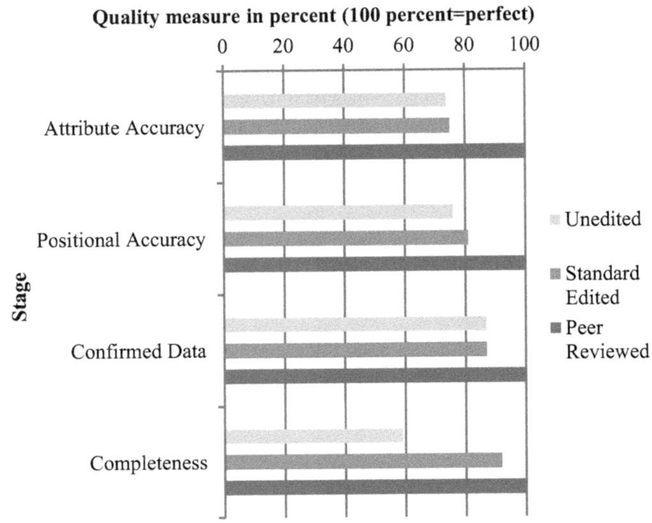

Fig. 6.2 Fiscal year 2012 Colorado pilot project quality study results

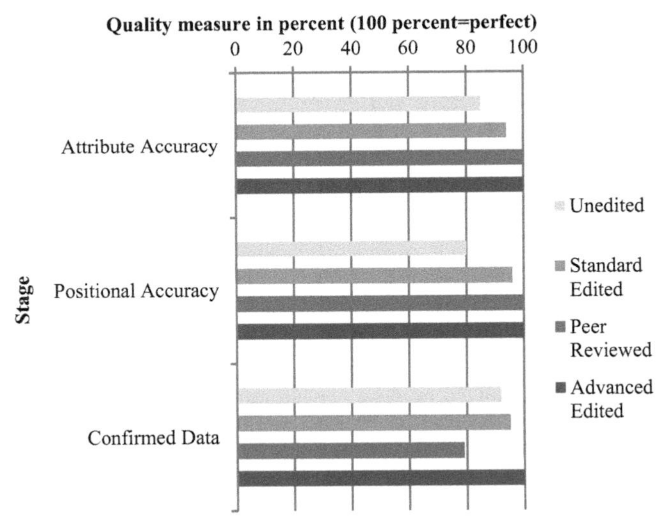

Fig. 6.3 Fiscal year 2014 national quality study results

Figures 6.2 and 6.3 show a different representation of selected quality measures from Tables 6.1, 6.2, 6.3, 6.4 and 6.5. The figures illustrate the consistent pattern, namely that for almost all quality measures, the volunteers improve the baseline data and quality increases through hierarchical stages. Confirmed data show the percentage of points analyzed that were correctly placed on the map.

6.4 Analysis and Discussion

Data quality assessment and monitoring are a top priority for the TNMCorps project. The purpose of conducting a third quality study was to validate the quality assessment methodology used for the project. As evidenced in the literature, the hierarchical editing process is a common approach to data quality assurance (Goodchild and Li 2012). The results from our quality study confirm the effectiveness of a tiered-editing approach. As volunteers progress in the hierarchy, positional and attribute accuracy improve, which improves the overall quality of the baseline data. An in-depth data quality analysis has helped identify potential strengths and weaknesses within the project. As discussed by Hunter et al. (2012), data quality concerns arise when data review processes and volunteer communication stagnate. The effectiveness of the hierarchical editing approach is an evident strength in the TNMCorps project. Improvement of the data quality over time speaks to the dedication of TNMCorps volunteers to citizen science. Daily quality monitoring provides the opportunity to communicate and collaborate with volunteers through frequent user feedback. Dynamic data quality review processes and individual user feedback are two strengths that set TNMCorps apart among other citizen science initiatives. Some weaknesses were also identified. Data contributed to TNMCorps is not cross-referenced with an authoritative dataset. One of the purposes of TNMCorps is to collect and develop authoritative datasets in collaboration with The National Map and USGS topographic maps. As mentioned by Bordogna et al. (2014), referencing an authoritative dataset is one of the primary approaches to data quality management. Because a complete authoritative dataset does not exist at this time for cross-referencing purposes, the hierarchical editing structure was applied to ensure the highest data quality possible. The majority of TNMCorps data are contributed by a small number of volunteers, which is a trend commonly referred to as "participation inequality" (Brovelli et al. 2016). Participation inequality is present in the project, but cannot be seen as an evident strength or weakness. It is crucial to note that contributions from Advanced Editors are significantly more accurate than baseline data. A potential setback to having a large portion of data contributed by a small volunteer base is the loss of key contributors over time. However, the number of core volunteers has continued to grow and sustain data contributions thus far due in part to active recruitment efforts via social media and http://www.volunteer.gov. TNMCorps engages volunteers through mapping challenges, volunteer badge recognition, and Advanced Editor recruitment.

Future research is needed to assess different facets of data quality and volunteer motivation. Areas of future research include topics such as the following: how to increase the commitment and recruitment of Advanced Editors to expand our user base, how to continue to establish trust with volunteers, how to assess the role of trust in the improvement of data quality over time, and the deployment of a mobile application for TNMCorps. Through continued research and the dedication of volunteers, TNMCorps can be a sustainable citizen science project providing a needed resource in support of TNM and US Topo maps.

6.5 Conclusions

This analysis of the FY12 Colorado pilot project and the FY14 nationwide expansion of the VGI project validate the data collection model. For all structure feature types, volunteer involvement has been shown to improve positional accuracy, attribute accuracy, and reduce errors of commission. Errors of omission are more difficult to study and quantify, but the study of post offices in the FY12 Colorado pilot project study provides some evidence that the volunteer model improves completeness as well.

The cornerstone of this model is a Wikipedia-like hierarchy, where editors and reviewers collaborate to improve attribute information and locational accuracy of baseline data to increase data quality. The FY12 Colorado pilot project and FY14 national data quality studies demonstrated that volunteer edits improve our baseline structures data even from Standard Editors and further review by Peer Reviewers. Advanced Editors improve the data further, and sample-based inspection by USGS personnel can monitor these processes.

References

Bordogna G, Carrara P, Criscuolo L, Pepe M, Rampini A (2014) A linguistic decision making approach to assess the quality of volunteer geographic information for citizen science. Inf Sci 258:312–327. doi:10.1016/j.ins.2013.07.013

Brovelli MA, Minghini M, Zamboni G (2016) Public participation in GIS via mobile applications. ISPRS J Photogramm Remote Sens 114:306–315. doi:10.1016/j.isprsjprs.2015.04.002

Craglia M, Shanley L (2015) Data democracy – increased supply of geospatial information and expanded participatory processes in the production of data. Int J Digital Earth 8:1–15. doi:10.1080/17538947.2015.1008214

Goodchild MF (2007) Citizens as sensors: the world of volunteered geography. Geojournal 69(4):211–221. doi:10.1007/s10708-007-9111-y

Goodchild MF, Li L (2012) Assuring the quality of volunteered geographic information. Spat Stat 1:110–120. doi:10.1016/j.spasta.2012.03.002

Hunter J, Alabri A, van Ingen C (2012) Assessing the quality and trustworthiness of citizen science data. Concurr Comput: Pract Exp 25(4):454–466. doi:10.1002/cpe.2923

Moore L (2011) US topo — a new national map series. Dir Mag, May 16. http://www.directionsmag.com/entry/us-topo-a-new-national-map-series/178707

Neis P, Zielstra D (2014) Recent developments and future trends in volunteered geographic information research: the case of OpenStreetMap. Future Internet 6(1):76–106. doi:10.3390/fi6010076

Poore BS, Wolf EB, Korris EM, Walter JL, Matthews GD (2012) Structures data collection for the national map using volunteered geographic information, Open-file report 2012–1209. U.S. Geological Survey, Reston

U.S. Geological Survey (USGS) (2012) *Guidelines for contributing structures data to the national map*. U.S. Geological Survey, Reston

Wolf EB, Matthews GD, McNinch K, Poore BS (2011) OpenStreetMap collaborative prototype, phase one, Open-file report 2011–1136. U.S. Geological Survey, Reston

Software

FME (2012) Windows. Safe software, 2012

ArcGIS Data Reviewer 2010 Windows. ESRI

Chapter 7
On Reliability of Routes Computed Based on Crowdsourced Points of Interest

Monir H. Sharker, Jessica G. Benner, and Hassan A. Karimi

Abstract Today's routing services provide routes that meet different needs and preferences of different users. To compute desired (optimal) routes, these services must support databases that contain accurate origin and destination locations, a high accuracy road network database, and an optimization algorithm. Origin and destination (O/D) locations are commonly collected by professional, commercial, and crowd sources via manual or automatic (geocoding) approaches. The routes computed for the same pairs of locations (O/D) obtained from different sources may be different. Considering the increased interest in collecting points of interest (POIs) through crowdsourcing, in this chapter, we address this research question: Are the routes computed using crowdsourced POIs (as O/D) reliable? To address this question, we conducted experiments where routes (shortest and fastest) computed using crowdsourced POIs (e.g., through OpenStreetMap) were compared with the routes computed using POIs obtained from professional and commercial sources. Metrics including route length, travel time, Euclidian distance, and number of road segments were used in the comparisons. The results reveal that, in general, though there are no significant differences between the routes, differences usually occur at or near origins and destinations.

Keywords Crowdsourced POI • Routing • OpenStreetMap • OSM reliability • Volunteered geographic information

7.1 Introduction

With the advent of Internet-based and mobile user devices, forums and social networking sites with many different options for user interaction, it has become increasingly easy to share location information. Today, roads, areas, and points of interest (POIs), among other objects of interest, are being collected and/or updated

---

M.H. Sharker (✉) • J.G. Benner • H.A. Karimi
Geoinformatics Laboratory, School of Information Sciences, University of Pittsburgh, Pittsburgh, PA, USA
e-mail: mhs37@pitt.edu

M. Leitner, J. Jokar Arsanjani (eds.), *Citizen Empowered Mapping*,
Geotechnologies and the Environment 18, DOI 10.1007/978-3-319-51629-5_7

through crowdsourcing services such as OpenStreetMap (OSM) (Benner and Karimi 2013). Originally intended for supporting collection of street data, OSM has grown to include many new features which facilitate collection of not only street data but also POIs as different land use features like school, hospital, airport, hotel, park (Goetz and Zipf 2012, Jokar Arsanjani et al. 2015). One issue regarding OSM, as well as all crowdsourcing services, is the reliability of the information collected in it (Girres and Touya 2010; Haklay 2010; Neis et al. 2012; Hochmair and Zielstra 2013). Considering the volunteered nature of collecting and sharing POIs, and other data, in crowdsourcing services, some POIs may be marked on the wrong side of the street or at the wrong intersection. Such errors in accurately pinpointing locations of POIs for finding routes could lead to unexpected consequences such as loss of time backtracking to the real location, extra cost (e.g., fuel consumption) to travel around blocks, or loss of driver's temperament which may affect his/her driving behavior.

Crowdsourced POIs are important sources of origin and destination (O/D) locations to find routes. While shortest or fastest are most common, other criteria include least intersections, least number of turns, least fuel cost, and least air pollution exposure (Sharker and Karimi 2014). Location accuracy of POI (O/D) pairs plays an important role in computing routes while other factors are quality of the underlying road network database and the performance of the optimization algorithm. However, since people use different devices and assumptions in collecting POIs, location accuracy of crowdsourced POIs is uncertain and inconsistent. Considering the increased interest in the use of OSM, and similar crowdsourced databases, for routing, the research question that is addressed in this chapter is: Are the routes computed using crowdsourced POIs (as O/D) reliable? The hypothesis is that *the differences in POI locations among crowdsourced data and authoritative and/or commercial data do not significantly impact the computed routes*. To address this research question, route reliability is measured based on metrics such as length (both Euclidian distance and route length), travel time, and number of road segments. The main contribution of this chapter is the method of examining and analysing the reliability of routes computed based on crowdsourced POIs as O/D pairs. The rest of the chapter is organized as follows. Section 7.2 provides background information on crowdsourcing, routing, and entity matching in geospatial databases. The method used and the experiments conducted are explained in Sect. 7.3. Section 7.4 provides the results of comparison and analysis followed by Sect. 7.5 that concludes the chapter and presents future research directions.

7.2 Background

In this section, first a discussion of crowdsourcing, specifically the OSM service, is provided. Next, the routing process and previous/related work on routing using OSM are overviewed. Finally, the process of geospatial entity resolution, a process commonly used to match referenced locations, is explained.

7.2.1 OpenStreetMap

OSM is a mapping service that has grown from a small start-up project in 2004 (Gyford 2004) to one of the largest sources of open geographic data in the world by 2016. Any user can add data to the map or modify the data added by other users. The dataset is generated as users: (1) digitize features or edit the locations and attributes of existing features, (2) bulk import complete datasets such as geographic names information system (GNIS) or TIGER line files in the US, and (3) upload their personal GPS trajectories. Users can add nodes, ways and relations to the database and use them to represent real-world features like buildings, street segments, turn restrictions on certain streets, land use, among other features. For more information about the OSM data model, refer to the OSM Wiki entry on Elements.[1] POIs in OSM can be represented as both nodes and closed ways depending on the type of POI being represented. Collected POIs in OSM are used in different applications and services including routing services.

Several studies have investigated the quality of OSM data. Haklay (2010) found that OSM data is fairly accurate (within 6 m on average) when compared to the UK Ordinance Survey data. Girres and Touya (2010) found an average positional difference of 6.65 m and small geometric differences between polygon features (e.g., lakes) in France. Ziestra and Zipf (2010) noted that the OSM street network in Germany had inadequate support for car navigation applications and conclude that while OSM offers a large amount of data, it is not an adequate alternative to TeleAtlas due to its lack of coverage in rural areas. Neis et al. (2012) focused on the expansion of the street network and route network for car navigation in OSM Germany and found that, when compared to Tom Tom (TeleAtlas), OSM included more pedestrian level data compared to data for car navigation. However, the quality of OSM data is expected to improve over time which may address the abovementioned issues.

7.2.2 Routing Using OSM

OSM data supports trip planning options for travellers in such services as OpenTrip-Planner, skobbler, and Garmin. According to the OSM Routing Wiki[2] entry in April 2016, there are at least nine routing services utilizing OSM data. The attributes of the road networks, such as one-way streets and speed limits on road segments (part of a road that is stored in the road network database as a unique component), in OSM's database are discussed in the OSM Routing wiki entry and the OSM Tags for Routing wiki entry. The literature, as of writing this article in April 2016,

---

[1] OSM Wiki "Elements": http://wiki.openstreetmap.org/wiki/Elements

[2] OSM Wiki "Routing": http://wiki.openstreetmap.org/wiki/Routing

does not include any work related to our works in this chapter. However, the work of Schmitz et al. (2008) is worth mentioning. They used OpenRouteService.org (ORS), a route service that utilizes OSM data, to test the fitness of OSM data for routing in Germany and several other European countries. They counted number of failed route requests sent to ORS over three time periods and found that the decrease in failed requests indicated the suitability of OSM data for routing over time. The authors identified the OSM data problems as unrecognized junctions and inconsistent attribution, and OSM's strengths as data richness and speed of data correction. Their work is focused on a specific routing service to assess the overall quality of OSM data, including road network quality, which is different than our work. They evaluate OSM data quality based on number of successful/unsuccessful routes, whereas our work is focused on assessing the quality and reliability of OSM POIs for routing.

7.2.3 Geospatial Entity Resolution

In order to compare the routes computed based on crowdsourced POIs with the routes computed based on POIs obtained through other approaches, a matching method is needed. One such method is "Geospatial entity resolution" which determines a set of "true" locations from multiple data sources (Kang et al. 2007). More specifically, the geospatial entity resolution method determines whether two location references, in different datasets, represent the same real-world location (Sehgal et al. 2006). Several approaches have been applied in a real-world setting. The most common method relies solely on spatial coordinates and matches objects using a one-sided nearest join (Beeri et al. 2004). Other methods include string-matching algorithms and feature-based matching using location features (i.e., attributes) (Sehgal et al. 2006). A newer development is the use of interactive visualization tools to aid in the matching process (Kang et al. 2007).

There are several issues when dealing with entity resolution in the geospatial domain such as data collectors using different terms and language, and data collectors' use of varying scales and geometries (points vs. areas) to represent the real-world entities (Sehgal et al. 2006). POI attributes such as type (e.g., school, hospital) can be used to increase the confidence of a match among entities (Sehgal et al. 2006). Geospatial databases generally include attributes alongside the POI geometry, thus, the method employed in Sehgal et al. (2006) is used in this work.

7.3 Method and Experiments

This section details the method used and the experiments conducted in this work. First, an overview of our method is presented which includes a description of each data source. Next, the POI matching and route computation methods are discussed.

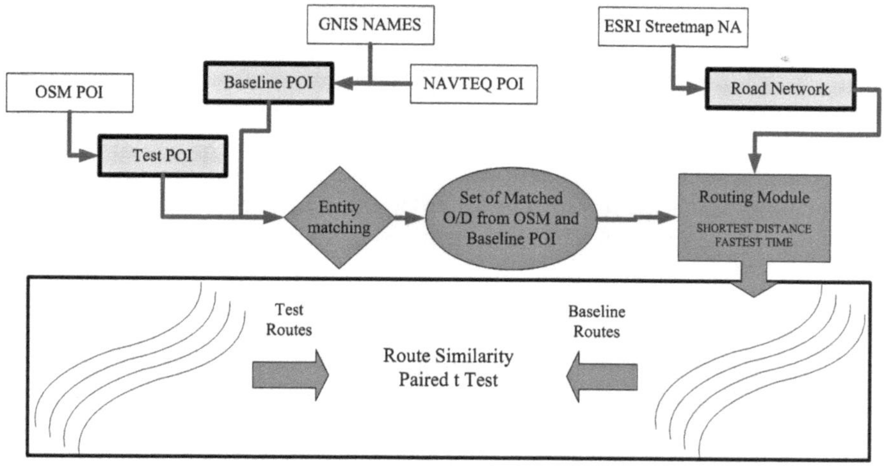

Fig. 7.1 Method of reliability measures comparing routes computed using test POIs and baseline POIs

Finally, the similarity of computed routes is explained. Figure 7.1 illustrates the method used in the experiments and will be discussed in the next section.

7.3.1 Method and Data Sources

The study was designed to compare routes computed between the same set of O/D pairs obtained from different sources of POIs. The experiments were focused on routes computed for car navigation. Two sets of POI were used in the method: a set of crowdsourced POIs and a set of baseline POIs. The first set was used to test the reliability of routes computed using crowdsourced POIs as O/D pairs. The crowdsourced POIs utilized in the experiments were obtained from OSM. The second set, collected by professionals and commercial entities, was used as a baseline to compare routes computed using POIs from the first set. We used two different sets of POIs as baseline: one maintained by the U.S. Board of Geographic Names and another collected by HERE (formerly NAVTEQ), a commercial navigation service provider. For comparison, we utilized the geospatial entity matching method (Sehgal et al. 2006) to identify matched pairs between the test and baseline datasets. Once matched, Euclidean distance between each pair was computed and recorded.

Another dataset needed for computing and comparing routes for car navigation is a road network. We selected the North America Detailed Streets dataset provided by the Environmental Systems Research Institute (ESRI). We chose this dataset, among other options such as the street networks provided by OSM and NAVTEQ, based on the criteria that it is independent of both the test and the baseline datasets and it contains all the required parameters for route computation. Different road network datasets, available both publicly and commercially, have different positional errors and/or generalizations (Frizzelle et al. 2009) that may influence route computation. Routing between POIs may be influenced by the road network data characteristics if the road network data and the POI data are obtained from the same data source. On the other hand, using one road network (independent of any POI data sources) to compute routes helps reduce the margin of error because the influence, if any, from the road network dataset is consistent across all the computed routes. In the experiments, we computed routes based on shortest distance and fastest time criteria for each matched pair of POIs resulting in two sets of routes: one from test POIs and another from baseline POIs. Route similarity measures (for reliability assessment), for both shortest and fastest routes, were recorded for the following metrics: route length, travel time, number of segments, and Euclidean distance (between each pair of O/D in each route).

The road network data was obtained from ESRI for Chicago Metro Area as a test case. Comparable test and baseline POIs in the area are randomly chosen from the sets of matched pairs. All datasets were processed in the same coordinate system and projection; World Geodetic System 1984 (WGS84), Universal Transverse Mercator Zone 16 N. All datasets were clipped to the administrative boundary of the Chicago Metropolitan Area. This boundary was extracted from the Urban Areas dataset in the U.S. National Atlas. Given the limited availability of NAVTEQ data, only POIs within the city of Chicago were obtained. However, the method used in the study is not impacted by the choice of geographic area. The remainder of this section discusses the specific datasets used in the experiments.

7.3.1.1 Baseline POIs

The Geographic Names Information System (GNIS) is a national standard governing geospatial nomenclature designed by the U.S. Geological Survey and managed by the U.S. Board on Geographic Names (BGN). The scale of the GNIS dataset is at 1:24,000 and the absolute horizontal positional accuracy complies with the National Map Accuracy Standard of +/−12.2 m relevant to either a 7.5 min USGS topographic map or a USFS map (USGS 1999). On the BGN website[3], the GNIS is described as the "official repository of domestic geographic names and data". The state of Illinois dataset was obtained from 'States, Territories, Associated Areas of the United States' data category on the U.S. Board from the Geographic Names website. In both 2007 and 2009, the GNIS dataset of Domestic Names was imported into OSM's database. Even though it has been shown that contributors have made modifications to many imported GNIS points (Hochmair and Zielstra 2013), any influence from the GNIS data imports from OSM's database was eliminated by filtering out POIs with any of the following keys,

[3]BGN Website: http://geonames.usgs.gov/domestic/index.html

associated with GNIS, from the Test POI (OSM) dataset: gnis:Class, gnis:County, gnis:County_num, gnis:ST_alpha, gnis:ST_num, ele, gnis:county_id, gnis:created, and gnis:state_id.

The premium POIs from NAVTEQ is a comprehensive database for commercial navigation systems and services. In the experiments, NAVTEQ Premium data for the city of Chicago, IL, included in the Premium Data package, was utilized. The scale of the selected POIs in the Premium Data package is at 1:95,000 and the absolute horizontal accuracy of POI are +/− 5 m relevant to the street centreline (NAVTEQ 2012).

7.3.1.2 Test POIs

We tested the reliability of the routes computed based on POIs from OSM's database stored as points. The POI dataset for the Chicago Metropolitan Area was extracted from a 2014 extract available on the Geofabrik webpage[4] for North America using the osmosis tool.[5]

7.3.1.3 Road Network Data

The road network data for the Chicago Metropolitan Area, with length, speed limit, and directionality as minimum required attributes, was extracted from the North American Detailed Streets dataset, maintained by ESRI. The travel time on each road segment of the extracted network was computed by using segment length and segment speed limit attributes. This database, which is independent of both POI data sources (test and baseline POIs), was used as the reference road network for route computation.

7.3.1.4 POI Matching

We ensured that the POIs (O/D pairs) in both test and baseline sets correspond to the same entities. For this, we used geometric information, i.e., *location (x, y)*, and semantic information, i.e., *semantic (N, T),* where N is entity name and T is POI type. Table 7.1 shows a summary of available POIs in the study area. The two POI baseline datasets, GNIS and NAVTEQ, support semantic information for each POI, however, in the case of OSM, only 1717 out of 2470 POIs (70%) support semantic information N.

The POI datasets were filtered in three steps. In the first step, the POI *type (T)* was used to filter out a subset of POIs to be used in the study; the chosen *T* are school, post office, hospital, library, and fire station. These five POI types were chosen as

[4]Geofabrik North America: http://download.geofabrik.de/north-america.html

[5]osmosis tool: http://wiki.openstreetmap.org/wiki/Osmosis

Table 7.1 Number of records in each POI dataset that include geometry and semantic information

POI dataset	Location (x,y)	Type (T)	Name (N)
OSM	2470	2470	1717
GNIS	6206	6206	6206
NAVTEQ	18,180	18,180	18,180

Table 7.2 Matched POI details in Chicago Metro Area. + this value (−1) due to the triad

	OSM		GNIS		NAVTEQ	
POI Type	Total	Matched (22)+	Total	Matched (12)	Total	Matched (11)
Fire station	16	8	342	8	–	0
Post office	24	4	58	1	65	3
Library	13	2	80	1	93	1
Hospital (Dept.)	3	2	150	2	78	1
School	15	6	2568	0	1118	6

a representative sample of typical route destinations. In the second step, the POI *location (x,y)* was used to filter out POIs; for each *T*, nearby POIs of the same *T* are chosen as candidates. In the third step, each candidate POI is further evaluated based on its *name (N)*. Table 7.2 shows the matching summary found for each dataset. Matches were found for *T* (fire station, post office, library, and hospital) between OSM and GNIS, and *T* (post office, library, hospital, and school) between OSM and NAVTEQ.

Figure 7.2 shows the sets of matched pairs within Chicago Metropolitan Area. All except one matched pair between OSM and GNIS fall outside of the municipality boundaries of Chicago while all OSM and NAVTEQ pairs fall within the city. We computed the Euclidean distance between all matched POIs and Table 7.3 shows the summary statistics for the results. The range of distances between matched OSM and GNIS POIs is larger with the smallest distance being 2.81 m and the largest distance 505.59 meters. The range of distances between OSM and NAVTEQ POIs is smaller with a minimum of 21.01 m and a maximum of 139.94 m. On average, the Euclidean distance between matched OSM and GNIS POIs is larger than those matched in OSM and NAVTEQ. Figure 7.3 shows the Euclidian distance for each matched pair in NAVTEQ (Fig. 7.3a) and GNIS (Fig. 7.3b). The NAVTEQ pairs show larger distances between POIs than POIs in GNIS pairs. The GNIS matched pairs show small Euclidean distances for the majority of POIs but also include two extreme cases (#118 and # 130) that are significantly larger than the other cases. Both POIs were checked and their locations were confirmed by using OSM History Viewer[6] and visually inspected on the map to ensure that correct matching was performed. The metadata available in the OSM History viewer confirmed that the POIs in question were added to OSM along with the location information as reported and were not the result of a processing error in our study.

[6]OSM History Viewer site: http://osmhv.openstreetmap.de/index.jsp

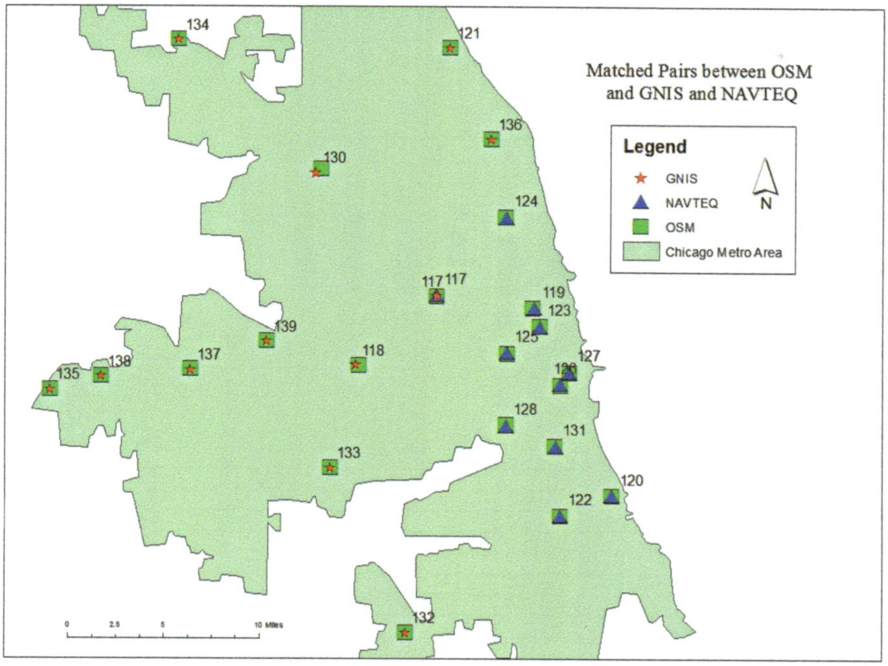

Fig. 7.2 Map of matched POI pairs

Table 7.3 Summary of Euclidean distance between matched POIs

	GNIS	NAVTEQ
	Distance (m)	
Min	2.81	21.01
Max	505.59	139.94
Average	74.49	60.21

7.3.1.5 Route Computation

The sets of matched POIs were utilized for route computation, OSM–GNIS and OSM–NAVTEQ. For routing, each POI was used as origin to all other POIs as destinations; n(n − 1) shortest and n(n − 1) fastest routes. A total of 132 routes were computed for OSM–GNIS O/D pairs and 110 routes for OSM–NAVTEQ O/D pairs. For each computed route, total distance, Euclidean distance, travel time, and number of road segments were recorded. Figure 7.4 shows the computed routes and the differences in both shortest and fastest routes over test and baseline POI datasets. Fig. 7.4a shows the shortest and fastest routes computed between OSM and GNIS POIs and Fig. 7.4b shows the shortest and fastest routes computed between OSM and NAVTEQ POIs.

Since routes computed based on POIs from OSM with GNIS and OSM with NAVTEQ were non-overlapping, the descriptive statistics were summarized in two

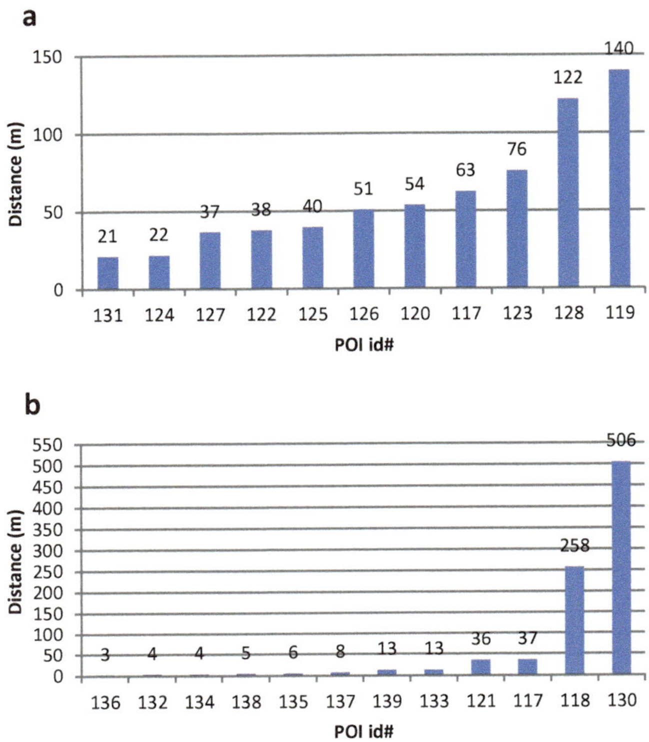

Fig. 7.3 Euclidean distance between individual matched POI pairs: (**a**) OSM and NAVTEQ; (**b**) OSM and GNIS

different tables. Table 7.4 shows a summary of 132 shortest routes that were computed for matched POIs in OSM and in GNIS. The computed routes, for OSM POIs, covered different route lengths ranging from 5.34 km to 70.72 km where the range of travel time was 6.34–80.39 min. The range for number of segments in the computed routes was 71–824. Table 7.5 shows the descriptive statistics of 110 shortest routes that were computed for matched POIs in OSM and in NAVTEQ. The computed routes, for OSM POIs, covered route lengths ranging from 1.77 to 27.02 km where the range of travel time was 1.88–29.64 min. The range for number of segments in the computed routes was 24–394. On average, compared to the shortest routes between GNIS POIs, corresponding shortest routes between OSM POIs are longer, faster, and have more segments. Conversely, compared to the shortest routes between NAVTEQ POIs, even though the shortest routes between OSM POIs are also longer, they take more time to travel and have equal number of segments.

Similarly for fastest routes, the descriptive statistics for OSM routes vs GNIS routes and OSM routes vs NAVTEQ routes are shown in Table 7.6. and Table 7.7, respectively. The ranges of route length, travel time, and number of segments in

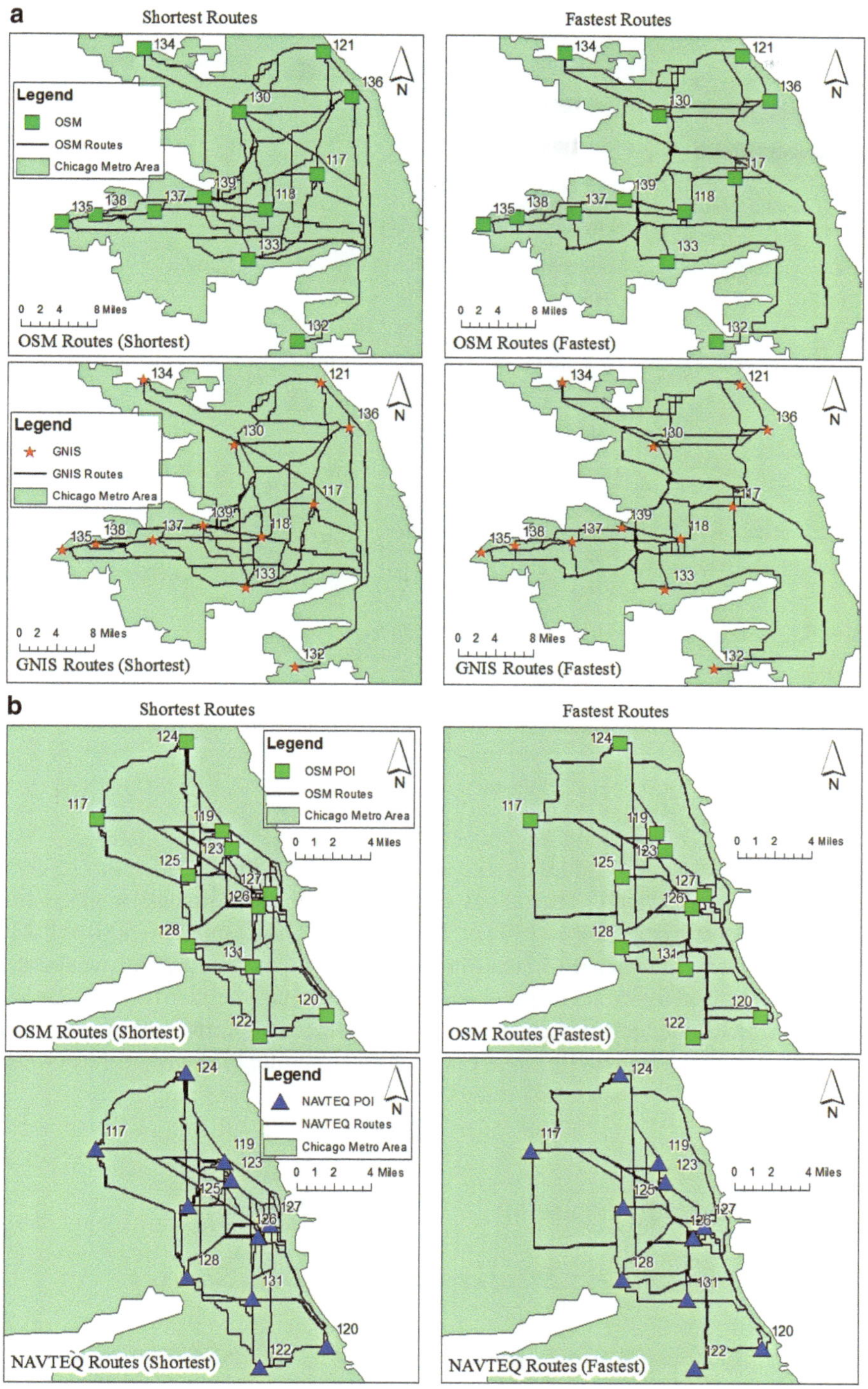

Fig. 7.4 Computed routes: (**a**) OSM/GNIS shortest and fastest routes; (**b**) OSM/NAVTEQ shortest and fastest routes

Table 7.4 Average measures for OSM and GNIS shortest routes

Average measures	Route length (m)	Travel time (s)	Number of segments
OSM	32,081	2163	374
GNIS	32,058	2169	372
Mean difference	−22	5	−1

Table 7.5 Average measures for OSM and NAVTEQ shortest routes

Average measures	Route length (m)	Travel time (s)	Number of segments
OSM	11,945	797	166
NAVTEQ	11,935	794	166
Mean difference	−9	−3	0

Table 7.6 Average measures for OSM and GNIS fastest routes

Average measures	Route length (m)	Travel time (s)	Number of segments
OSM	37,938	1848	402
GNIS	37,966	1849	403
Mean difference	28	1	<1

Table 7.7 Average measures for OSM and NAVTEQ fastest routes

Average measures	Route length (m)	Travel time (s)	Number of segments
OSM	13,003	649	176
NAVTEQ	12,908	644	174
Mean difference	−95	−5	−2

OSM and GNIS fastest routes (Table 7.6.) are 5.8–85 km, 6–66 min, 75–999, and 74–1189, respectively. The ranges of the same set of parameters in OSM and NAVTEQ fastest routes (Table 7.7) are 1.7–30 km, 1.88–22 min, 24–390, and 23–488, respectively. In the case of fastest routes, OSM routes, on average, are shorter, faster and include fewer segments than GNIS routes. On the contrary, the fastest routes in OSM are longer, slower and include more segments than those in NAVTEQ routes. For both routing criteria, NAVTEQ routes are shorter than OSM routes, while OSM routes are faster than GNIS routes.

7.4 Results and Discussion

7.4.1 Route Similarity Measures Between Data Sources

To test the reliability of using crowdsourced POIs in routing, a statistical significance test ($\alpha = 0.05$) was conducted on the computed parameters (total distance, Euclidean distance, travel time, and number of road segments) for all routes. The

Table 7.8 Statistical significance test results between OSM and baselines for shortest routes

Shortest routes	GNIS vs. OSM			NAVTEQ vs. OSM		
Similarity measure	t-Stat	P(T<=t) One-tail	P(T<=t) Two-tail	t-Stat	P(T<=t) One-tail	P(T<=t) Two-tail
Route length	−0.011	0.495	0.991	0.011	0.495	0.991
Travel time	0.041	0.484	0.968	0.049	0.48	0.961
Number of segments	−0.048	0.481	0.961	0.006	0.497	0.995

Table 7.9 Statistical significance test results between OSM and baselines for fastest routes

Fastest routes	GNIS vs OSM			NAVTEQ vs OSM		
Similarity measure	t-Stat	P(T<=t) One-tail	P(T<=t) Two-tail	t-Stat	P(T<=t) One-tail	P(T<=t) Two-tail
Route length	0.011	0.495	0.991	−0.107	0.457	0.914
Travel time	0.011	0.495	0.990	−0.128	0.448	0.897
Number of segments	0.027	0.488	0.977	−0.158	0.437	0.874

null hypothesis is that H_0: *there is no difference in routes computed using POIs from OSM compared to the routes computed using POIs from the baseline*. To test this hypothesis, the route differences were investigated in terms of mean route length, travel time, number of segments, and Euclidian distance.

The reliability of crowdsourced POIs, when used as O/D pairs, in routing is examined by similarity measures between OSM routes and GNIS routes and between OSM routes and NAVTEQ routes. Results of the statistical tests comparing OSM with GNIS and OSM with NAVTEQ are shown in Table 7.8 for shortest routes. Results of the same significance test comparing the fastest routes between OSM with GNIS and OSM with NAVTEQ are shown in Table 7.9. The comparison tests were performed on each of the parameters by using paired t-test for significance level of 5% ($\alpha = 0.05$). P-values for both one-tail and two-tail tests are reported. The result illustrates that none of the parameters are statistically significant ($p > 0.05$) for neither GNIS and OSM comparison nor NAVTEQ and OSM comparison. The differences in means are zero, that is, there are no statistically significant differences in the means of parameters computed for routes using POIs from OSM and POIs from GNIS or NAVTEQ. Based on these results, we can infer that the null hypothesis is not rejected, which means that OSM routes are not significantly different from GNIS and NAVTEQ routes.

A detailed pairwise comparison was performed for all route parameters. Figure 7.5 is a visualization of the similarity between computed routes discussed earlier. The figure shows one example of a pairwise comparison for route lengths of OSM and GNIS routes. The vertical axis reflects the length of each route which is represented along the horizontal axis. In the figure, solid lines represent GNIS routes and dashed lines represent OSM routes. We show only one sample graph, instead of 16 (4 parameters, 2 pair of comparisons: OSM-GNIS and OSM-NAVTEQ, 2

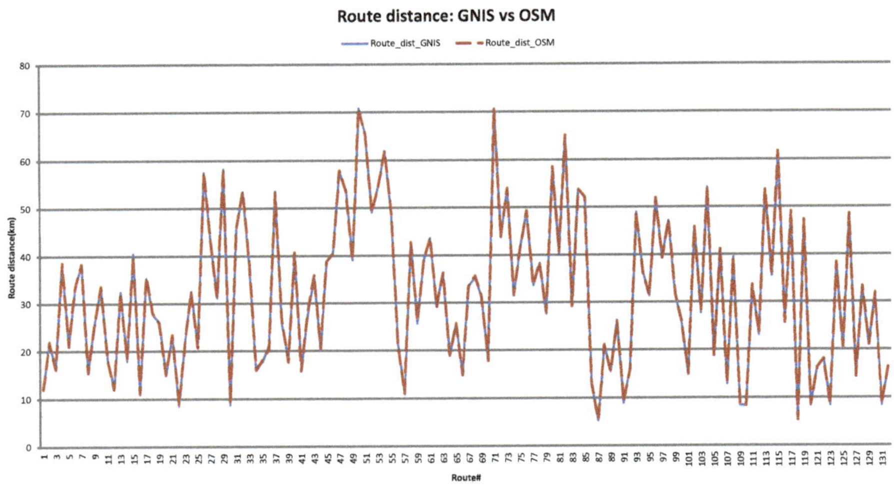

Fig. 7.5 Pairwise comparison of route length obtained for POIs from GNIS vs. OSM

preferences: shortest and fastest; $4 \times 2 \times 2 = 16$), because our results for this comparison show a similar pattern for all route parameters and for all route pairs. The main observation from this comparison is that the lines are mostly overlapping which means that the parameters of GNIS-OSM routes and NAVTEQ-OSM routes are similar.

The results of the significance test using average measures of route similarity indicates that there is no significant difference between routes computed between crowdsourced POIs and routes computed between the same POIs collected from professional and commercial sources. However, upon further examination at/around the origins and destinations of the computed routes, important differences are observed related to route length, number of segments, geocoding and whether a POI is an origin or destination in the route. These differences are important because they could lead the user to a wrong O/D location such as wrong side of the road, wrong intersection, rear side of a building (instead of front), just to name a few. These topics and a brief discussion of the limitations of this study are discussed in this section.

7.4.2 Distance and Segments

For each of the above comparative analyses and significance tests, the effects of route length (long, medium, short) on differences in number of segments between OSM routes and GNIS/NAVTEQ routes are shown in Fig. 7.6. This finding indicates that route length may not have an effect on the difference in number of segments. This means that differences in number of segments between short OSM routes, short GNIS routes and short NAVTEQ routes may even be higher than those computed

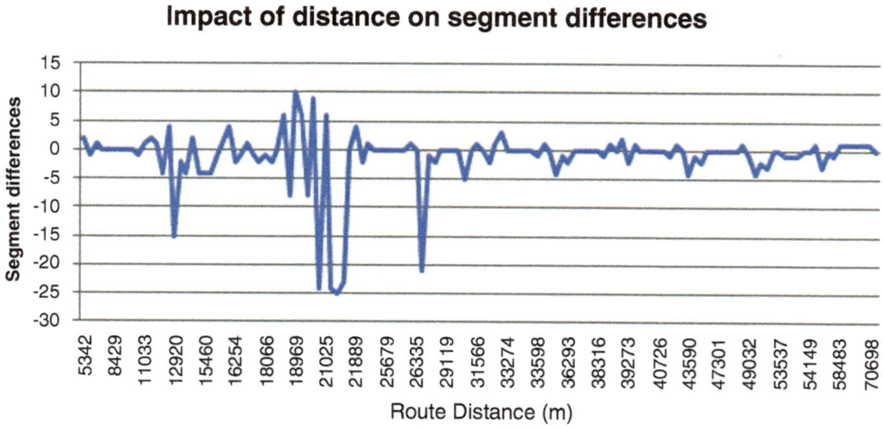

Fig. 7.6 Distance effects on route segment differences

for longer routes. For example, a 35.89 km long OSM route has 25 more segments than its corresponding GNIS route. Conversely, in a longer 70.73 km OSM route, there is exactly the same number of segments as in its corresponding GNIS route.

An analysis of Euclidean Distances between each pair of POI showed several data points with extreme values (possible outliers). It is logical that the number of segments within the pairs of routes that include these possible outliers as O/D would be larger in one route than the other. Figure 7.7 shows the differences in number of segments for each computed route (horizontal axis) and compares them by using the two routing criteria. Figure 7.7a shows the impact of POI's 130 and 118 (POIs with large Euclidian distances between them) with the majority of routes including those POIs reporting more than 10-segment difference between the shortest routes. Figure 7.7b shows the large differences between the fastest routes for POI's 120 and 127. Since these POIs did not report high Euclidian distances between them, other factors are suspected to play a role in the differences in number of segments. From these two cases, it becomes clear that the route length and number of segments parameters, while similar in the average case, exhibit unexplained behaviours in individual cases. This indicates the need for more study to identify new parameters that can account for variation between routes.

7.4.3 Geocoding

One reason for different POI locations is the method used to collect them. POIs can be collected manually or automatically. OSM and GNIS POIs are collected manually. OSM contributors use a combination of consumer grade GPS receivers on smartphones and digitization using an OSM Editor to add POIs to the database. GNIS POIs are collected via digitization of U.S. Geological Survey Topographic

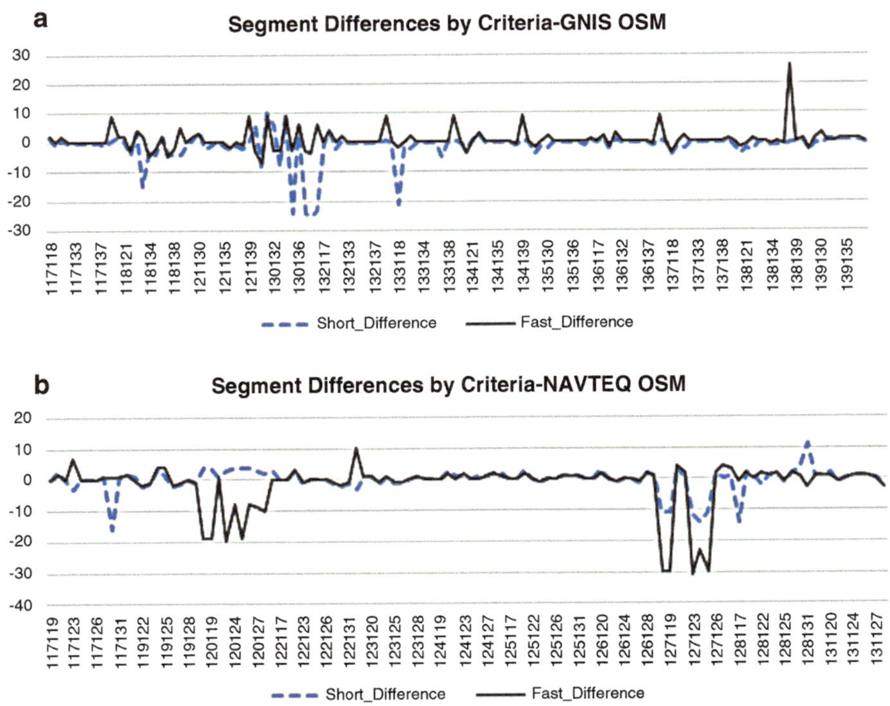

Fig. 7.7 Differences in number of segments for individual routes for (**a**) GNIS and OSM routes and (**b**) NAVTEQ and OSM routes

maps and some other Federal maps. NAVTEQ POIs are collected automatically through a geocoding algorithm. Reasons for locations difference are the assumptions and logics used in different geocoding algorithms (Karimi et al. 2004); for details on differences between different geocoding approaches and algorithms see Roongpiboonsopit and Karimi (2010a), b, and Karimi et al. (2011). Another difference is whether a POI is marked along the street or on the rooftop of a building.

These differences mean that the same POI could be closer to one segment in one case and closer to another segment in another case. Figure 7.8 depicts a triad of POIs that were matched from all three POI datasets overlaid on a set of buildings footprints created by the City of Chicago. Both OSM and GNIS show POI locations on the roof of the building while the NAVTEQ POI locations are along the street. OSM POIs are at the center of a square block of street segments, thus equidistant to each of those four segments. This means that each segment has an equal chance to be chosen as the start (origin) or end (destination) segment. Conversely, the NAVTEQ POI locations are close to the street segments making them closer to the actual start (origin) or end (destination) segments. Assuming that Fig. 7.8 is representative of the pattern of geocoding (rooftop vs. street) for the three datasets and knowing that NAVTEQ's database is designed for routing, it may be the case that OSM POIs are

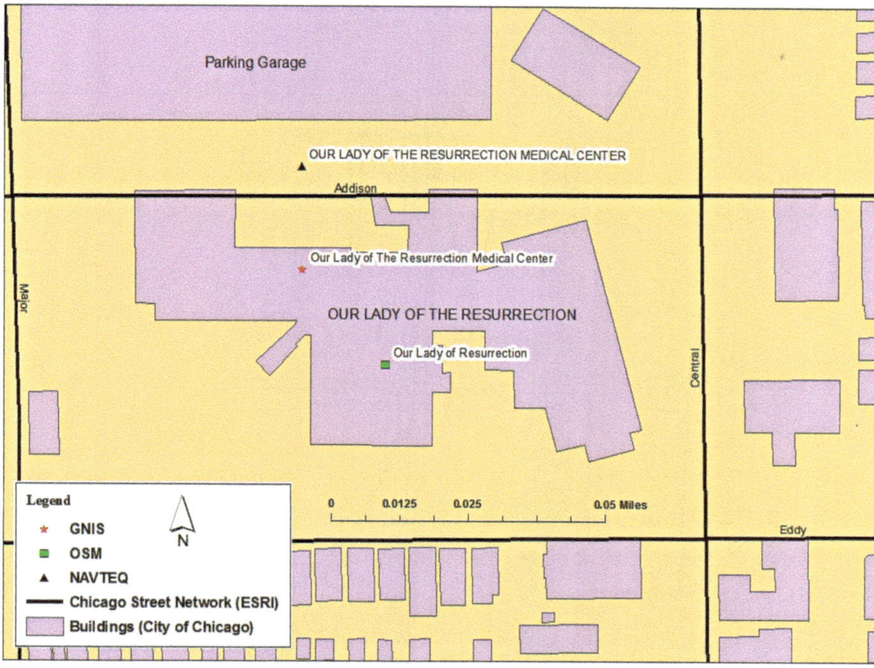

Fig. 7.8 Matched POIs illustrating impact on computed routes

collected along the rooftops. By removing the possible outliers (POI ID # 128, 119 in Fig. 7.3a and POI ID # 118, 130 in Fig. 7.3b) a different pattern of closeness emerges. Without the outliers, the average Euclidean distance between OSM and GNIS POI is now smaller (12.9 m) than the average distance between OSM and NAVTEQ (44.7 m). This may indicate that OSM POIs are farther from streets and less suitable for routing.

7.4.4 End Points

One important observation is that the differences, even though insignificant, occur usually at the two ends of a route (near origin and destination). This is because locations of POIs do not coincide in different datasets and the routes are computed using the closest road segment to each POI. Figure 7.9 shows four shortest routes: two based on OSM and GNIS data and two based on OSM and NAVTEQ data. The inset of each of the four maps shows the full extent of the route. In all cases, the majority of the routes overlap except at the end points. Each map is a large-scale view of the end points of the route. These views show the large differences between the segments that are selected along the end of a route.

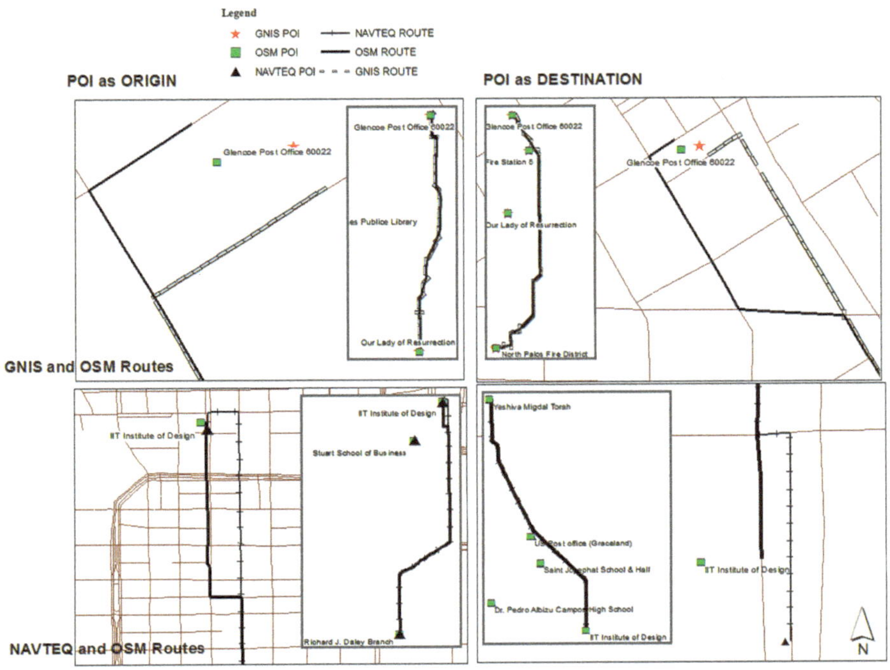

Fig. 7.9 Comparison of route end points

Finally, there are two limitations to this work. First, the use of commercial POI (i.e., NAVTEQ) was constrained by the availability of the dataset, which limited the study areas. Second, the set of POI that could be matched between the POI sources was small. The number of data points in OSM that supported semantic information was too low to support a higher match rate. The lack of semantic information is a significant issue with crowdsourced POIs. Choosing a study area with a higher level of semantic information, assuming commercial data is available in that area, is one potential approach to overcome this issue.

7.5 Conclusions and Future Research

We studied the reliability of crowdsourced POIs for routing. The results indicate that shorter routes may have larger differences in number of segments than longer routes. This may be explained by a higher availability of alternative routes when travel distance is short. The finding of the study indicates that crowdsourced POIs may be reliably used for routing with the exception of some differences near the end points in some cases. In short, it is shown that on average routes generated from POIs in OSM are no different from professionally collected POIs (e.g.,

GNIS) and commercial datasets collected automatically (e.g., NAVTEQ). However, several geospatial database issues such as, POIs representing multiple geometries (nodes and areas) (Sehgal, Getoor, and Viechnicki 2006) and lack of consistent and complete attributes (POI type, name) (Schmitz, Zipf and Neis 2008) still exist. The support for semantic information required for matching in OSM was very low (10.6%). Also, if a user wants to search for a particular destination and the POI in OSM has the data point in the database but there is insufficient data to identify it as the desired POI, then it cannot be used to generate a route to that destination. Potential errors in computed routes using crowdsourced POIs as O/D pairs could be due to poor accuracy of particular POI locations since the OSM data is open for change and can be changed by anybody, anytime. A route may start from a different road segment close to origin and end on a different road segment close to destination since it is evident from the study that most of the differences occur at the two ends of the computed route.

There are several future research directions for this study. First, extending the study to include other modes of travel such as walking, for different travellers, including those with special needs (e.g., visually impaired and wheelchair users), and bicycling other than driving. Second, investigating the impact of using additional parameters, such as amount of overlaps in comparable routes, similarity measures using metrics such as Hausdorff distance, and incorporating user preferences, which may provide new insight into the reliability of OSM's POIs. Third, a formal model of similarity that captures the essential parameters, including endpoint factors, to compare route similarities would be a useful tool for future studies of similarity. Lastly, an investigation of how the POI location (e.g., whether it is on the border of the building, close to the border, or at the center) affects the routing start/end points is the next immediate step to understand the impact of POIs on routing.

References

Benner JG, Karimi HA (2013) Geo-crowdsourcing. In: Karimi HA (ed) Advanced location-based technologies and services. CRC Press, Boca Raton, pp 1–28

Beeri C, Kanza Y, Safra E, Sagiv Y (2004, August) Object fusion in geographic information systems. In: Proceedings of the thirtieth international conference on very large data bases, vol 30. VLDB Endowment, pp 816–827

Frizzelle BG, Evenson KR, Rodriguez DA, Laraia BA (2009) The importance of accurate road data for spatial applications in public health: customizing a road network. Int J Health Geogr 8(1)1

Girres JF, Touya G (2010) Quality assessment of the French openstreetmap dataset. Transactions in GIS 14(4):435–459

Goetz M Zipf A (2012) Openstreetmap in 3D – detailed insights on the current situation in Germany. In: Gensel J, Josselin D, Vandenbroucke D (eds) Multidisciplinary research on geographical information in Europe and beyond

Gyford P (2004) Euro Foo camp: steve coast – opentextbook & openstreetmap. In: Blog of Phil Gyford. http://www.gyford.com/phil/writing/2004/08/21/euro_foo_camp_st.php

Haklay M (2010) How good is volunteered geographical information? A comparative study of openstreetmap and ordinance survey datasets. Environ Plann B: Plann Des 37:682–703

Hochmair H, Zielstra D (2013) Development and completeness of points of interest in free and proprietary data sets: a Florida case study. GI Forum, 1–10 Salzburg, Austria, 3–5 July 2013.

Jokar Arsanjani J, Mooney P, Zipf A, Schauss A (2015) Quality assessment of the contributed land use information from OpenStreetMap versus authoritative datasets. In: Jokar Arsanjani J, Zipf A, Mooney P, Helbich M (eds) OpenStreetMap in GIScience: experiences, research, applications. Springer, Cham, pp 37–58. ISBN:978-3-319-14279-1

Kang H, Sehgal V, Getoor L (2007) GeoDDupe: a novel interface for interactive entity resolution in geospatial data. In: 2007 11th international conference information visualization (IV '07), July. Ieee 489–496 doi:10.1109/IV.2007.55

Karimi HA, Durcik M, Rasdorf W (2004) Evaluation of uncertainties associated with geocoding techniques. J Comput Aided Civ Infrastruct Eng 19(3)170–185

Karimi HA, Sharker MH, Roongpiboonsopit D (2011) Geocoding recommender: an algorithm to recommend optimal online geocoding services for applications. Trans GIS 15(6)869–886

NAVTEQ (2012) NAVSTREETS street data reference manual v4.4 1 April 2012 Nokia Location & Commerce, NAVTEQ, Chicago, Illinois, 1702 pp

Neis P, Zielstra D, Zipf A (2012) The street network evolution of crowdsourced maps: open-StreetMap in Germany 2007–2011. Future Internet 4:1–21

Roongpiboonsopit D, Karimi HA (2010a) Comparative evaluation and analysis of online geocoding services. Int J Geogr Inf Sci 24(7)1081–1110

Roongpiboonsopit D, Karimi HA (2010b) Quality assessment of online street and rooftop geocoding services. Cartogr Geogr Inf Sci (CaGIS) J 37(4)

Schmitz S, Zipf A, Neis P (2008) New applications based on collaborative geodata – the case of routing. In: XXVIII INCA international congress on collaborative mapping and SpaceTechnology. Gandhinagar, Gujarat, India

Sehgal V, Getoor L, Viechnicki PD (2006) Entity resolution in geospatial data integration. In: Proceedings of the 14th annual ACM international symposium on advances in geographic information systems – GIS '06, vol 4. ACM Press, Arlington, p 83. doi:10.1145/1183471.1183486

Sharker M, Karimi HA (2014) Computing least air pollution exposure routes. Int J Geogr Inf Sci 28(2)343–362

USGS (United States Geological Survey). (1999). National map accuracy standards, USGS Fact Sheet 171–99, November 1999, Available: https://pubs.usgs.gov/fs/1999/0171/report.pdf

Zielstra D, Zipf A (2010) A comparative study of proprietary geodata and volunteered geographic information for Germany. AGILE 2010. Guimarães, Portugal

Chapter 8
A Comparison of Volunteered Geographic Information (VGI) Collected in Rural Areas to VGI Collected in Urban and Suburban Areas of the United States

Kari J. Craun and Ming Chih-Hung

Abstract Volunteered Geographic Information (VGI) is being collected worldwide and may be a source for authoritative data providers such as national mapping organizations. In order to evaluate the usability of these data as part of an authoritative dataset, it is first necessary to understand the quality and reliability of the data. Several studies have been conducted in Europe to compare a volunteer-provided dataset, OpenStreetMap (OSM), to authoritative data sources. The methodology used in these studies was the basis for studying OSM data over rural, suburban, and urban areas in three regions of the United States. The methodology was adapted to compare the volunteer-provided data in OSM to TIGER data from the U.S. Census Bureau which was used as the baseline data to initially populate OSM in the United States. The results showed that road network lengths in all areas studied were increased by volunteers. The increases were greater in more densely populated areas. The density of the OSM road network was generally found to be higher than the density of the baseline TIGER dataset, especially in urban areas. The types of features collected by volunteers were similar to the baseline dataset, but, showed increased percent feature content for pedestrian transportation features. These results are consistent with previous studies which compared OSM data to authoritative data sources.

Keywords Volunteered geographic information • Authoritative data • OpenStreetMap • National mapping organizations • TIGER

K.J. Craun (✉)
USGS, 1400 Independence Road, Rolla, MO, 65401, USA
e-mail: kcraun@usgs.gov

M. Chih-Hung
Northwest Missouri State University, 1333 Garrett-Strong, Maryville, MO 64468

M. Leitner, J. Jokar Arsanjani (eds.), *Citizen Empowered Mapping*,
Geotechnologies and the Environment 18, DOI 10.1007/978-3-319-51629-5_8

8.1 Introduction

Historically, detailed and accurate maps have been difficult to create. Surveying and cartography were fields occupied strictly by professionals who spent years learning and perfecting their use of tools and techniques. The complexity of map-making is evidenced by the following paragraph describing mapping methods used by the U.S. Geological Survey (USGS) in the mid-late 1800s,

> By far the greater amount of the field work was accomplished with the gradienter, a small transit, having a small telescope with striding level, 3-in. vertical and 3-in. horizontal circles both reading to minutes, mounted on a planetable tripod. With it the topographer, from primary and secondary triangulation stations that usually were on the highest peaks, read many horizontal and vertical angles to peaks, summits, ends of ridges, forks of creeks, and other salient features. (Evans and Frey 2009)

While the level of complexity of authoritative, accurate maps can be very high, tools are now available for the general population to easily record information about their location and store and/or transmit that information to others as part of a map. For example, it is possible for a person using a cell phone with embedded Global Positioning Systems (GPS) technology to take a picture of an object and place it on a map using one of many web-based systems designed to share that information. It is also possible to 'geotag' (provide a geographic position of, e.g., latitude, longitude) a photo along with other information and place it on a map background. Hand-held GPS receivers specifically designed for location-based navigation and data collection by non-professionals are also readily available and inexpensive. While the tools of the professional mapping community are still complex, there are now devices that many people routinely carry with them on a daily basis (cell phones, laptop computers, handheld GPS devices) that can be used to perform geospatial data capture. This proliferation of low-cost locational device technology, along with easy access to web mapping and social media (for sharing) applications provides the basis for the growth in crowdsourced mapping or 'volunteered geographic information (VGI).' (Goodchild 2007)

VGI is a term coined by Michael Goodchild to refer to a phenomenon of, '. . . large numbers of private citizens, often with little in the way of formal qualifications . . . working toward . . . creation of geographic information.'(Goodchild 2007) Contributions from volunteers can range in content from base map information to citizen science observations such as bird counts or weather observations. The common component in VGI is that it has a spatial component; in other words, the information is associated with a particular place. The information or observation, after being collected by a volunteer, is usually provided to some central organizing entity or an independent group of volunteers that combines information from multiple volunteers into a broader context often using a map display for visualization. As VGI has gained in popularity and as the volume of this type of data has increased, many national mapping organizations have begun to consider its use as an additional source for developing and maintaining authoritative data sources (Coleman et al. 2009). Coleman et al. (2009) explored this idea and outlined several important questions to be addressed by mapping organizations prior to incorporating volunteer-derived data as a source. Among those questions is, "How do organizations attract new volunteer produsers? How do they keep existing volunteers 'engaged' – or is it assumed they will cycle in and out?" (Coleman et al. 2009) These questions begin to focus on the idea of sustainability of VGI data sources in maintaining an authoritative data source. Authoritative data sources are viewed as '. . . current, reliable, and trusted . . . ' (Ponzio 2004) This implies a sustained level of maintenance (currency), completeness (reliable, trusted), and quality (accuracy). The questions posed by Coleman et al. (2009) for national mapping organizations relate to the appropriateness of the use of VGI to contribute to a dataset that is considered to be an authoritative data source.

Several recent studies have examined questions regarding the completeness of mapping data collected by volunteers compared to more traditional authoritative and commercial datasets. Haklay (2010), Girres and Touya (2010), Zielstra and Zipf (2010), Zielstra and Hochmair (2011), Jokar and Vaz (2015), and Neis et al. (2012) all compared OpenSteetMap data (collected by volunteers) with data from more traditional sources. Results generally indicated that data collected by volunteers approached a level of completeness similar to more traditional sources in highly populated, relatively affluent areas. Haklay et al. (2010) also studied the quality of VGI as it related to the number of volunteers contributing data in a particular area. These results generally support the potential for VGI to be used as part of an authoritative dataset.

8.2 Research Objective

The objective of this research is to examine the relative amounts of VGI collected in various geographic areas in the United States compared to a common baseline dataset. Specifically, the research quantifies and describes the number and types of edits done in OpenStreetMap from 2007 through 2011 in selected urban, suburban, and rural areas of the United States compared to the baseline TIGER data used to initially populate the OpenStreetMap database. Additional OSM data from 2015 to 2016 was incorporated to understand trends over time regarding the quantities of features being collected by volunteers. By comparing VGI collected in urban, suburban, and rural areas, the research will provide an increased understanding of the effect of population density on the amount and type of features collected by volunteers. This work parallels the European studies done by Haklay (2010), Girres and Touya (2010), Zielstra and Zipf (2010), and Neis et al. (2012) and geographically expands the work done in the United States by Zielstra and Hochmair (2011). The intent of the work is to begin to systematically analyze VGI in the United States. It is not meant to be a comprehensive analysis of national datasets such as those completed in the previously referenced publications, although the results could indicate whether further work toward such a comprehensive analysis is warranted. The research also helps the geospatial community and national mapping organizations better understand the volume and type of data being collected by volunteers and how it may contribute to the development of authoritative data sources.

8.3 Study Areas

In order to understand whether population density in an area impacts volunteer data collection, study areas included urban areas (population density approximately 1000 persons/square mile minimum); suburban areas (population density approximately 300–500 persons/square mile); and rural areas (population density less than 100 persons/square mile) in the United States. The eight counties and one city that were included in the study are listed in Table 8.1 and shown in Fig. 8.1. The study areas included in the urban areas meet the population density criteria of 1000 persons/square mile; those included as suburban meet the suburban criteria of 300–500 persons/square mile; and the rural criteria of less than 100 persons/square mile. The median household income was included as a consideration in selecting these study areas in order to ensure comparisons were made between areas with similar socioeconomic status. This is a consideration because Haklay (2010) had shown that socioeconomic factors significantly affect contributions by volunteers. By comparing areas of similar socioeconomic status, the differences in volunteer contributions

Table 8.1 Geographic study areas, including socioeconomic and demographic information

Area/County	2010 Population	Population density (persons per square mile)	Median household income, 2009
East Coast			
Washington, DC	601,723	9856.5	$58,906
Washington County, MD	147,430	322.1	$48,883
Washington County, VA	54,876	97.8	$40,638
Midwest			
Jackson County (Kansas City metro area), MO	674,158	1115.3	$45,798
Clay County, MO	221,939	558.6	$57,983
Phelps County, MO	45,156	67.2	$38,126
West Coast			
King County, WA	1,931,249	912.9	$67,706
Snohomish County, WA	713,335	341.8	$64,677
Mason County, WA	60,699	63.3	$48,104

U.S. Census Bureau (2011)

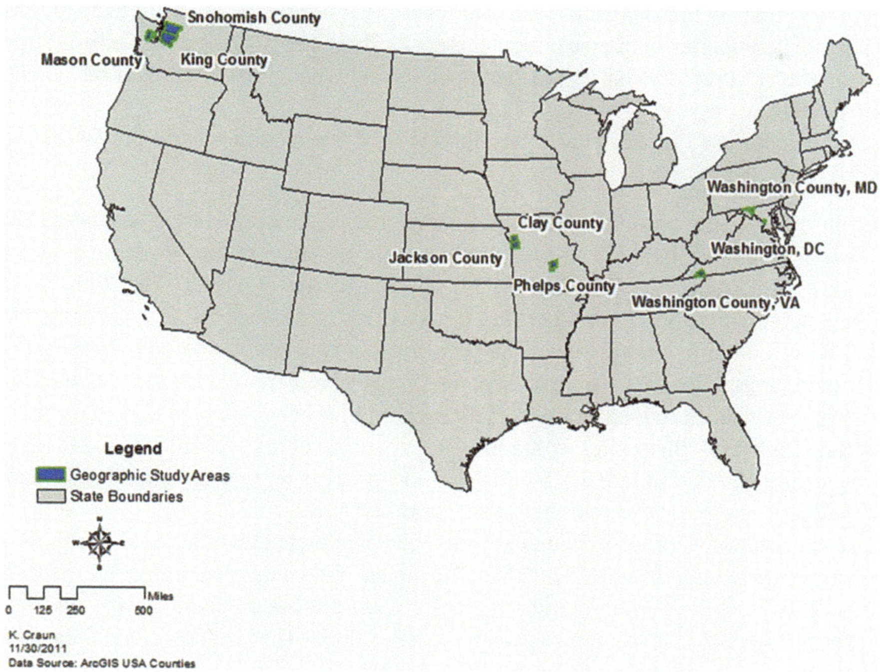

Fig. 8.1 Map of geographic study areas

due to this variable were limited. Geographic areas of interest were also included from both the east and west coasts and the midwest to begin to understand whether there are significant regional differences in volunteer contributions across the United States.

8.4 Data Sources

OSM data is a focus of studies on VGI because it is one of the largest sources of this type of data and because it is freely available under an Open Data Commons Open Database License (ODbL) (OpenStreetMap 2013). When OSM was founded in 2004, much of the geospatial data in Europe, particularly in the United Kingdom, was licensed and thus not freely available to everyone. OSM, then, essentially started with a blank map for volunteers to complete in many places. In the United States, the situation was somewhat different. There was a relatively complete dataset provided by the United States Census Bureau which was used as a starting point in OSM. Thus, volunteers did not start with a blank map, they started with the Census data (TIGER) and then provided edits and additions to that baseline. This baseline dataset makes it possible to compare a snapshot of the OSM data at a point in time to the baseline data. The changes introduced by volunteers (OSM snapshot) compared to the baseline dataset (TIGER) can provide some information about where volunteers have provided updates and what types of updates have been provided.

OSM data used in this study was obtained by downloading from the following web site in 2011:

(http://downloads.cloudmade.com/americas/northern_america/united_states#downloads_breadcrumbs). These files are available by state in zip file format under an Open Data Commons Open Database License (ODbL). The files were extracted from the OpenStreetMap database on December 13, 2011. Each state extract contains several files. The data used in this study was contained at the 'statename.shapefiles.zip' link. Separate Shapefiles were included for the various feature types contained in OSM. The specific Shapefiles used in this study were the 'highway' Shapefiles containing all roads features. Additional OSM data was downloaded in late 2015/early 2016 to provide an update to the initial data studied. These data were downloaded from the site:

(http://download.geofabrik.de/index.html). The data were downloaded as shape files in zip file format by state. The TIGER/Line baseline data were uploaded to OSM in 2007 and 2008. The details of this process are documented on the wiki at: (http://wiki.openstreetmap.org/wiki/TIGER_2005). The TIGER/LINE files used to initially populate the OSM database for the United States were 2006 second edition TIGER files. These were downloaded from:

(http://www2.census.gov/geo/tiger/tiger2006se/). The files at this site are organized by state and then by county (designated by FIPS codes for each county). Note that because these files were not made available by the Census Bureau in Shapefile format prior to 2007, the files were converted in order to more easily compare them with the OSM data in Shapefile format. Freeware software, TGRtoSHP, available from the University of Tennessee at Knoxville (tnatlas3.geog.utk.edu/freeware) was used for this conversion. Road features in the TIGER data were selected for conversion as allowed for by the TGRtoSHP software.

8.5 Conceptual Framework and Methodology

In order to understand the role of VGI as a method of geospatial data collection and particularly its role as a potential authoritative data source for a national mapping organization, several important questions must be answered regarding whether these data meet the criteria of being '. . . current, reliable, and trusted...' (Ponzio 2004). In order to further characterize the quality of potential authoritative data sources, Haklay (2010) and others had evaluated several aspects of VGI data quality, primarily for OSM data in Europe. In addition to evaluating data characteristics, prior research also evaluated characteristics of volunteer populations themselves, such as population density and socioeconomic status. The analysis of these characteristics has begun to help the mapping community understand whether VGI may serve as a reliable source of information across broad geographic areas where volunteer populations vary in terms of their numbers (population density) and ability and interest in collecting VGI (possibly related to socioeconomic status). Projections of what will happen in the future with volunteer data collections have been and will continue to be drawn from an understanding of: the data themselves; the characteristics of the volunteer populations collecting data; and trends in VGI collection.

The work described in this paper analyzed data and volunteer populations in the United States. Similar to the work done in Europe, VGI was compared to an authoritative data source. Unlike the situation in Europe, however, OSM was populated initially using authoritative data (Census TIGER) in the United States and volunteers have edited and added to that baseline. So, rather than looking at data collected solely by volunteers, the OSM dataset analyzed in this study represents the baseline authoritative data source plus volunteer-created changes to that baseline. Contributions to OSM within the United States were analyzed based on content of the data (road length), geographic area of collection, and types of features collected.

8.5.1 Methodology Overview

This study consisted of two phases. The Phase 1 analysis involved comparing the line densities and the road network lengths in the two datasets in each of the study areas. Road network lengths were also compared for data collected through late 2015/early 2016. Phase 2 methodology analyzed the types of features present in the TIGER and the OSM data. In order to understand how these feature types related to each other, a crosswalk was developed between TIGER and OSM feature types.

8.5.1.1 Phase 1 Analysis

The first steps in the process after download of the data involve preprocessing the baseline data (TIGER files) and the OSM (VGI) data for analysis and comparison. These preprocessing steps generally involved clipping the data to county boundaries for the study areas and ensuring the datum/coordinate systems (WGS-84/geographic coordinates), as well as the units of length (km) for road features were consistent. This consistency of units and length was very important because one of the primary comparisons for Phase 1 of the study was performed on the road network lengths in both the TIGER/Line and OSM data for two time periods (2011 and 2015/2016). These comparisons were done using two methods for the TIGER compared to the 2011 OSM data. First, the county-based data was further subdivided into 5 km square tiles. Each tile for both the OSM and TIGER data was analyzed using the 'line density' concept (km per square km). In other words, for the linear features (roads), the number of km of roads was calculated within each square km. For each 5 km tile, the TIGER line density was subtracted from the OSM line density. This allowed for an easy visual understanding of where the TIGER data was denser than the OSM data and vice versa. The road network lengths for the TIGER data and for the OSM data from the two time periods were also accumulated and compared for each county as a whole.

8.5.1.2 Phase 2 Analysis

Two types of analyses were conducted to help answer the question, 'what types of data are volunteers collecting and how does that compare to the TIGER (baseline authoritative) data?' The first type of analysis compared the percentage of primary feature types (percent of total number of features) in the TIGER data (based on Census Feature Class Codes, CFCC) to the OSM feature types in the same study area (by county). For this analysis, statistics were generated for the 'Type' field in the OSM attribute table and for the 'CFCC' field in the TIGER attribute table. The CFCC codes were then translated into feature type names based on descriptions in U.S. Census Bureau (2007). A table comparing TIGER feature types and OSM feature types was also created to allow for easier comparison between the two types of data. This table was created by using the feature definitions for OSM documented at:

(http://wiki.openstreetmap.org/wiki/Map_Features) (OpenStreetMap Wiki 2013) and the feature definitions as defined by CFCC codes in U.S. Census Bureau (2007).

In the final analysis step, the feature types unique to the OSM data were analyzed. Because these feature types are contained in only the OSM data, they originate with volunteers. These volunteer-added features were identified and then analyzed and portrayed by feature type as a percent of total features added by the OSM volunteers. The percent of each feature type is calculated by dividing the feature count for each type into the total number of features (based on count) for each county.

8.6 Results

8.6.1 Results from Phase 1

The results of the Phase 1 comparison of road length differences are summarized in Figs. 8.2 and 8.3. Figure 8.2 shows a comparison of total road network length in kilometers for each county for the TIGER baseline data versus the OSM 2011 and OSM 2015/16 data. In all cases, the road network length was higher for the OSM data than for the TIGER data. In addition, the road network lengths were increased for the OSM data over the time period from 2011 to 2015/16. Figure 8.3 shows the

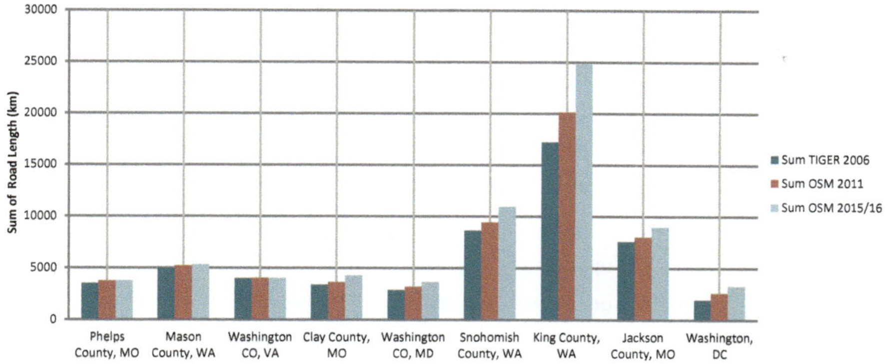

Fig. 8.2 Road network feature length (km) summary for TIGER baseline; OSM 2011; OSM 2015/16

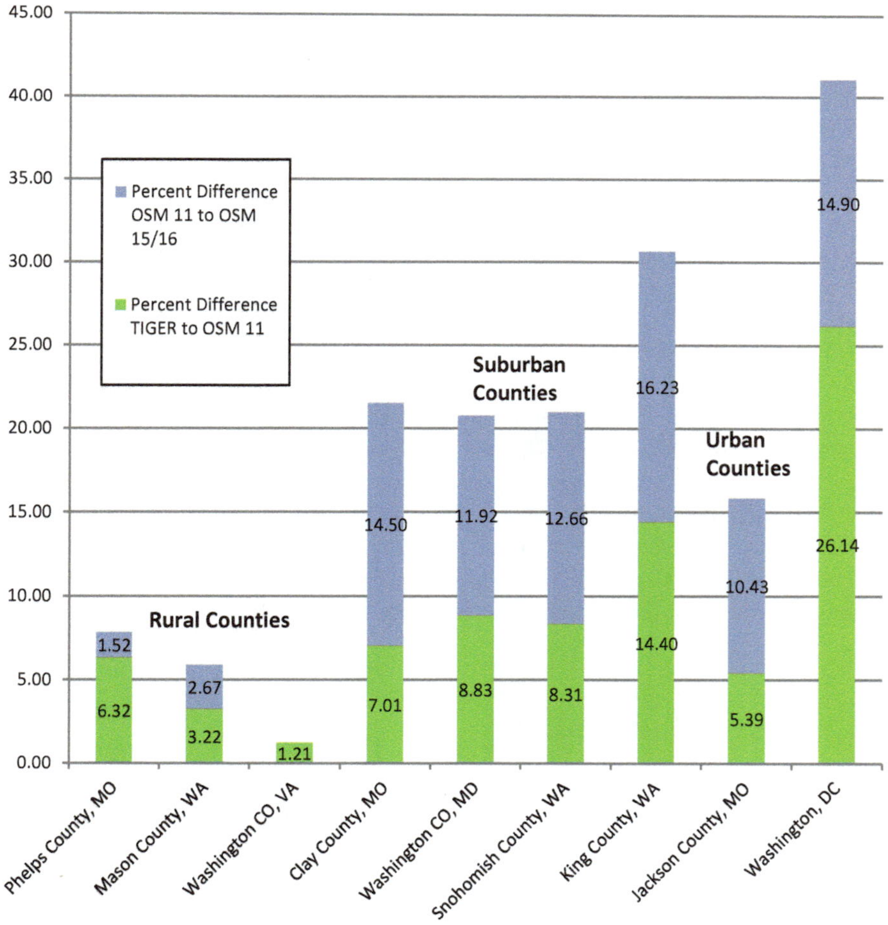

Fig. 8.3 Percent difference between TIGER baseline and OSM 2011 and OSM 2015/16 road network lengths

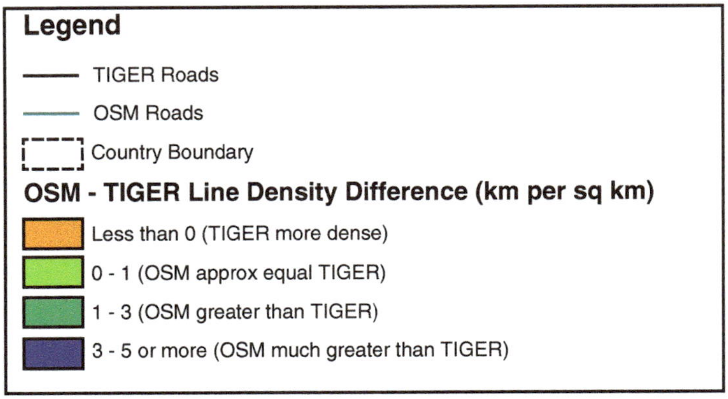

Fig. 8.4 Legend for line density comparison maps in Figs. 8.5, 8.6 and 8.7

percentage increases were higher in the suburban and urban counties than for the rural counties in all cases for both the comparison between the TIGER baseline and OSM 2011 data and between the OSM 2011 and OSM 2015/16 datasets.

The second part of the Phase 1 analysis involved comparing the line density of OSM data vs. TIGER data in each county. Figure 8.4 contains a legend which is applicable for all of the line density comparison maps. Figures 8.5, 8.6 and 8.7 show where TIGER data was denser, less dense, and approximately equal to the OSM data in terms of line density (km of linear features per square kilometers of area). In general, the gold on these maps show areas where the TIGER data is denser; the green areas indicate where the two data sources are approximately equal in terms of density; and the blue areas indicate where OSM data are denser. The darker blue shade indicates a higher density difference between the OSM and TIGER data.

In general, for all of the counties studied, more urban areas tended to have a greater density of OSM data. TIGER data was denser only in the rural and suburban counties of Missouri and the density differences in these areas were very small and likely represent differences in the way the data are collected or stored rather than true differences in feature length. In all areas, the trend seemed to be that the areas of fewer lines, generally, tended to be more equal in terms of OSM and TIGER road density. Where linear features were denser, OSM tended to have greater line density.

8.6.2 *Results from Phase 2*

Phase 2 of the data analysis examined the feature types in the TIGER and OSM datasets. In the TIGER data, feature types are designated by Census Feature Class Codes (CFCC). In order to adequately compare TIGER and OSM features, it was necessary to develop a crosswalk of CFCC and OSM feature types. Table 8.2 contains a general feature type in column 1, then, the corresponding OSM feature

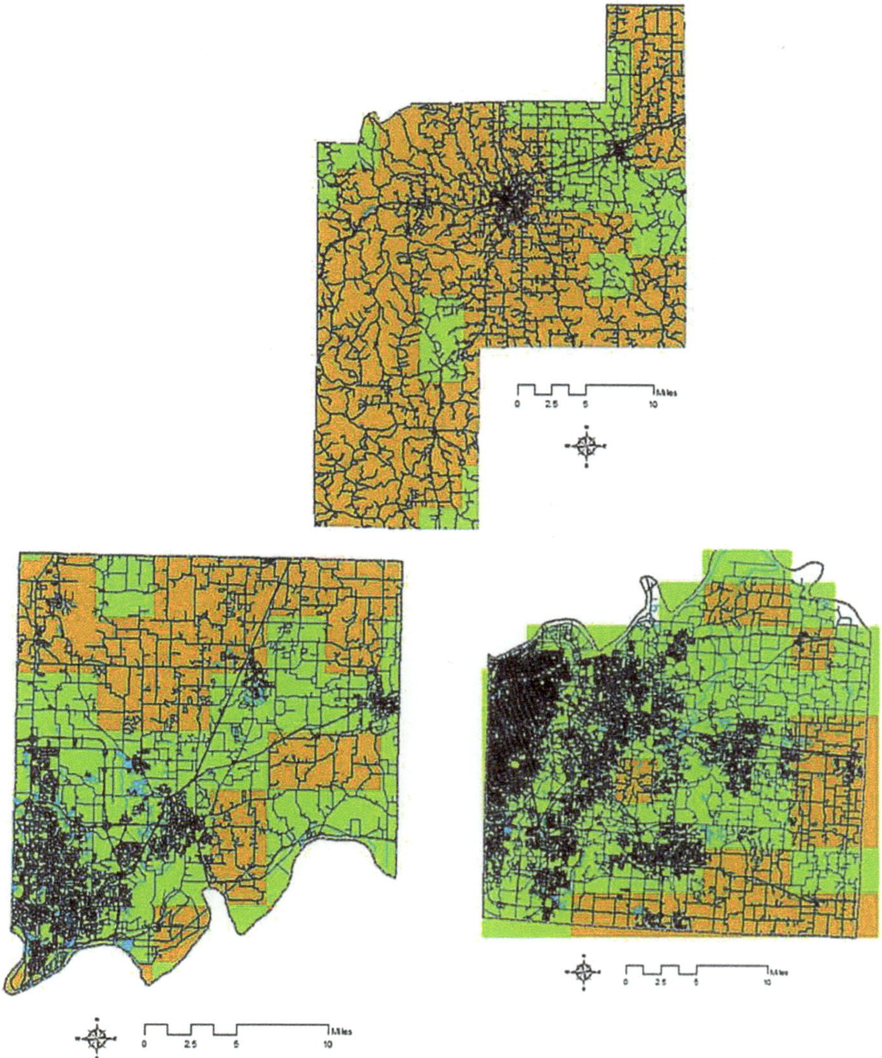

Fig. 8.5 OSM and TIGER road density comparison for Phelps County (*top*); Clay County (*bottom left*); and Jackson County (*bottom right*) in Missouri

type and definition in columns 2 and 3. Corresponding TIGER CFCC codes and definitions are shown in columns 4 and 5. Note that the features included in the table are based on the OSM features contained in the data in this study. There are OSM and TIGER features not included in the table that are outside the scope of this study. Figures 8.8, 8.9 and 8.10 show the comparison of the top ten types of TIGER features in each county (grouped by state or region) to the top ten types of OSM

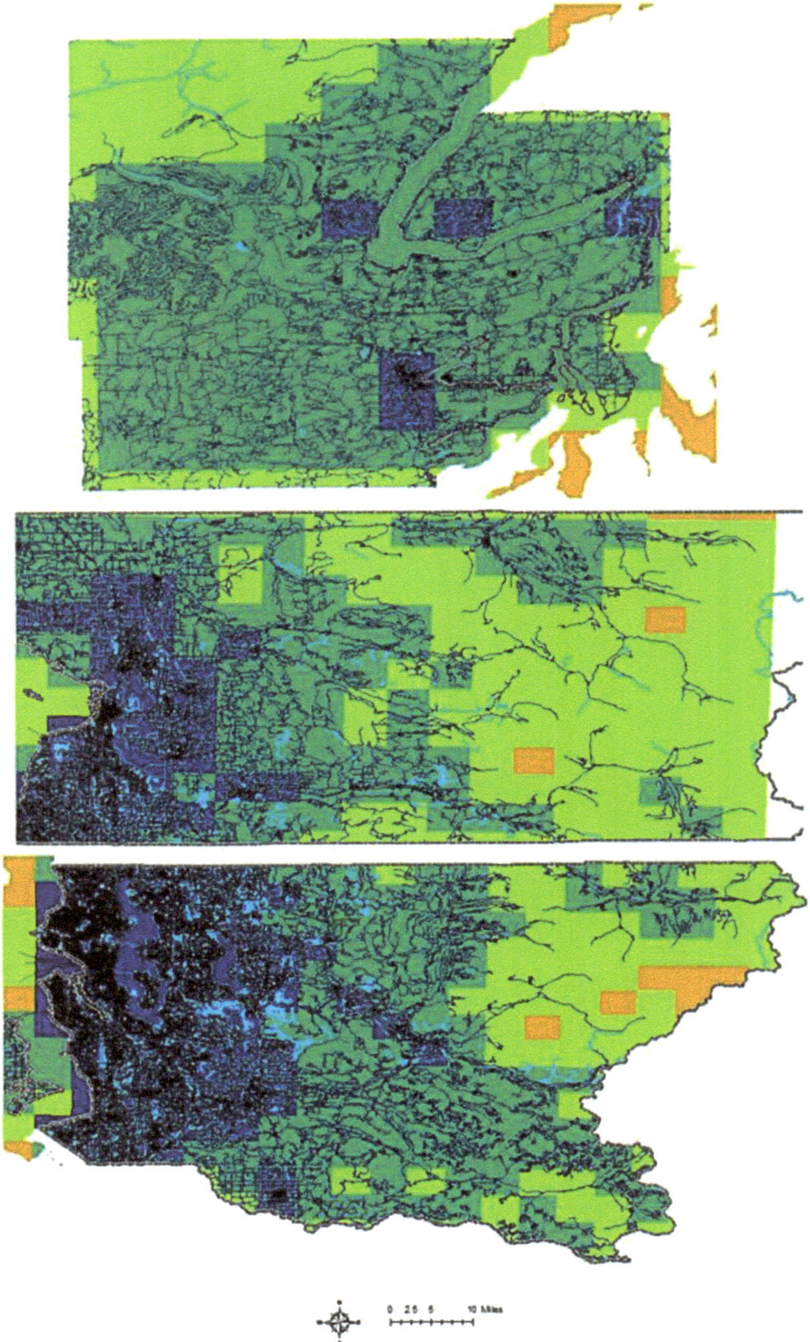

Fig. 8.6 OSM and TIGER road density comparison for Mason County (*top*); Snohomish County (*middle*); and King County (*bottom*) in Washington State

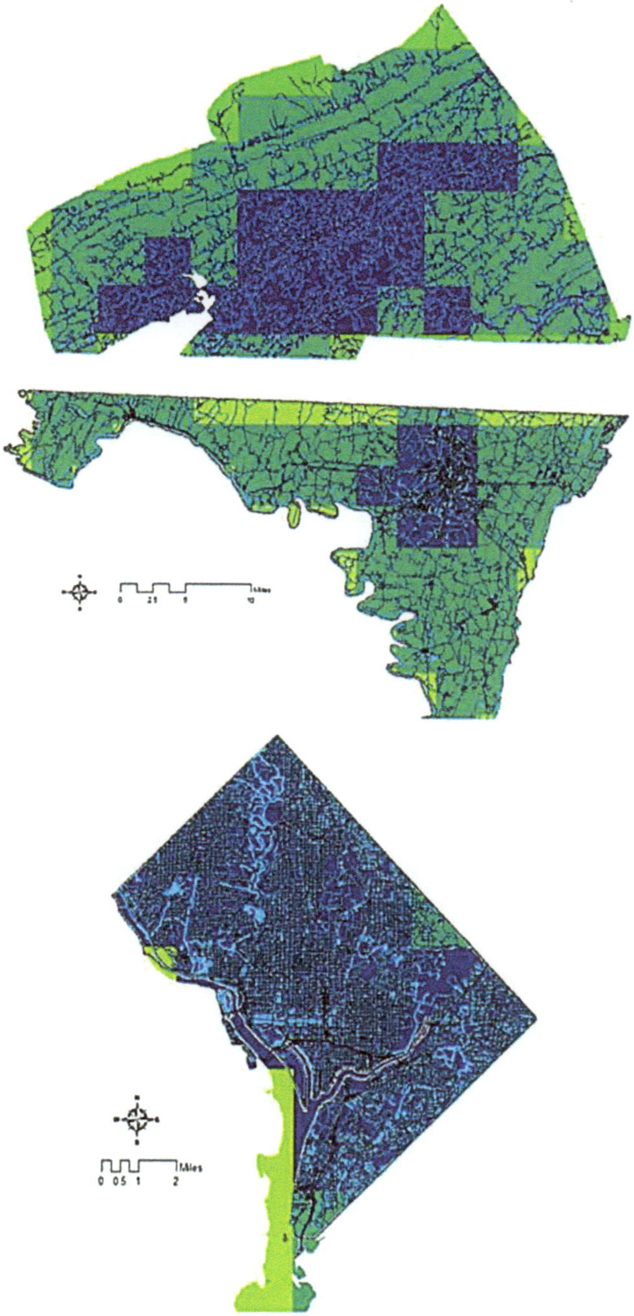

Fig. 8.7 OSM and TIGER road density comparison for Washington County, VA (*top*); Washington County, MD (*middle*), and Washington, DC (*bottom*)

Table 8.2 OSM to census CFCC code crosswalk

Feature type (general)	OpenStreet Map feature type: name	OpenStreetMap feature definition	Census CFCC code(s)	Census feature definition
Separated, major highway	Highway: motorway	A restricted access major divided highway, normally with two or more running lanes plus emergency hard shoulder. Equivalent to the Freeway, Autobahn, etc.	A15	Primary road with limited access or interstate highway, separated
Access ramp	Highway: motor-way_link	The link roads (sliproads/ramps) leading to/from a motorway from/to a motorway or lower class highway. Normally with the same motorway restrictions.	A63	Access ramp, the portion of a road that forms a cloverleaf or limited access interchange
Interstate highway, generally unseparated	Highway: trunk	Important roads that aren't motorways. Typically maintained by central, not local government. Need not necessarily be a divided highway. In the UK, all green signed A roads are, in OSM, classed as 'trunk'.	A11	Primary road with limited access or interstate highway, unseparated
Primary highway, unseparated	Highway: primary	Administrative classification in the UK, generally linking larger towns.	A21	Primary road without limited access, US highways, unseparated
Secondary highway, unseparated or separated	Highway: secondary	Administrative classification in the UK, generally linking smaller towns and villages	A31 OR A35	Secondary and connecting road, state and county highways, unseparated OR separated
Tertiary road, unseparated	Highway: tertiary	A "C" road in the UK. Generally for use on roads wider than 4 metres (13′) in width, and for faster/wider minor roads that aren't A or B roads. In the UK, they tend to have dashed lines down the middle, whereas unclassified roads don't.	A31	Secondary and connecting road, state and county highways, unseparated

(continued)

Table 8.2 (continued)

Feature type (general)	OpenStreet Map feature type: name	OpenStreetMap feature definition	Census CFCC code(s)	Census feature definition
Minor road, unseparated, not residential	Highway: unclassified	To be used for minor roads in the public road network which are not residential and of a lower classification than tertiary. Please do not use this as a marker for roads where the classification is unknown, for which highway=road should be used. Use highway=residential for minor roads lined with housing. See highway=service for access roads	No comparable feature	No comparable feature
Service road	Highway: service	For access roads to, or within an industrial estate, camp site, business park, car park etc. Can be used in conjunction with service=* to indicate the type of usage and with access=* to indicate who can use it and in what circumstances.	A74 OR A73	Private road or drive for service vehicles, usually privately owned and unnamed. Primary type of use is for access to oil rigs, farms, or ranches OR Alley, road for service vehicles, usually unnamed, located at the rear of buildings and property.
Pedestrian walkway	Highway: footway	For designated footpaths; i.e., mainly/exclusively for pedestrians. This includes walking tracks and gravel paths. If bicycles are allowed as well, you can indicate this by adding a bicycle=yes tag. Should not be used for paths where the primary or intended usage is unknown. Use highway=pedestrian for pedestrianised roads in shopping or residential areas and highway=track if it is usable by agricultural or similar vehicles.	A71	Walkway or trail for pedestrians, usually unnamed

(continued)

Table 8.2 (continued)

Feature type (general)	OpenStreet Map feature type: name	OpenStreetMap feature definition	Census CFCC code(s)	Census feature definition
Unpaved, rough track or trail	Highway: track	Roads for agricultural or forestry uses etc. Often rough with unpaved/unsealed surfaces. Use track type=* for tagging to describe the surface.	A51	Vehicular trail, road passable only by 4WD vehicle, unseparated
Stairway, pedestrian	Highway: steps	For flights of steps (stairs) on footways. Use with step_count=* to indicate the number of steps	A72	Stairway, pedestrian
Local, residential street	Highway: residential	Roads which are primarily lined with housing, but which are of a lowest classification than tertiary and which are not living streets. Using abutters=residential together with tertiary, secondary etc. for more major roads which are lined with housing.	A41 OR A45	Local, neighborhood, and rural road, city street, unseparated OR separated
Bicycle path	Highway: cycleway	For designated cycleways; i.e., mainly/exclusively for bicycles. Add foot=*only if default-access-restrictions do not apply.	No comparable feature	No comparable feature
Non-specific use path (e.g., trail)	Highway: path	A non-specific or shared-use path. Probably better to use highway=footway for paths mainly for walkers,highway=cycleway for one also usable by cyclists, highway=bridleway for ones available to horses as well as walkers and highway=track for ones which is passable by agriculture or similar vehicle	A51 OR A72	Vehicular trail, road passable only by 4WD vehicle, unseparated OR Pedestrian walkway

features in the same county. The feature names, while different, may be translated or crosswalked based on the information in Table 8.2.

Generally, in all of the TIGER to OSM feature type comparisons, the highest percentage of total features are 'residential' in the OSM data which compares to the feature type 'local, neighborhood, and rural road, city street, unseparated OR

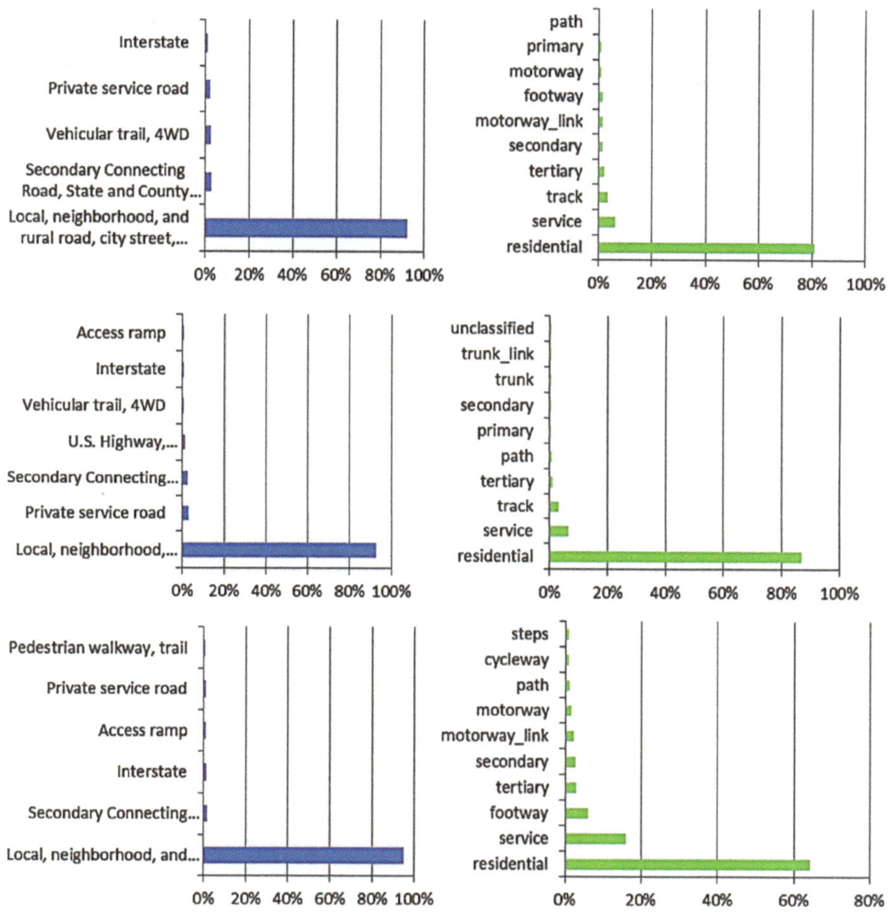

Fig. 8.8 Comparison of TIGER (*left, blue*) to OSM (*right, green*) percent of total feature types in Washington State, including Snohomish County (*top*); Mason County (*middle*); King County (*bottom*)

separated' in the TIGER data. This feature type represents more than 80% of all features in six out of the nine counties studied for TIGER data. The percent of total features for OSM is lower for the comparable feature type ('residential'), but, this may be due to differences in definitions of feature types between the two datasets. For example, OSM contains a higher percentage of service road features than the TIGER data and it may be because some of these features are included in the 'local, neighborhood, and rural road...' feature type in the TIGER data. One notable difference between the TIGER and OSM feature types occurs in the Washington, DC and King County, WA datasets. In both of these cases, there is a much higher percentage of 'footway' features in the OSM data than 'walkway' features in the TIGER data. In Washington, DC, approximately 10% of road features are classified

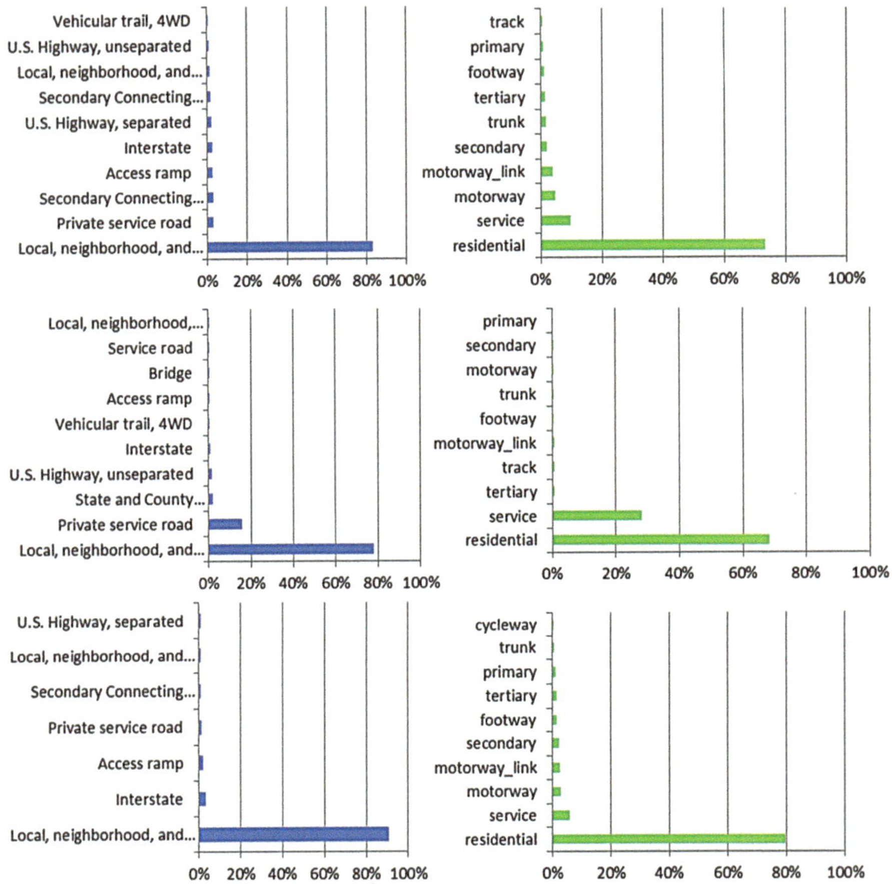

Fig. 8.9 Comparison of TIGER (*left, blue*) to OSM (*right, green*) percent of total feature types in Missouri, including Clay County (*top*); Phelps County (*middle*); Jackson County (*bottom*)

as 'footway.' The comparable feature in the TIGER dataset, 'walkway or trail for pedestrians' does not show up as one of the top feature category types, meaning that the percent of total is less than 1%. In King County, WA footway features represent approximately 6% of the total number of features in the OSM data. Again, in the TIGER data, the comparable feature type is less than 1%. The diversity of feature types also appears to be higher in the OSM data in Washington, DC and King County, WA, meaning there are more different feature types comprising a higher percentage of the total. In the Washington, DC OSM dataset, there are six feature types with 5% or higher of the percent total features. In the King County, WA OSM dataset, there are three feature types with 5% or higher of the percent total. Only in the Washington County, MD TIGER dataset do we have three feature types with 5% or higher of the percent total. In the other TIGER datasets, two or fewer feature types comprise 5% or more of the total features. The final step in the

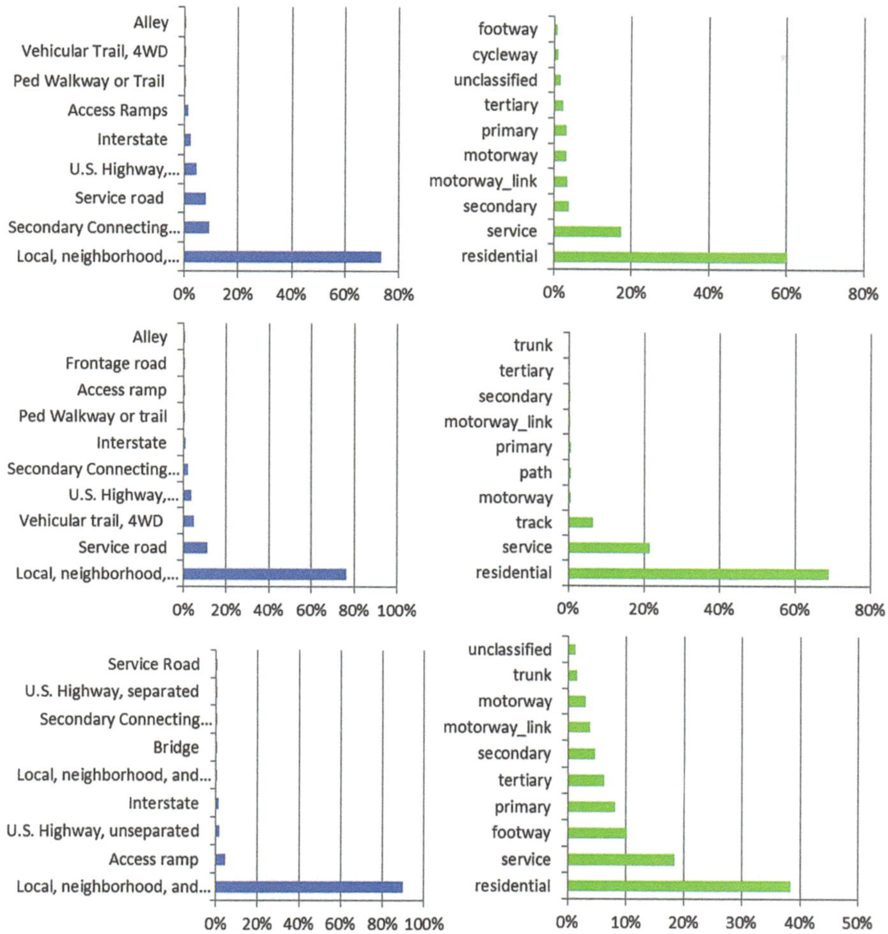

Fig. 8.10 Comparison of TIGER (left, blue) to OSM (*right, green*) percent of total feature types in including Washington County, MD (*top*); Washington County, VA (*middle*); Washington, DC (*bottom*)

Phase 2 analysis looks specifically at the types of features added to the OSM data that are *not included* in the TIGER data. These are the features added to the baseline dataset (TIGER) by volunteers to create the resultant OSM dataset. Figures 8.11, 8.12, and 8.13 include charts showing the feature types added for urban, suburban, and rural counties, respectively.

The highest percentages of feature type additions are residential roads. These types of features represent a range of 36% (Washington, DC) to over 80% (Mason County, WA) of the additions in OSM. There are also a fairly large number of service road additions, including a high of 28% of feature additions in Phelps County, MO and 18% and 19% in King County, WA and Washington, DC, respectively.

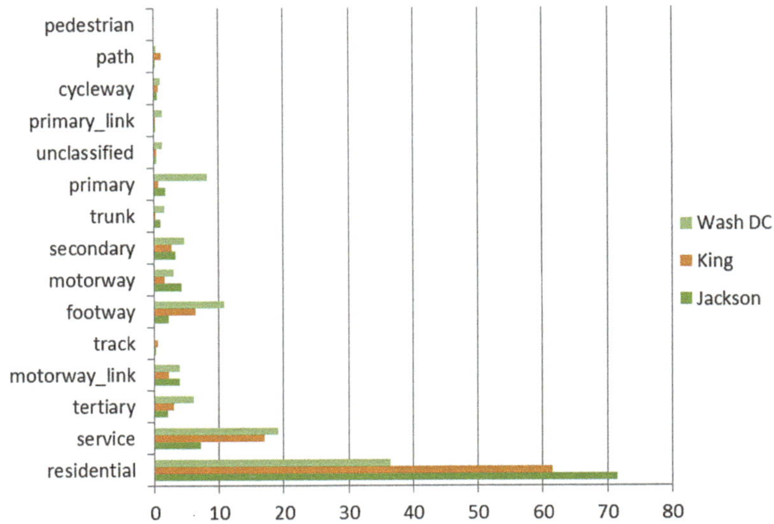

Fig. 8.11 OSM feature type additions in urban counties (percent of total additions)

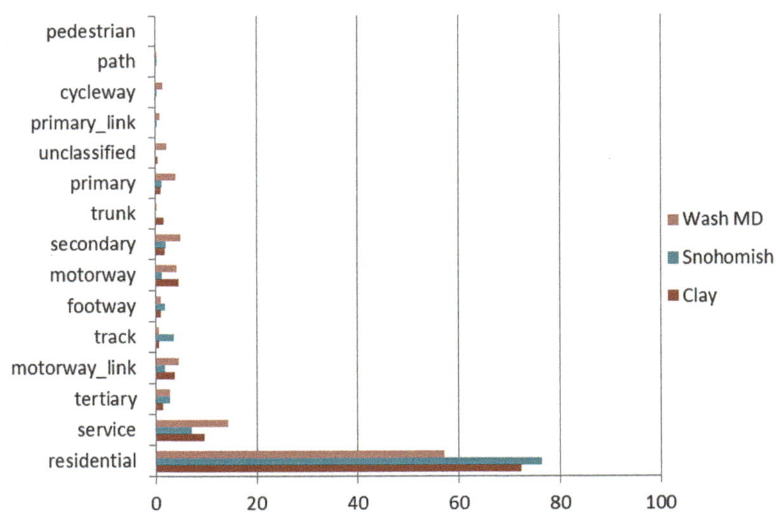

Fig. 8.12 OSM feature type additions in suburban counties (percent of total additions)

In Phelps County, MO these types of features may denote private roads leading to farms or ranches that were not included in the TIGER data. In King County, WA and Washington, DC, these types of features are more likely to be alleys or service entrances to businesses. The 'track' feature is added in relatively large percentages in the rural counties of Mason, WA and Washington, VA. This is logical given this type of feature is defined in Table 8.2 as, 'Vehicular trail, road passable only by

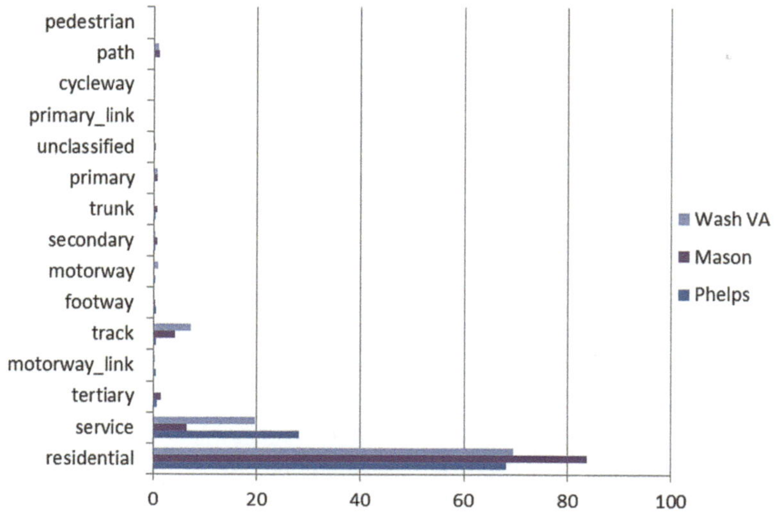

Fig. 8.13 OSM feature type additions in rural counties (percent of total additions)

4WD vehicle, unseparated' and would be more likely found in a rural area. And, finally, the 'footway' feature type accounts for more than 10% of the feature types added to OSM in Washington, DC during this time period. Again, this seems logical given that there are typically more pedestrian-only paths in urban areas than in other regions.

8.7 Discussion

8.7.1 Implications of Phase 1 Results

The Phase 1 analysis examined the road network feature length in eight counties and one city in urban, suburban, and rural areas distributed in three regions of the United States. The lengths of the OSM data from 2011 and the baseline TIGER dataset used to initially populate the OSM database were compared. The rationale for doing this comparison is that the differences between these datasets represent the contributions of volunteers contributing to the OSM database after its initial population. The results indeed indicate there has been a substantial increase in the road network feature length in all areas studied. The average percent difference between the road network feature lengths ranged from more than a 30% difference in Washington, DC to a low of an approximately 15% difference in Jackson and Clay Counties in Missouri. In all cases the road network feature lengths were higher in the OSM dataset. In addition, the increases in feature length in the OSM dataset were larger for urban areas, next larger for suburban areas, and lowest in rural areas.

The only exception is in Missouri where the Clay County suburban OSM dataset had grown slightly more than the Jackson County urban OSM dataset. Both showed an increase, on average, of 15% in road network feature length over the baseline TIGER dataset.

The Phase 1 analysis also included a comparison of the line density between the TIGER baseline and OSM datasets in the nine areas studied. Line density difference maps were created to show where one dataset was more/less dense than the others. Generally, these maps show that OSM data is more dense in the center of urban/suburban/rural populated areas and less dense (or more comparable to the TIGER baseline dataset) in the areas outside of those core populated areas. The only areas where the TIGER baseline dataset was denser than the OSM dataset were in Phelps and Clay counties in Missouri and a very small area in Jackson County, Missouri. These areas also showed the lowest increase in overall road network feature length of OSM data over the TIGER dataset. Overall, the results of the Phase 1 analysis provide an understanding of where volunteers are collecting data in various regions of the country. The results also confirm that correlations noted in Europe between population density and VGI collection quantities hold true in the United States. Finally, these results begin to help us understand if there are regional differences between the eastern, midwestern, and western United States, in terms of volunteer data collection activity.

8.7.2 *Implications of Phase 2 Results*

The Phase 2 analysis examined and compared the types of features being collected by volunteers in the OSM dataset to those present in the baseline TIGER dataset. It was first necessary to develop a crosswalk of feature types so that an 'apples to apples' feature type comparison could be performed. Once this crosswalk was established, the top feature types in each dataset (OSM and TIGER) were identified as a percentage of the total number of features. Very generally, the types of features collected in the two datasets are similar. The largest overall feature type, in terms of percent of total is the 'local, neighborhood, and rural road, city street. . . ' category in the TIGER dataset and the 'residential' feature type in the OSM dataset. This is true in all nine areas studied. What is notable about the feature types is that one begins to see the differences in the TIGER and OSM datasets emerging in the U.S. that have been identified in Europe. Neis et al. (2012) particularly noted the increase in service roads, pedestrian walkways (highway: footway in OSM), and tracks or 4WD vehicular trails in OSM vs. other data sources. In other words, features that are non-car navigation features show an increase in OSM compared to other data sources, such as the commercial data sources (Neis et al. 2012). In the datasets included in this study, this trend is particularly evident in Washington, DC where the feature 'footway' in the OSM dataset constitutes about 10% of the total features whereas a comparable feature in the baseline TIGER dataset represents less than 1% of the total features. This is also true to some extent in King County, WA. Overall, 'service roads' also comprise a much larger percent of the total features in the OSM datasets than in the TIGER baseline dataset. It is somewhat unclear if this is because of a true difference in the data being collected by volunteers or if there is a difference in the categorization of this type of feature in the TIGER vs. OSM data. However, in the second part of the Phase 2 analysis, the results confirm that a large number of service road features are being collected by volunteers. In most cases, this is the second largest feature type being collected by volunteers with the largest number of features collected by volunteers falling into the residential or local road feature type.

The Phase 2 results, again, help characterize the type of data being collected by volunteers and help understand whether there are differences in the data being collected in urban, suburban and rural areas and in the varying regions of the country studied. These results also offer some comparison to the research done in Europe to analyze OSM data. Overall, the results from the work done here, although not as comprehensive in some ways as the work done in Europe, appear to be consistent with the results found in European studies of OSM data.

8.8 Conclusions

The results of the study offer a beginning in terms of understanding and characterizing the quantity and types of VGI being collected in the United States in order to populate the OSM database. This characterization shows that: (1) in all areas studied, volunteers have increased the road network feature length over the baseline TIGER dataset; (2) increases in road network feature lengths are larger in urban areas (areas with higher population density) than in suburban and rural areas; (3) the density of the OSM road network is generally higher than the density of the TIGER dataset, particularly in urban areas; (4) the feature type with the highest percentage in both the OSM and TIGER datasets is the residential road feature type (called, 'local, neighborhood, and rural road'...in the TIGER dataset); (5) residential roads are also the most common features added by volunteers, however, service roads constitute a large percent of the additions; (6) non-vehicle navigation features, including pedestrian footways and tracks or 4WD vehicular trails appear to constitute a larger percentage of features in OSM than in the TIGER baseline dataset.

While these results are not sufficient to fully characterize a dataset in order to evaluate its suitability as an authoritative dataset, they do offer positive indications that further study is warranted given the amount and types of data being collected in all areas studied. The results are also consistent with many of the findings of previous research conducted in Europe and the few studies that have been done in the U.S. to date. Given the lack of research done in the U.S. on the data being collected in OSM, this work fills a niche in terms of providing a starting point for understanding VGI contributions to OSM in the U.S. and for predicting the potential longer term sustainability of VGI as a data source. In addition, there are differences between the overall context for OSM in the U.S. vs. in Europe that also makes this research unique. OSM was initially populated in the U.S. in the 2006–2007 timeframe with the TIGER data from the U.S. Census Bureau. In many parts of Europe, there was no baseline data; so, all of the data in OSM was collected by volunteers (Zielstra et al. 2013). This offers an advantage in terms of analysis of VGI because it allows us to look at the data at any point in time and compare it back to the baseline TIGER dataset. It also has disadvantages because characteristics of the data have their origins in a mix of volunteer and non-volunteer sources. Thus, it is necessary to first separate out the VGI portion of the data from the original TIGER dataset in order to fully understand the characterization.

8.9 Future Work

As discussed above, this study only begins to characterize the OSM data in the U.S. There is a tremendous amount of potential future work to further characterize this information and evaluate the potential of VGI more generally as an authoritative data source. Included here are a few suggestions for future work.

A first suggested action to extend this research would be to update the OSM data in the U.S. to a current snapshot to compare to the TIGER baseline dataset. One could then look at longer term trends in the same geographic areas studied. Further expansion of this research could include other geographic areas and/or take a nation-wide look at some of the statistics, including road network lengths in OSM vs. the baseline TIGER dataset. A map highlighting areas of relative concentration of OSM data collection throughout the U.S. might be useful in pointing out where more detailed information could or should be collected in order to understand whether there are regional or local patterns in the types of features being collected.

Second, look at data quality in a more comprehensive way, particularly positional and attribute accuracy; completeness; and currency. The challenge in the U.S. is, of course, that there is no public domain transportation data source that is nationally consistent and available for use as a reference. In other words, there is nothing really to compare the OSM data to unless a researcher would be able to gain access to a commercial transportation dataset for use in a comparison, such as TomTom's MultiNet dataset. This type of data is commonly used by GPS devices for navigation and TomTom self-reports 100% coverage of the street network in the United States (TomTom 2013).

Third, take a closer look at volunteer activity. In other words, rather than focusing on the data, focus on the characteristics of the individuals providing the data. For example, try to understand *where* volunteers are working most and try to understand the incentives (*why* are volunteers working) in those particular areas.

Finally, look at other VGI projects in order to determine whether the trends in OSM data development is consistent with those other projects. For example, examine other projects such as the USGS National Map Corps, Wikimapia, or other VGI projects to evaluate the characteristics and quality of both the data being collected and the characteristics of the volunteers.

References

Coleman DJ, Georgiadou Y, Labonte J (2009) Volunteered geographic information: the nature and motivation of produsers. Int J Spatial Data Infrastruct Res 4:332–358. doi:10.2902/1725-0463.2009.04.art16

Evans RT, Frye HM (2009) History of the topographic branch (division): U.S. geological survey circular 1341. 196

Girres J, Touya G (2010) Quality assessment of the French OpenStreetMap dataset. Trans GIS 14(4)435–459. doi:10.1111/j.1467-9671.2010.01203.x

Goodchild MF (2007) Citizens as sensors: the world of volunteered geography. GeoJournal 69:211–221. doi:10.1007/s10708-007-9111-y

Haklay M (2010) How good is volunteered geographical information? A comparative study of OpenStreetMap and ordnance survey datasets. Environ Plann B Plann Des 37:682–703. doi:10.1068/b35097

Haklay M, Basiouka S, Antoniou V, Ather A (2010) How many volunteers does it take to map an area well? The validity of Linus' law to volunteered geographic information. Cartogr J 47(4)315–322. doi:10.1179/000870410X12911304958827

Jokar J, Vaz E (2015) An assessment of a collaborative mapping approach for exploring land use patterns for several European metropolises. Int J Appl Earth Obs Geoinf 35:329–337. doi:10.1016/j.jag.2014.09.009

Neis P, Zielstra D, Zipf A (2012) The street network evolution of crowdsourced maps: OpenStreetMap in Germany 2007–2011. Future Internet 4:1–21. doi:10.3390/fi4010001

OpenStreetMap (2013) Copyright and license. http://www.openstreetmap.org/copyright. Accessed 19 June 2016

OpenStreetMap Wiki (2013) Map features OpenStreetMap Wiki. http://wiki.openstreetmap.org/wiki/Map_Features. Accessed 19 June 2016

Ponzio FJ (2004) Authoritative data source (ADS) framework and ADS maturity model. In: Proceedings of the ninth International Conference on Information Quality (ICIQ-04). Massachusetts Institute of Technology, Cambridge, pp 346–357

TomTom (2013) MultiNet coverage. http://www.tomtom.com/en_gb/licensing/products/maps/multinet/#tab:tab2. Accessed 6 Jan 2014

U.S. Census Bureau (2007) Technical documentation: 2006 second edition TIGER/Line® technical documentation. pp 88–93

U.S. Census Bureau (2011) State and county quickfacts. www.census.gov/quickfacts. Accessed 19 June 2016

Zielstra D, Hochmair H (2011) Digital street data: free versus proprietary. GIM Int 25(7). http://www.gim-international.com/issues/articles/id1739-Digital_Street_Data.html. Accessed 19 June 2016

Zielstra D, Zipf A (2010) A comparative study of proprietary geodata and volunteered geographic information for Germany. http://koenigstuhl.geog.uni-heidelberg.de/publications/2010/Zielstra/AGILE2010_Zielstra_Zipf_final5.pdf .Accessed 19 June 2016

Zielstra D, Hochmair H, Neis P (2013) Assessing the effect of data imports on the completeness of OpenStreetMap – a United States case study. Trans GIS 17(3):316. doi:http://dx.doi.org/10.1111/tgis.12037

Part III
Environmental Monitoring and Perception

Chapter 9
Identifying Frostquakes in Central Canada and Neighbouring Regions in the United States with Social Media

Andrew C. W. Leung, William A. Gough, and Yehong Shi

Abstract Following the ice storm of December 2013 in southern Ontario, the general public heard noises that resembled falling trees and reported these occurrences on social media. These were identified as a rare phenomenon called cryoseism, or more commonly known as frostquakes. These occurrences became the first large-scale documented frostquakes in Canada. Using meteorological metrics, we were able to forecast two subsequent frostquake events in January 2014 that coincided with reports on social media. In total, six more episodes of frostquakes as well as their locations were identified in January and February of 2014. Results showed that in central Canada, frostquake occurrences ranged from Windsor, Ontario to the west to Montreal, Quebec to the east and from Niagara Falls, Ontario to the south to North Bay, Ontario to the north. In the United States, the reports came from states bordering the Great Lakes and the New England areas. Two frostquake clusters were identified, one in and around the Greater Toronto Area and the other in eastern Wisconsin. Frostquakes were most frequently heard at nighttime. We critically assess the use of social media as an observation network including the possibility of false positives and population bias. This study demonstrates that rare phenomena such as frostquakes can be identified and assessed using data gathered through social media.

Keywords Frostquake • Cryoseism • Social media • Crowdsourcing • Collaborative mapping

---

A.C.W. Leung (✉) • W.A. Gough • Y. Shi
Department of Physical & Environmental Sciences, University of Toronto Scarborough, 1265 Military Trail, Scarborough, ON, Canada
e-mail: andrewc.leung@mail.utoronto.ca

M. Leitner, J. Jokar Arsanjani (eds.), *Citizen Empowered Mapping*, Geotechnologies and the Environment 18, DOI 10.1007/978-3-319-51629-5_9

9.1 Introduction

Frostquakes, also known as cryoseism, are relatively rare weather phenomenon. They occur after sudden freezing of the ground under specialized conditions and are characterized by a "boom" or "cracking" noise that resembles falling trees. Sometimes, a small tremor is also reported (Lacroix 1980). Occurrences are infrequent (Nikonov 2010; Barosh 2000) and recurrences can be delayed by decades or longer.

In southern Ontario, frostquakes were first heard on the night of December 24, 2013, just after the ice storm of 2013 (December 20–22, 2013). The general public reported these noises on various social media platforms such as Twitter, Facebook, and online discussion boards. Some described the noise as similar to someone banging their fist against the wall or a gunshot (Allen 1993). Many individuals reported on social media that they were asleep and were woken up by the noise. A number of them mentioned that their pets became startled when the noise began. According to media outlets, some people called the police believing that someone was firing a shotgun or that their house was being broken into. Meteorologists from local media in Toronto, Ontario identified that the noises were likely the result of frostquakes and elaborated on the antecedent processes that gave rise to them. Seismic events were quickly ruled out as seismic stations in Canada did not find any seismic waves in this area that night. Meteors were also ruled out.

After reading the term online or hearing it in the news, the public appeared to be looking for more information and turned to Wikipedia (Fig. 9.1). Prior to December 25, 2013, the page on cryoseism received about 300 views on average per day (searches for the terms "ice quake" and "frostquake" on Wikipedia are redirected to

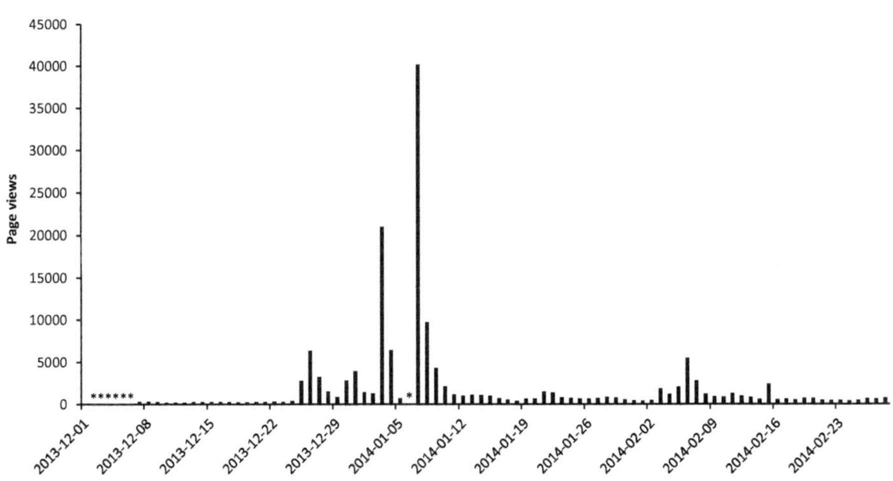

Fig. 9.1 Number of non-mobile page views on cryoseism article on Wikipedia. * indicates dates with missing page view data

the cryoseism page). Just after the first wave of frostquakes in southern Ontario, the number of views on cryoseism spiked up to 2772 views on December 25 and then 6363 views on December 26. After that, visits to this page subsided but still were well above the average prior to the first wave. When frostquakes returned during the night of January 2–3, 2014, the news media once again covered the event in newspaper, on TV, and online. Because of the attention generated by these various publications, the Wikipedia page drew over 21,000 views on January 3. The public became conscious of the noise so that when frostquakes occurred once again on January 6–7 the cryoseism entry on Wikipedia also drew over 40,000 views, the highest daily visit in that page's history. These page views are considered to be a conservative number since the traffic statistics software did not include mobile views into the total, which accounts for roughly 30% of all page views (Heilman and West 2015).

Prior to 2013, the only frostquake officially reported in Canada occurred near a seismic station in Sadowa, Ontario (44.8°N, 79.2°W) on January 18, 2000 and the occurrence was recorded on a fortuitously closely located seismometer (Natural Resources Canada 2016). Eight probable frostquakes from 1870 to 1898 in New Brunswick were identified by Burke (2004).

A search of the scientific literature produced scant results. Lacroix (1980) examined frostquakes and their intensity in the New England area up to 1979. He also identified that frostquakes were frequent in January. Barosh (2000) reported additional frostquakes in New England area, including the damage caused by related cracking. Fujita and Sleep (1991) confirmed three frostquakes and four probable ones from 1872 to 1922 in Michigan. Allen (1993) monitored and recorded frostquake activities with seismographs in Sebago Lake region of Maine during the winter of 1990–1991. Burke (2004) found three likely frostquake events in eastern Maine on top of the eight events in New Brunswick. Nikonov (2010) examined events spanning 1803–1908 in Eastern Europe. He identified three critical factors were required for the formation of frostquakes: moist soil, low to no snow cover, and a sudden drop in temperature that exceeded −20 °C.

A major reason why frostquakes appear so infrequently in the scientific literature is their relative infrequency and difficulty in detection, it is a largely unstudied phenomenon. While networks have been set up to detect earthquakes, frostquakes are too localized and infrequent to be effectively monitored in a similar fashion. Past occurrences rely on anecdotal information such as journals or newspaper reports (Burke 2004), the social media of the time. With the advent of the internet and the contemporary social media, frostquakes have the potential to be reported more readily and thus researched. We note that gathering data through social media has been used in natural science and earth science research in the past (Hyvärinen and Saltikoff 2010; Ogden 2013). For example, US Geological Survey used Twitter to improve its earthquake monitoring response time (Earle et al. 2011). In Europe, forest fires are usually detected by remote sensing but also augmented by citizens contributing volunteered geographic information (VGI) in the forms of blogs, tweets, and photos (De Longueville et al. 2010). Other climatology-related observations that benefitted from VGI include assessing the availability of outdoor skating rinks due to warmer winters in Canada (Robertson et al. 2015), flooding and storm surges caused by Hurricane Sandy in New York in 2012 as well as tornado damage in Oklahoma in 2013 (Middleton et al. 2014). However, this is the first time that frostquake data are gathered using VGI.

In this work, the questions we seek to answer are as follows:

What were the climate conditions of the January 18, 2000 Sadowa frostquake?
What was the geographical range and associated climate conditions of frostquakes in Canada and US during the winter of 2013–2014?
What role can social media play in detecting frostquakes?

9.2 Methods

Two approaches were taken to investigate the frostquakes in Ontario during the winter of 2013–2014, climate data analysis and social media reporting. We have limited our analysis to begin with the frostquakes that occurred on January 2/3, 2014. While frostquakes were first heard on the night of December 24, 2013, we are not including reports from that night for several reasons. First, in southern Ontario, over one million houses lost electricity as a result of the ice storm that was a precursor to the formation of frostquakes. Power was not restored to some homes until a week later. There would be inherent bias of under-reporting or no reporting towards those who lost power since they could not go online to report their observations. Second, because of the ice storm, many trees and branches had fallen. These noises that appeared to be coming from frostquake could actually be trees falling down under the weight of the ice and could have been mistakenly identified as frostquakes or vice versa. This could lead to false positive reports. Third and finally, the term frostquake was not familiar to Canadians. Only a few Canadian TV media outlets in the Toronto area ran online stories on December 25–26 about the booming sound by mentioning the term frostquake. The term was not publicized until another round of frostquakes on January 2, 2014 as noted above.

9.2.1 Climate Data Analysis

We analyze local weather data using the criteria established by Nikonov (2010). These metrics included saturated soils, low to no snow cover and a rapid drop of temperature to below −20 °C. We did this first for the January 18, 2000 frostquake reported at Sadowa, Ontario and then for the frostquakes that occurred during the winter of 2013–2014.

For the January 18, 2000 frostquake event we used Muskoka Airport (44°58′ N, 79°18′ W) weather station data for climatological analysis. The weather station is approximately 27 km away from Sadowa. We examined the daily temperature (minimum, mean, and maximum), precipitation (rain and snow), and snow on ground.

For the winter of 2013–2014 we used data from fifteen Canadian weather stations from Environment Canada's Climate Archives (Fig. 9.2a) and data from six American weather stations from National Ocean and Atmospheric Administration's Climate Data Online (Fig. 9.2b). Station selection was based on the spatial range of frostquake reports and the number of reports mentioned in that area. Similar to the event in 2000, we used daily temperature, precipitation, and snow on ground.

Precipitation is reported using "trace" as a measure and we sought to quantify this. For Canadian stations, trace amount of daily rainfall (<0.2 mm/day) was given a value of 0.1 mm/day and trace amount of daily snowfall (<0.2 cm/day) was given a value of 0.07 mm/day (Mekis and Vincent 2011). The adjustment for trace amounts of precipitation in United States stations was more problematic due to different measuring equipment and different definition of trace precipitation that stems from its use of imperial units. Unlike Canada, US weather stations use standard rain gauges to measure the amount of snowfall (Doesken and Judson 1997). In Yang et al. (1998), trace amount of daily rainfall (<0.01 in/day; 0.254 mm/day) was assigned the same value (0.1 mm/day) as Canadian weather stations. For trace amount of daily snowfall (<0.1 in/day; 0.254 cm/day), Sugiura et al. (2003) suggested an assigned value equal to a quarter of its measuring limit, which is 0.025 in/day or 0.0635 mm/day. This assigned trace daily snowfall value for American weather stations was almost identical to the value given to Canadian weather stations.

Handling trace snow depth was more problematic. For the US, substantive inconsistencies exist in terms of how trace amounts of snow depth was interpreted and recorded among airport weather stations and volunteer stations (Doesken and Judson 1997). In addition, while all Canadian airport and volunteer weather observers record snow depth at or around 6 am daily, some US volunteer stations record snow depth in the early evening while US airport weather stations report snow depth at midnight. In both Canada and US, snow depth is rounded to the nearest whole unit of measurement (cm for Canada, inch for US). Thus, a snow depth of 0.5–1.0 cm is reported as 1 cm. For Canadian stations snow depth below 0.5 cm is described as "trace". Since snow depths of 0.1–0.4 cm were considered equally likely to occur, the average value of 0.25 cm was assigned. This approach is identical in principle with that used by Mekis and Vincent (2011). Similarly, for US stations, trace snow depth between 0.1 in and 0.4 in was given a value of 0.25 in, which is equal to 0.635 cm.

9.2.2 Social Media

The first approach was to analyze social media reports, particularly from Twitter and produce maps of frostquake reports for each frostquake episode during the winter of 2013–2014. To identify frostquake events on Twitter, the search terms "frostquake", "cryoseism", and "ice quake" were used. All frostquake reports from January 2 to February 28, 2014 were examined. Location was identified as the city mentioned in an individual's tweet or post if given. Otherwise, the city location that

a

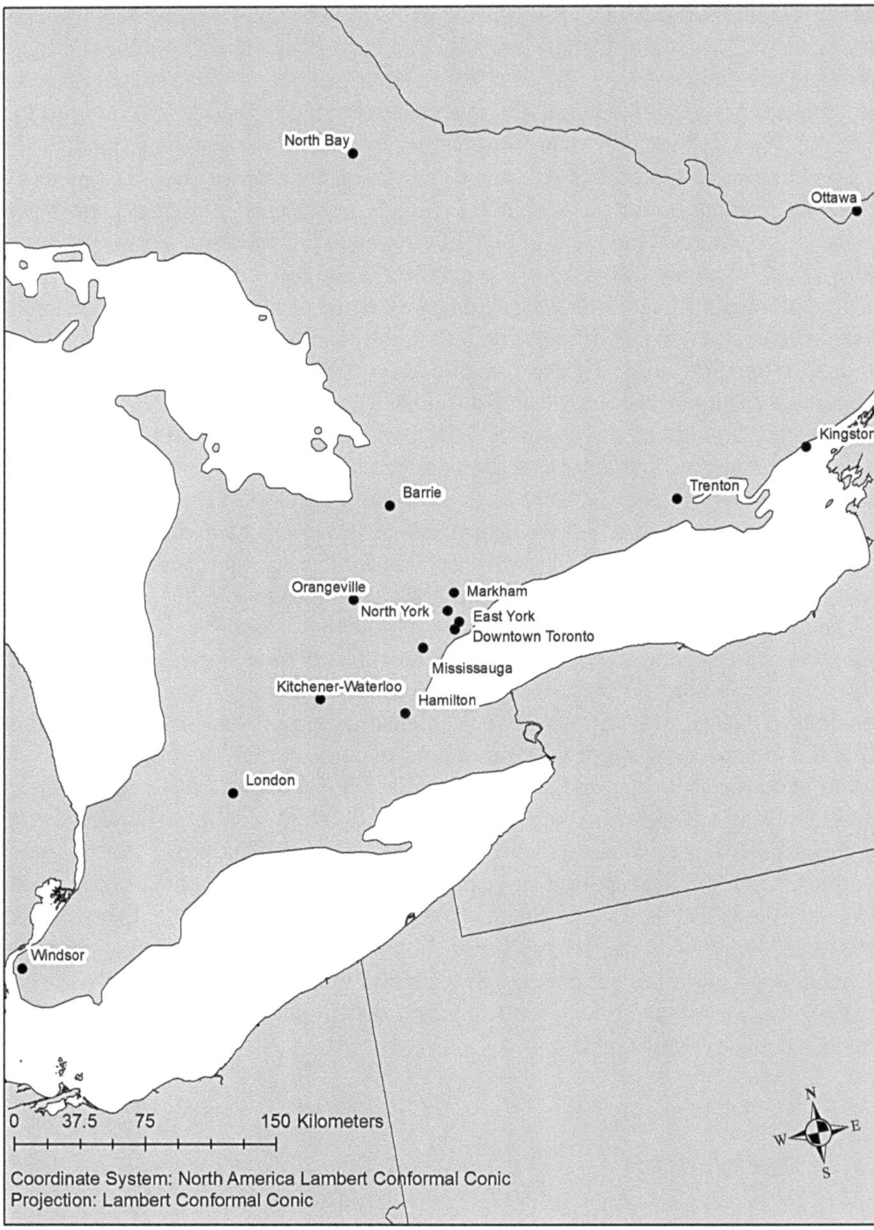

Fig. 9.2 Weather stations in (**a**) Canada and (**b**) US chosen for temperature and snow depth analysis

b

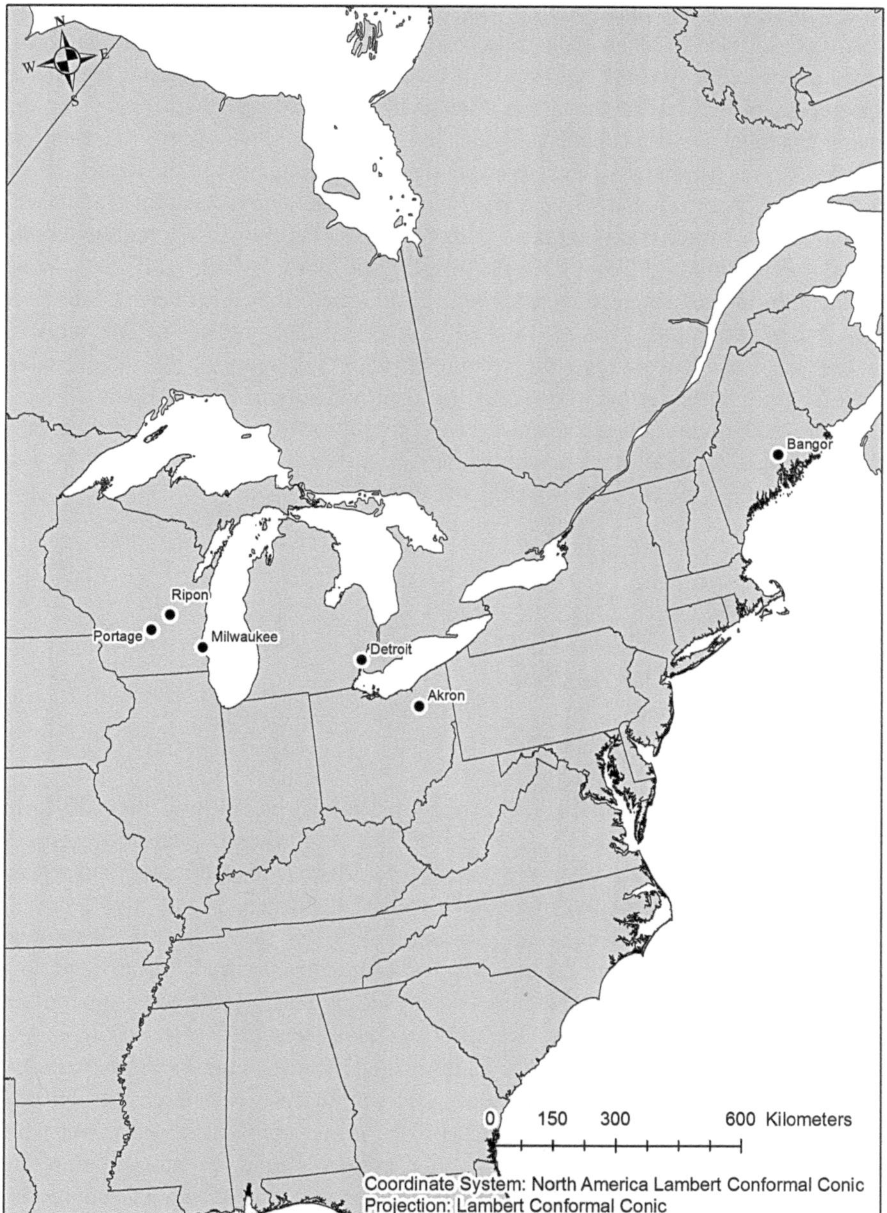

Fig. 9.2 (continued)

an individual associated with on their user profile was assumed to be the location where the frostquake occurred. For date and time, the timestamp of the tweet or post was assumed to be the time of occurrence if the user mentioned that they just heard the noise just prior to posting. If this was not case and the individual specified the approximate time of the noise, then that time was used as a proxy for the actual frostquake occurrence. Finally, if the individual did not specify the time the frostquake was heard, only the date was assigned to the location of the report. Those that did not specify location were not included in the study. Additional locations were obtained from a user-generated online Google Map (https://www.google.com/maps/d/viewer?mid=zId7WwTT0PPk.kmYXHjIndA-w), which solicited social media users to collaboratively mark when and where they heard the frostquake. This online map's URL was also linked to multiple news media's online versions of the story and encouraged the readers to add their reports. Results from all crowdsourced information were sorted by date and grouped by individual towns and cities. The reports were cleaned by examining obvious plotting errors on the user-generated Google Map. Locations were removed if the points were plotted in the middle of a large waterbody (e.g. Lake Ontario).

9.3 Results

9.3.1 Climate Data Analysis

9.3.1.1 January 18, 2000

The only frostquake officially confirmed by seismic record was the one from Sadowa, Ontario on January 18, 2000 at 6:55 pm. According to Natural Resources Canada (2016), that night was very cold and 12 frostquakes were recorded within a 2-h period. Coincidentally, individuals from Skowhegan, Maine also reported frostquakes around the same time, on January 14–15, 2000 (Maine Geological Survey 2016). Upon examining the weather conditions for both locations at that time, both had above 0 °C temperature two days prior to a quick drop in temperature. The temperature drop on January 16, 2000 was quite large, from 0 to −25 °C in one day. On the day of the frostquake, Sadowa had 8 cm of snow cover on the day of frostquake while Skohegan had none. The way that water entered the soil was also different between these locations. At Skowhegan, rain was recorded two days prior to the frostquake. But at Sadowa, a rain event occurred seven days before the frostquake occurred. We believed that the increase in soil moisture was caused by melting of the snow cover on the ground, as the snow cover reduced from 11 cm to 8 cm. Therefore, it appeared that the saturated soils required for frostquakes could be the result of either rainfall or melting snow on the ground. Since the events in Sadowa and Skowhegan happened within the same week, we believe that the spatial variability is mainly caused by the particular temperature, rainfall, and snow depth at the respective locations.

9.3.1.2 Winter of 2013–2014

Temperature graphs for Canadian and American weather stations are shown in Fig. 9.3. Thawing followed by a quick drop in temperature was observed in the following periods at Canadian stations: December 19–22, December 26–29, January 3–6, January 9–14, January 16–17, and February 18–23. Similar observations at the American stations were found on December 26–29, January 9–15, and February 17–23.

Snow depths are presented in Fig. 9.4. Most stations in southern Ontario had less than 30 cm of snow on the ground from December 2013 to January 2014. All of the American stations saw a decrease in snow cover after January 6–12. On average, almost all of these American stations had less than 20 cm of snow depth on average.

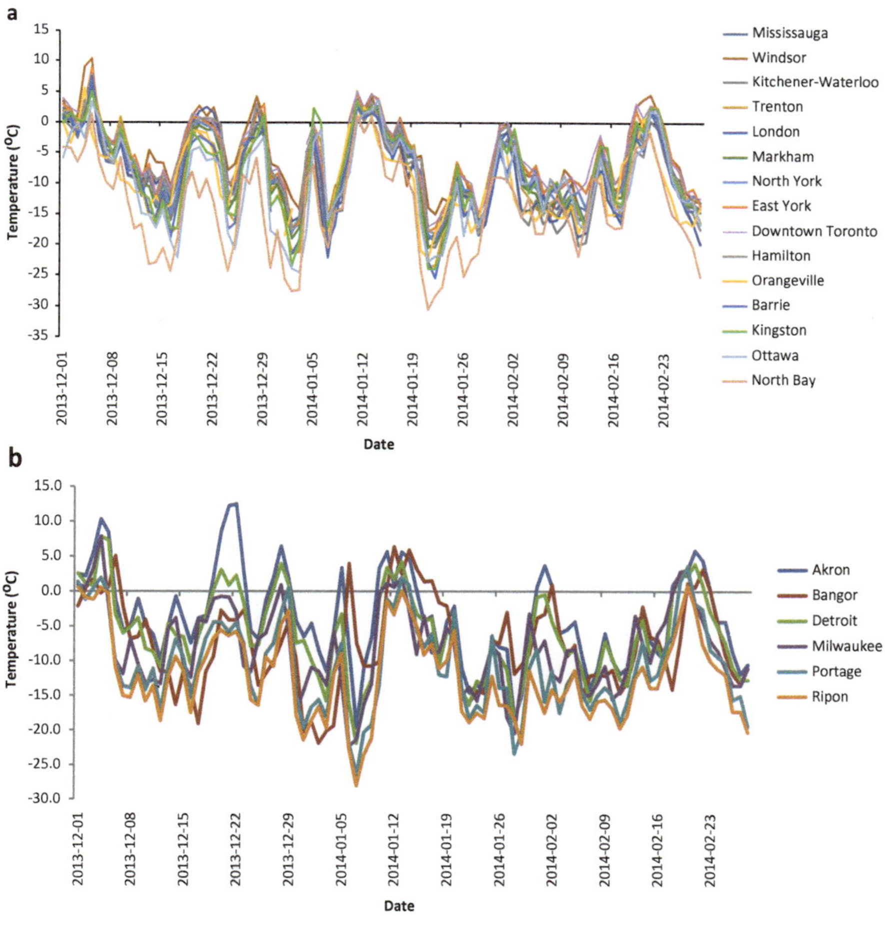

Fig. 9.3 Temperature at weather stations in (**a**) Canada and (**b**) US during the winter of 2013–2014

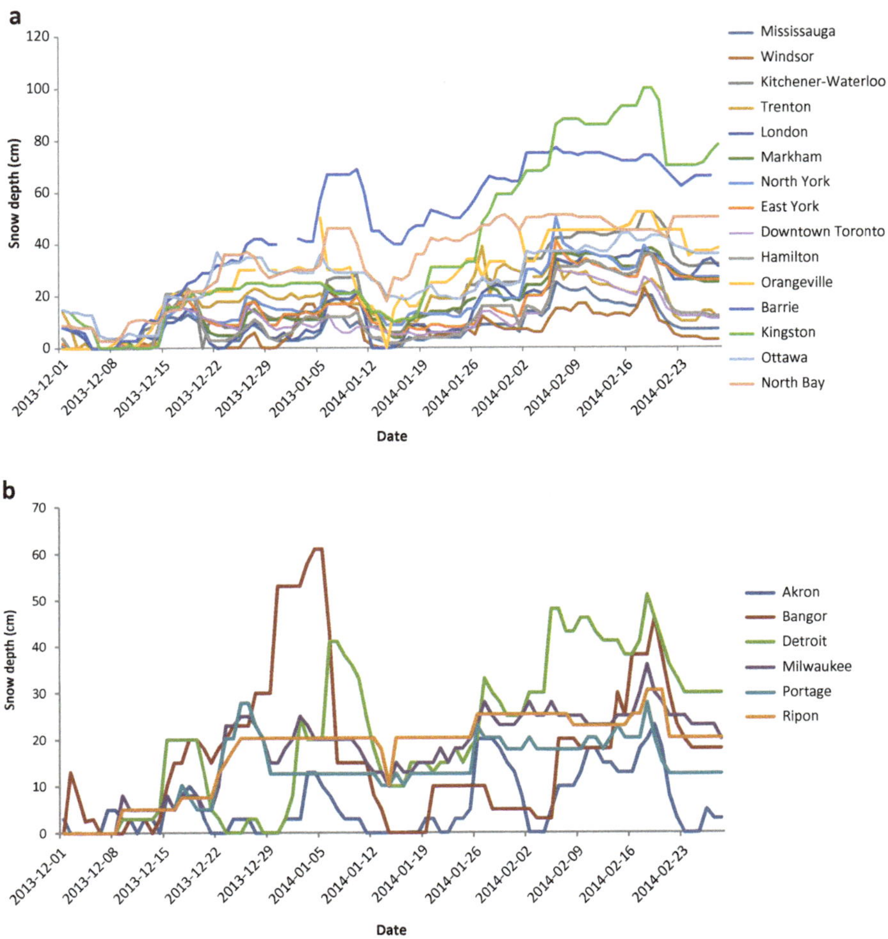

Fig. 9.4 Snow depth at weather stations in (**a**) Canada and (**b**) US

9.3.2 *Social Media*

Overall, there were 2301 frostquake reports recorded through social media (Table 9.1). We have generated maps of frostquake reports for three events with the highest number of reports: January 2/3, January 6/7 and January 20–22 (Fig. 9.5). Over 2100 public reports were recorded from these three events alone and the majority were from Canada (Table 9.1). Spatial analysis showed that regions with high population density (Greater Toronto Area) also experienced the highest number of reports. For the January 2/3 event, most reports came from Toronto and Brampton, Ontario (Fig. 9.5a). Virtually all of the reports were from Ontario, though there were two reports from Wisconsin and one each from Indiana and New

Table 9.1 Breakdown of date ranges with frostquake and the number of reports within each range

	Number of reports		
Dates with frostquake	Canada	US	Total
January 2–3	878	4	882
January 6–7	824	158	982
January 13–15	5	0	5
January 20–22	261	10	271
January 23–25	7	11	18
January 26–29	9	16	25
February 1–3	7	18	25
February 5–7	4	32	36
February 8–12	6	6	12
February 17	0	1	1
February 22–23	4	2	6
February 25–28	27	11	38
Total	2032	269	2301

York State. For January 6/7 event, the highest number of reports came from Toronto and around the Green Lake area in Wisconsin (Fig. 9.5b). There were also reports from Montreal, Quebec ($n = 3$), Montague, Prince Edward Island ($n = 1$), and St. John's, Newfoundland and Labrador ($n = 1$). From the United States, we received multiple reports from Indiana, Ohio, Michigan, Vermont and Maine. We also had Colorado, Iowa and Virginia that did not experience frostquake prior to this event. For the January 20–22 event, Toronto and Newmarket, Ontario had the highest number of reports (Fig. 9.5c). We also had one frostquake reported in Minnesota, a new state for reporting. There were 55 reports not classified because the public only specified the location and did not include the time or date of the event.

Using all of the gathered reports from Fig. 9.5, we combined all the counts from each location on various dates to create a density report map (Fig. 9.6). The density report map identifies the actual cluster of reports by taking into consideration the higher population in urban areas. The density report values in each community is calculated by adding the total number of frostquakes reported on January 2/3, 6/7 and 20–22 of 2014 then dividing by the population of the community. The population of each community is based on Statistics Canada's 2011 Census and U.S. Census Bureau's 2010 Census data. In total, there were 236 communities that experienced frostquakes during those periods. We found two clusters of reports around the Toronto region and eastern Wisconsin region.

Based on Google Map reports, Twitter and Facebook posts, we created a temporal distribution of the timing of the frostquakes (Fig. 9.7). We found that the most common time when frostquakes were reported was at night. In Canada, most reported hearing frostquake during the overnight period, especially between 1am and 3am (Fig. 9.7a). In the US, most reported that they heard frostquakes between 7 pm and 11 pm (Fig. 9.7b).

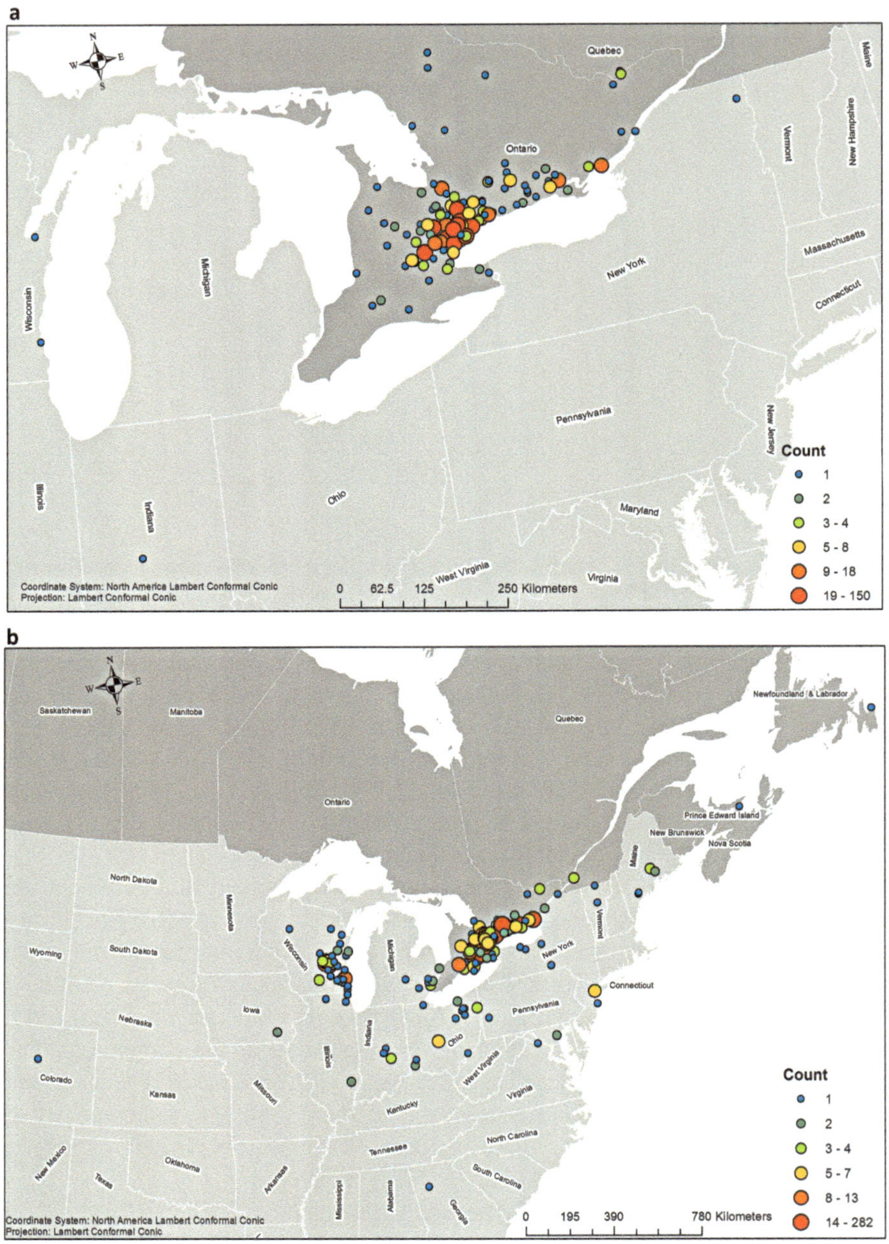

Fig. 9.5 Plots of reported frostquake locations on (**a**) January 2/3, (**b**) January 6/7, and (**c**) January 20–22 of 2014. *Dot sizes* and *colours* are scaled to the number of reports in each community. *Larger dots* represent more reports from a particular town or city

c

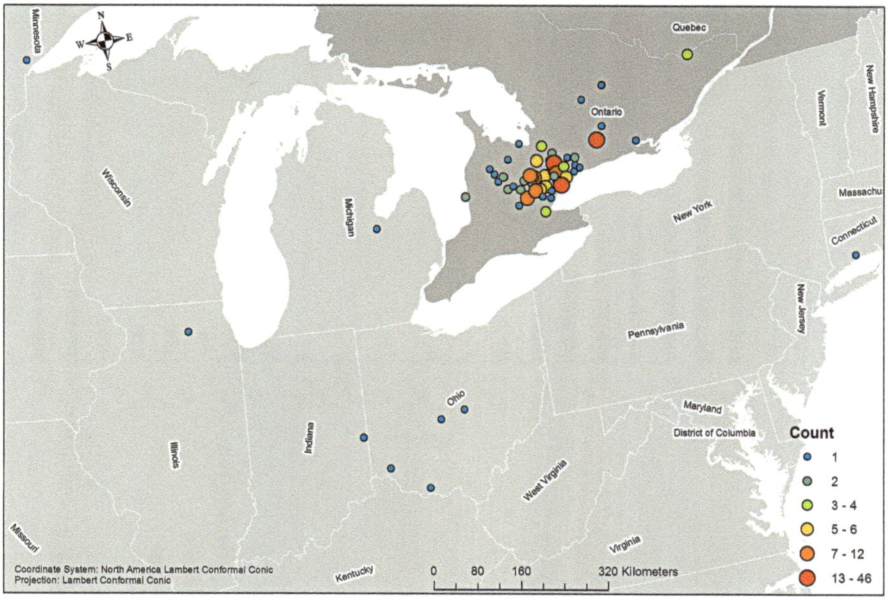

Fig. 9.5 (continued)

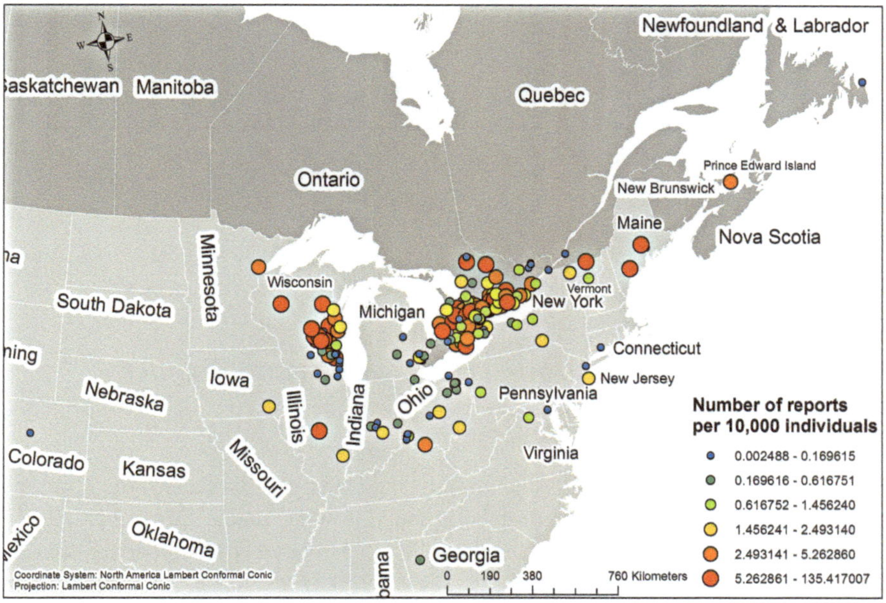

Fig. 9.6 Density report map for frostquakes. Individual values are classified by number of reports per 10,000 individuals in the community

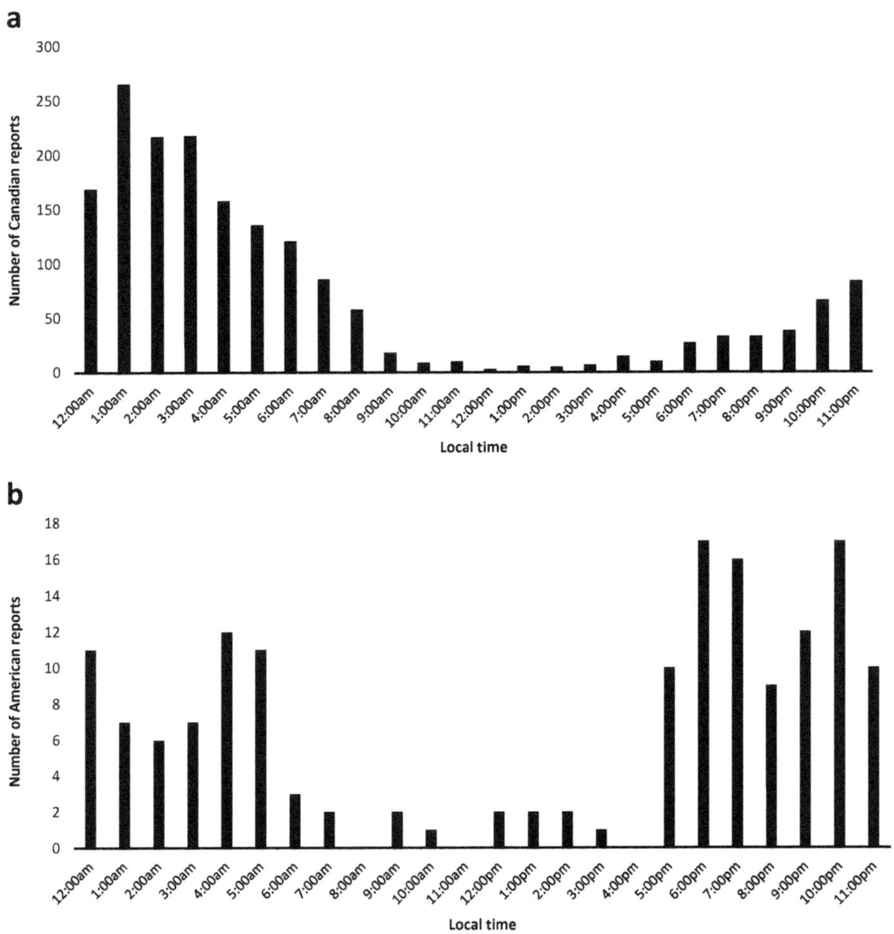

Fig. 9.7 The local time in (**a**) Canada and (**b**) US at which the public reported to have heard a frostquake

9.3.3 Coincidence of Frostquake Reporting and Weather Conditions

For the winter of 2013–2014 the climate data analysis indicated for southern Ontario, dramatic temperature drops for the following dates: December 19–22, December 26–29, January 3–6, January 9–14, January 17–20, January 24–27, February 15–19 and February 26–28 (Table 9.2). Frostquakes are reported in Table 9.1 for days where there was a large difference between maximum temperature and minimum temperature and often after temperature passed through the melting point of ice. Since Toronto had the most reports in the overall period,

Table 9.2 Daily maximum, minimum and the difference between maximum and minimum temperature for the Toronto City weather station in downtown Toronto

Date	Maximum temperature (°C)	Minimum temperature (°C)	ΔTemperature (°C)
1/1/2014	−8.4	−14.5	6.1
1/2/2014	**−14.3**	**−19.2**	**4.9**
1/3/2014	**−7.1**	**−22.3**	**15.2**
1/4/2014	0.3	−7.2	7.5
1/5/2014	1.4	−1.7	3.1
1/6/2014	**2.4**	**−15.8**	**18.2**
1/7/2014	**−15.8**	**−22.2**	**6.4**
1/8/2014	−7.7	−16.2	8.5
1/9/2014	−2.9	−11.8	8.9
1/10/2014	4.1	−4.0	8.1
1/11/2014	7.3	2.9	4.4
1/12/2014	3.4	1.3	2.1
1/13/2014	7.4	1.9	5.5
1/14/2014	5.3	1.3	4.0
1/15/2014	2.4	−3.5	5.9
1/16/2014	−0.4	−3.6	3.2
1/17/2014	2.3	−1.2	3.5
1/18/2014	−0.8	−6.9	6.1
1/19/2014	−1.9	−7.2	5.3
1/20/2014	**−1.8**	**−16.5**	**14.7**
1/21/2014	**−14.2**	**−19.8**	**5.6**
1/22/2014	**−12.1**	**−20.5**	**8.4**
1/23/2014	**−12.2**	**−17.4**	**5.2**
1/24/2014	**−6.8**	**−17.7**	**10.9**
1/25/2014	**−2.8**	**−13.8**	**11**
1/26/2014	**−3.8**	**−14.6**	**10.8**
1/27/2014	**−3.7**	**−16**	**12.3**
1/28/2014	**−11.6**	**−18.6**	**7.0**
1/29/2014	**−9.2**	**−16.5**	**7.3**
1/30/2014	−0.4	−10.6	10.2
1/31/2014	1.5	−2.5	4.0
2/1/2014	**1.4**	**−1.1**	**2.5**
2/2/2014	**0.9**	**−6.1**	**7.0**
2/3/2014	**−2.4**	**−9.7**	**7.3**
2/4/2014	−3.9	−10.5	6.6
2/5/2014	**−5.2**	**−11.7**	**6.5**
2/6/2014	**−6.5**	**−13.2**	**6.7**
2/7/2014	**−8.5**	**−13.9**	**5.4**
2/8/2014	**−8.6**	**−14.2**	**5.6**
2/9/2014	**−7.2**	**−12.1**	**4.9**
2/10/2014	**−5.7**	**−12.0**	**6.3**

(continued)

Table 9.2 (continued)

Date	Maximum temperature (°C)	Minimum temperature (°C)	ΔTemperature (°C)
2/11/2014	**−8.8**	**−13.6**	**4.8**
2/12/2014	**−5.2**	**−14.1**	**8.9**
2/13/2014	−1.6	−9.0	7.4
2/14/2014	0.7	−4.4	5.1
2/15/2014	−2.4	−11.9	9.5
2/16/2014	−6.9	−12.8	5.9
2/17/2014	−2.2	−14.0	11.8
2/18/2014	0.7	−4.2	4.9
2/19/2014	7.8	−1.4	9.2
2/20/2014	2.6	−1.0	3.6
2/21/2014	5.1	0.7	4.4
2/22/2014	4.3	−0.6	4.9
2/23/2014	1.4	−4.7	6.1
2/24/2014	−4.3	−8.8	4.5
2/25/2014	**−5.2**	**−10.4**	**5.2**
2/26/2014	**−8.2**	**−14.2**	**6.0**
2/27/2014	**−8.3**	**−16.0**	**7.7**
2/28/2014	**−6.9**	**−17.7**	**10.8**

Dates in bold indicate that over ten reports were recorded in those periods

we used the weather station located in downtown Toronto (Toronto City) as the representative for the region. For January's frostquakes, the dates that had most public reports (January 2/3, January 6/7, January 20–22) all had a large drop in temperature and the minimum temperature was below −20 °C. Other dates with frostquake events also had a considerable drop in temperature but the minimum temperature did not drop below −20 °C.

9.4 Discussion and Conclusions

9.4.1 Weather Conditions

Nikonov (2010) concluded that moist soil, low snow cover, and a sudden drop in temperature were the variables required for frostquakes to occur. However there is a paucity of observations of frostquakes to explore the nature of the precursors. Using social media reported frostquakes and coincident climate data, we have an opportunity to study frostquakes in more detail.

The weather station data showed that these conditions were met during our study period. The snow depth was shallow when the temperature went from below 0°C to above (Fig. 9.4). A few days later, thawing stopped when the temperature quickly dropped and in some locations, by up to 20 °C within 24 h (Fig. 9.3). During and shortly after the temperature plunged, the public started to hear the noise or feel the shaking from the frostquakes.

In our study, we speculate that these steps need to have a specific sequence and we have documented this using local weather data. For example, if the sudden drop in temperature that results in freezing occurred before the soil became moist or before the snow cover was reduced to low levels, the frostquake would not have happened because the ground would have sufficient space for the soil to freeze and therefore not cause any cracking noise. Therefore, the only realistic sequence would be a location first having low to no snow cover which allowed the soil to become moist. After that, the temperature drop must greatly exceed the insulation effect of the snow cover so that the temperature in soil will quickly drop and water molecules inside the soil will freeze and expand.

9.4.2 Social Media

It is a commonly used research method to utilize VGI data gathered from the general public. Very often, these studies provide online forms for the public to fill out in a structured manner. This ensures the completeness of the data often by asking the user to select from a list of pre-determined options. In contrast, data collection from social media often faces bigger challenges. The submitted information are considered "free flowing" because the user chooses the level of detail and in an unstructured manner that sometimes require follow-up prior to analyzing the data. The main benefits of using social media for data gathering is a larger sample size that does not require the public to fill out a web form to report their findings. Instead of waiting for the public to engage with the scientists, social media allows scientists to reach out and obtain data directly from the public. In our study, the spatial data production of frostquakes can be classified as "bottom-up, amateur, and asserted" (Cinnamon 2015) as almost all of the data points were generated by the general public on social media or Google Map rather than by authoritative experts.

Hyvärinen and Saltikoff (2010) listed the service provider's terms and conditions, data retention policy, privacy, and copyright as the biggest challenges for collecting meteorological observations from social media. Their study analyzed user-submitted photos that were uploaded to Flickr (an online photo depository) and identified meteorological events at a specific time and location from the photos. We believe that our study has fewer issues related to content removal policy, privacy, and copyright. We collected location data in near real-life time, which circumvented the issue with the deletion of older materials. Hyvärinen and Saltikoff (2010) indicated that Twitter only stored 1 week of tweets before deletion. We noted that Twitter has since modified its content removal policy to keep tweets in perpetuity unless deleted by the user. In addition, there are third-party providers who store tweets on a particular topic or keyword for future retrieval. However, we did find that some frostquake tweets were deleted by the user in rare instances (<0.1% of all reports) by comparing the differences between the search results from Twitter with the third-party's. Privacy was not a concern for our study. We aggregated all reports from a town or city into one location thus resulting in a coarse resolution of the locations. Even though some users enabled their GPS locations while tweeting, the coordinates were imprecise or sometimes obviously inaccurate and unreliable on Twitter. For example, on a few occasions, the coordinates identify the user's location to be in the middle of Lake Ontario, 500 m away from the shorelines of Toronto.

While this is a novel approach of using social media for scientific data collection, there are some potential issues with using Twitter and Facebook to gather data. It is difficult to pinpoint the exact time and location when the frostquake occurred. 20% of the reports we gathered did not specify approximate time of occurrence. The number of reports is dependent on the population of a city and citizens who use social media, thus skewing towards larger cities like Toronto. Another under reporting problem arose from those who heard the noise but decided not to report it on social media. Furthermore, there is an age disparity for social media users. A survey conducted in 2009 in the US found that 75% of young adults from age 18–24 had a social media accounts whereas only 7% of those aged 65 and above had an account (Lenhart 2009). Therefore, reports gathered from different social media were more likely to be coming from teenagers and young adults rather than older adults or seniors. Our study's age bias was somewhat lessened when mainstream media included the link to our Google Map reporting system in their online news stories. Privacy setting on social media accounts also suppressed some reports. While Twitter's tweets were set to public by default, Facebook's posts were private by default. 60% of teenagers reported that they set their profiles (along with their posts) to private and only visible to their friends (Madden et al. 2013). Hence, frostquake reports were less likely to be found on Facebook because of the users' privacy settings. On the other hand, false positives are not uncommon on social media (Hyvärinen and Saltikoff 2010) and this was particularly true for frostquake since the only identification was a banging noise and light tremor but usually no physical observation can be made and in some cases a report could be completely fictitious. Some obvious false positive frostquake reports were identified. Some users reported hearing frostquake noises from places such as San Diego which did not have frost on the ground at that time of the year. We also had a number of well-intentioned reports from the Pickering, Ontario area on the morning of January 21, 2014. The volunteers later corrected themselves after discovering that the shaking and noise were the result of a nearby waste water treatment plant explosion and not by frostquake. These reports were removed from the analysis. A few plausible locations such as Atlanta, Georgia and Denver, Colorado were kept after examining the climatological conditions on January 6–7, 2014 and found the conditions to be possible for frostquake to have occurred (Fig. 9.5b).

We believe that the nature of the reports on social media between Canada and the US were different. The reported frostquake time in the US tends to be in the evening period where most people were still awake (Fig. 9.7b). They might have heard the noise and decided to mention it on the internet. On the other hand, the majority of Canadian reported frostquakes took place during the overnight period after most people went to bed (Fig. 9.7a). It is likely in this case that the public was woken up by the noise and decided to share their experience. In a number of reports, the public said that their pets were woken up by the sound or felt the jolt and the pets woke up their owners. A lot of Canadians mentioned in their tweets that they were surprised by the sound and in some cases they were delighted to finally hear it after their friends who shared similar experience with them. There were several explanations which explained why more frostquakes were heard at night than during the day. Night time temperature tends to be colder and a quick temperature drop appears to be a requirement for frostquakes to occur. Another reason to explain the temporal difference is that it is quieter at night and people are less active, which makes it easier for the frostquake to be heard or felt. On a greater temporal scale, we found more frostquakes happening in January than in February (Table 9.1) and this appeared to be in agreement with Lacroix (1980).

On social media, the public stated that the sound appeared to be coming from the roof even though the cause of the noise was the expansion of ice within the soil. At the time of their reports, these individuals were in various types of buildings (detached houses, apartments). It is unclear how the vibration sound resonated through different building materials (e.g. wood, concrete) and propagate to upper-level floors in apartment buildings. In addition, the frostquakes in 2013–2014 occurred in highly populated areas. This was noticeably different from reports from New Brunswick and New England where most of the people who heard the frostquakes were living in farm houses in rural communities (Allen 1993; Barosh 2000; Burke 2004).

9.4.3 *Frostquake Clusters*

Given that urban centres like Toronto and Montreal have large populations, their total number of reports are not unexpectedly high when compared to suburban and rural areas (Fig. 9.5). However, we did not observe a cluster of reports around Montreal and that reports from Montreal area only appeared on January 6/7, 2014 (Fig. 9.5b). The density map showing a cluster in Toronto (Fig. 9.6) is not surprising given that there were extensive media coverage and social media presence. It was also the first area that received prominent attention given that many people believed to have experienced it just after the ice storm of 2013. The spike in Wikipedia traffic to cryoseism article (Fig. 9.1) on December 26, 2013 and January 3, 2014 were very likely to be coming from people in the Toronto cluster because there were only four reports coming from the US up until January 3, 2014 (Table 9.1). The Toronto cluster consists of four of the top ten largest municipalities in Canada (Brampton, Hamilton, Mississauga, and Toronto) plus a number of suburban towns and rural villages. While most people heard or felt the frostquake were living in detached homes, some of them, especially in downtown Toronto, heard it inside their apartments. Despite the large population size, Hamilton and Mississauga were placed 28th and 30th percentile respectively on the density map while Toronto was placed at 57th percentile and Brampton was placed at 73rd percentile. There were a mix of low and high density reports from other communities within the Toronto cluster. However, the density map showed that Montreal had very low number of reports per 10,000 individuals.

The Wisconsin cluster, in contrast, is very different from the Toronto cluster. Most of the communities in the Wisconsin clusters had very small population but high density of reports. In fact, the top nine communities with the highest density reports were all from the Wisconsin cluster and seven of the nine communities had multiple reports of frostquakes. Yet, none of these nine communities had a population over 1500 and practically everyone lives in detached housing in these rural villages and towns. One thing in common between the Toronto and Wisconsin clusters was that both areas were reported extensively by mainstream media. In Wisconsin's case, the attention was drawn to The Weather Network (2014)'s video of a Wisconsin farmer discovering a crack about 30 m long and 20 cm deep after hearing booming noise which was attributed to frostquake. The video caused heightened awareness of this phenomenon and was the subject of local discourse. It led to a positive feedback loop of more awareness leading to more reports, and that in turn led to even greater awareness. We found that similar cracks were also present in two instances in Maine and Massachusetts (Allen 1993; Barosh 2000). In Allen (1993)'s study, the frostquakes were heard and felt between 7 and 9 pm, which is consistent with our temporal analysis of when frostquakes were most likely to be occur in United States (Fig. 9.7b). The timing was quite similar to the event in Sadowa, Ontario at 6:55 pm (Natural Resources Canada 2016) and the event in Rothesay, New Brunswick in 1884 from 9:30 pm to 10:30 pm (Burke 2004). However, we found that the Canadian events in this study took place much later in the night (Fig. 9.7a). Also, we were unable to explain why the traffic to Wikipedia's croyseism article did not show a noticeable spike after the third major event on January 20–22, 2014 (Fig. 9.1).

Using Fig. 9.6 to compare with existing literature, we were able to get a snapshot of which provinces and states were new to frostquakes. In Canada, there were no frostquakes reported in Quebec, Prince Edward Island and Newfoundland and Labrador prior to this study. Likewise, in United States, the states that did not have reported frostquakes before January 2014 were Colorado, Georgia, Illinois, Indiana, Iowa, Pennsylvania and Virginia. Almost all of the Canadian provinces and US states which experienced the first frostquake took place on January 6/7.

Our research demonstrated that yet another type of rare weather phenomenon like frostquake can be monitored through social media. We linked the VGI reports of frostquakes to concurrent weather conditions. Through the use of social media, we collected the greatest number of frostquake locations to date and identified two frostquake clusters. We also found three Canadian provinces and seven US states that experienced its first ever reported frostquake during our study period.

Acknowledgements We thank Ashley King for creating the initial user-generated Google Map that served as the collaborative VGI mapping platform for the public to report their observations.

References

Allen RP (1993) A study of cryoseisms ("Frostquakes") in the Sebago lake region, Maine. In: Symposium on the application of geophysics to engineering and environmental problems 1993, pp 415–430. doi:10.4133/1.2922017

Barosh PJ (2000) Frostquakes in New England. Eng Geol 56(2000)389–394. doi:10.1016/S0013-7952(99)00092-7

Burke KBS (2004) Historical seismicity in the central highlands, Passamaquoddy Bay, and Moncton Regions of New Brunswick, Canada, 1817–1961. Seismol Res Lett 75(3)419–431. doi:10.1785/gssrl.75.3.419

Cinnamon J (2015) Deconstructing the binaries of spatial data production: towards hybridity. Can Geogr 59(1)35–51. doi:10.1111/cag.12119

De Longueville B, Annoni A, Schade S, Ostlaender N, Whitmore C (2010) Digital earth's nervous system for crisis events: real-time sensor web enablement of volunteered geographic information. Int J Digit Earth 3(3)242–259. doi:10.1080/17538947.2010.484869

Doesken NJ, Judson A (1997) A guide to the science, climatology, and measurement of snow in the United States. Colorado State University Department of Atmospheric Science, Fort Colins, CO

Earle PS, Bowden DC, Guy M (2011) Twitter earthquake detection: earthquake monitoring in a social world. Ann Geophys 54(6)708–715. doi:10.4401/ag-5364

Fujita K, Sleep NH (1991) A re-examination of the seismicity of Michigan. Tectonophysics 186(1–2)75–106. doi:10.1016/0040-1951(91)90386-7

Heilman JM, West AG (2015) Wikipedia and medicine: quantifying readership, editors, and the significance of natural language. J Med Internet Res 17(3)e62. doi:10.2196/jmir.4069

Hyvärinen O, Saltikoff E (2010) Social media as a source of meteorological observations. Mon Weather Rev 138(8)3175–3184. doi:10.1175/2010MWR3270.1

Lacroix AV (1980) A short note on cryoseisms. Earthquake Notes 51(1)15–20. doi:10.1785/gssrl.51.1.15

Lenhart A (2009) Adults and social networking websites. Pew Research Internet Project. Available via http://www.pewinternet.org/2009/01/14/adults-and-social-network-websites/. Accessed 20 May 2016

Madden M, Lenhart A, Cortesi S, Gasser U, Duggan M, Smith A, Beaton M (2013) Teens, social media, and privacy. Pew Research Internet Project. Available via http://www.pewinternet.org/2013/05/21/teens-social-media-and-privacy/. Accessed 20 May 2016

Maine Geological Survey (2016) Reports of earth shaking in Maine possibly due to cryoseisms. http://www.maine.gov/dacf/mgs/hazards/earthquakes/quake-cryolist.htm. Accessed 20 May 2016

Mekis E, Vincent LA (2011) An overview of the second generation adjusted daily precipitation dataset for trend analysis in Canada. Atmos-Ocean 49(2)163–177. doi:10.1080/07055900.2011.583910

Middleton SE, Middleton L, Modafferi S (2014) Real-time crisis mapping of natural disasters using social media. IEEE Intell Syst 29(2)9–17. doi:10.1109/MIS.2013.126

Natural Resources Canada (2016) Frequently Asked Questions about Earthquakes (FAQ). http://www.earthquakescanada.nrcan.gc.ca/info-gen/faq-eng.php. Accessed 20 May 2016

Nikonov AA (2010) Frost quakes as a particular class of seismic events: observations within the East-European platform. Izv Phys Solid Earth 46(3)257–273. doi:10.1134/S1069351310030079

Ogden LE (2013) Tags, blogs, tweets: social media as science tool? Bioscience 63(2)148. doi:10.1525/bio.2013.63.2.15

Robertson C, McLeman R, Lawrence H (2015) Winters too warm to skate? Citizen-science reported variability in availability of outdoor skating in Canada. Can Geogr 59(4)383–390. doi:10.1111/cag.1222

Sugiura K, Yang D, Ohata T (2003) Systematic error aspects of gauge-measured solid precipitation in the Arctic, Barrow, Alaska. Geophys Res Lett 30(4)1192–1195. doi:10.1029/2002GL015547
The Weather Network (2014) Wisconsin: Ice quake causes property damage. http://www.theweathernetwork.com/news/articles/wisconsin-ice-quake-causes-property-damage/19445. Accessed 20 May 2016
Yang D, Goodison BE, Ishida S, Benson CS (1998) Adjustment of daily precipitation data at 10 climate stations in Alaska: Application of world meteorological organization intercomparison results. Water Resour Res 34(2)241–256. doi:10.1029/97WR02681

Chapter 10
Structuring Volunteered Geographic Information Collection to Improve Information Processing Efficiency in Environmental Management

Mu-Ning Wang Brandeis and Timothy L. Nyerges

Abstract Incorporating volunteered geographic information (VGI) in environmental monitoring has been treated as a great way to enhance public participation and improve the coverage of data collection. This research assumed that structuring VGI would improve manager use of VGI within decision-making. A better structure includes a clearly pre-defined geographic location and/or an organized content that fits managers' needs. This research tested the effectiveness of using a spatial decision unit (SDU) and compared the usability of free-form VGI and highly structured VGI. By using a case comparison method, four cases from recreation management and invasive species control were compared to explore if structured VGI can increase the usability of VGI in environmental management. Results showed that using SDUs shortens the matching process between VGI and management interests, but potential negatives might arise as side effects. Different structured monitoring forms have different functions. A highly structured monitoring form provides data processing efficiency, whereas free-form participation provides flexibility for volunteers. Managers should clearly define VGI usage and use appropriately structured forms for their needs.

Keywords Structured public participation • Volunteer geographical information (VGI) • Spatial decision unit (SDU) • Environmental monitoring • Environmental management

M.-N.W. Brandeis (✉)
School of Environmental and Forest Science, University of Washington, Box 352100, Seattle, WA 98195-2100, USA
e-mail: wang0209@uw.edu

T.L. Nyerges
Geography Department, University of Washington, Box 353550, Seattle, WA 98195-3550, USA

M. Leitner, J. Jokar Arsanjani (eds.), *Citizen Empowered Mapping*, Geotechnologies and the Environment 18, DOI 10.1007/978-3-319-51629-5_10

10.1 Introduction

Incorporating volunteer geographic information (VGI) as part of environmental monitoring is considered a great way to balance between public participation needs (Smith 1982; Reed 2008; Wyk et al. 2008) and increasing data collection coverage at the same time when managerial resource is extremely limited (Johnson and Sieber 2013). Collaborative projects that involve volunteers and/or stakeholders include citizen science, community-based (participatory) monitoring, participatory geographical information system (PGIS), and etc. from early 1990s (Obermeyer 1998; Jankowski and Nyerges 2001; Schlossberg and Shuford 2005; Dunn 2007; Goodchild 2007; Silvertown 2009). With Web 2.0 technology, citizen-driven data collection has been improved through web-based mapping interfaces (Elwood et al. 2012).

Collecting appropriate environmental information involves multiple levels of difficulties (McNie 2007). Changing environments need consistent monitoring because diverse natural regimes require different management standards (Holling 1973, 1978; Gunderson et al. 2008). Inappropriate temporal and spatial scales of information contribute to inappropriate decisions-making (Steffen et al. 2006; Cash et al. 2006). Another concern comes from VGI. VGI does not typically include traditional measures of accuracy (Elwood et al. 2012) and the credibility is hard to measure (Grira et al. 2010; Johnson and Sieber 2013). Managers have struggled to find adequate information for decision-making from existing mismatched information and complex decision-making processes (Wyk et al. 2008).

A National Research Council (1983) report proposed using structured decision-making processes to ease bounded rationality limitation in decision-making (Simon 1979). Gregory and Keeney (2002) suggest that a complex problem should be decomposed into various elements or stages to reduce the complexity of decision-making and later showed that this method can be generally applied in either policy decision making or routine management decisions (Gregory et al. 2012). However, Brewer and Stern (2005) found that public environmental managers are not trained to handle complexity during decision making.

Among the many drawbacks about using information collected from volunteers or stakeholders are that data are unstructured and/or not being collected properly so managers cannot use it for decision-making. Given the findings about structure in decision making processes mentioned earlier, it would seem that transforming VGI into more valuable information for environmental monitoring and management requires a transition from unstructured VGI into structured VGI in the form of information with which managers are familiar. Conceptually, Olsson et al. (2004) suggested that connecting management tasks, responsible person(s), and needed information can increase efficiency in decision-making processes. Grira et al. (2010) indicated that data collection quality is another main element affecting the functionality and quality of VGI. Practically, Nyerges et al. (1997) suggested that even public-oriented geographic information should consider information storage, retrieval, organization, and visualization. In sum, studies show that structuring information for decision-making and organizing information based on these needs could help managers improve their decision-making.

This research compares four cases with different ways to structure VGI. Since the way to structure VGI is connected with practical obstacles encountered from environmental management, the following section will first describe the lack of information in environmental management and why structured VGI can be valuable to land managers. The VGI concept in this research includes any information that is collected from volunteers wherein the VGI contains geographic-related characteristics as part of its content. Then we describe interventions to structure VGI. The interventions include a better spatial decision unit that is linked to management target(s) and various degrees of structured forms to organize VGI content. To examine effects associated with interventions, four empirical case comparisons are conducted to cultivate insights for improving the usability of VGI. From these comparisons we develop findings about structuring VGI, and develop conclusions about those findings in the form of recommendations for moving forward with VGI in environmental monitoring about recreation and invasive species control.

10.2 Practical Obstacles in Environmental Management

Monitoring provides critical feedback for environmental decision-making (Gibbs et al. 1999) to adjust future decision-making (Lyons et al. 2008) from an adaptive management perspective (Walters 1986; Uychiaoco et al. 2005). To develop comparable information throughout different time periods, scientific protocols are used to organize field observations. Regular monitoring is costly and normally only the data sample is limited due to the cost consideration. Our research found that recreation and invasive species control are two environmental management fields that are in extreme lack of managerial resources. With limited resources, researchers and managers suggest that volunteer-based activities could be included in monitoring or partial management to address limited resource situations (Brudney 1999; Propst et al. 2003). Four cases (two each for recreation management and invasive species control) were selected as a result of opportunistic circumstances, but aim to provide an interesting diversity of institutional missions, resources, and scales as described below.

10.2.1 Recreation Management Obstacles

Recreation, as one type of public land management, provides an indirect method of using natural resources and increases benefits received by humans. With such huge areas to manage, staff resources within recreation management are very limited. Consequently, volunteers become a crucial work force (Brudney 1999) for parks and recreation.

The two case studies we worked with for trail recreation management are: (a) Lord Hill Regional Park (LHRP) in Snohomish County, Washington, and (b) the Puget Sound region managed by the Washington State Department of Natural Resources (WA-DNR-P). LHRP is a 1400-acre regional park with various types of trails managed by the Snohomish Park and Recreation Department (Snohomish PRD). This park shared two rangers with nine other parks for monitoring and maintenance. WA-DNR-P has approximately 1.8 million acres and 1100 miles of trails managed daily with 40–50 employees (Jordan Reeves, personal communication, November 13, 2012). Because park or recreation employees are in limited supply, patrolling all parks and trails are not possible every day. LHRP and WA-DNR-P have to rely on users' reports or volunteers to detect incidents that happen on trails.

In the absence of a formal park-oriented participation setting, enthusiastic volunteers initiated LHRP stewardship meetings during 2010–2011 to build a communication channel with the Snohomish PRD. From discussions at these meetings, discussions about trail incidents were the only VGI extracted from meeting minutes and agendas. To strengthen communication, this paper's lead author coordinated and led a service-learning project to update trail maps and enhance the reporting mechanism. This service-learning project is a quarter-based course (GEOG469/569 GIS Workshop) hosted by Geography Department, University of Washington. This course invites case proposals from local government, nonprofit organizations, and other research institutes that have a short-term GIS project to complete. The Snohomish PRD observed that the updated trail map could be used to develop web-based reporting mechanisms, and the lead author was invited to develop a structured form to ease obstacles related to reports that were mis-transferred within the Snohomish PRD.

WA-DNR-P initiated a volunteer program called Forest Watch to ameliorate the lack of staffing across parks. One volunteer coordinator is responsible for volunteer recruitment, training, and processing information. The current reporting mechanism was through paper-based forms. To enhance and extend participation, the WA-DNR-P invited the lead author to enhance its volunteer reporting content and procedure; consequently, our research team summarized work orders of the department's daily tasks and routine decisions to identify potential for automatically generating and compiling VGI. To motivate volunteers, the WA-DNR-P coordinated Forest Watch volunteers in performing occasional trail maintenance, although most trail maintenance is implemented by environmental non-profit organizations through grants.

10.2.2 Invasive Species Control Obstacles

Likewise, involving citizens in tracking invasive species is a critical part of environmental management because threats from invasive species (i.e., plants, pests, and animals) to current agriculture and ecosystem are substantial. Increasing the number of observations helps with early detection and rapid response (EDRR) of invasive species spread. EDRR and control and management (CM) are two main strategic goals out of four that were set up from The National Invasive Species Council (NISC) by Executive Order 13112 in 1999 (National Invasive Species Council 2005). To generate better performance for EDRR and CM, volunteers in collaboration with environmental non-profit or governmental organizations become a requirement for invasive species control and management. Threats engendered by invasive species (i.e., plants, pests, and animals) to current agriculture and ecosystems are substantial. Invasive species damage can cause 120–137 billion dollars in economic losses per year in the United States. Over half of the plant species in the United States were imported before the twentieth century and some of them are threatening the health of ecosystems and affecting native species populations (U.S. Fish & Wildlife Service, 2012). Increasing the number of observations from volunteers can facilitate early detection and rapid responses for management decisions (Goodchild 2007). This study involved comparing the regional King County Weed Watcher (KCWW) program with the nationwide Early Detection and Distribution Map System (EDDMapS) program.

The KCWW program provides education and outreach as part of the King County Noxious Weed Control Program. This program began in 2006 in a specific area (Middle Fork Snoqualmie Valley) and expanded to the Snoqualmie Pass Area and other nearby wilderness areas in 2010. Beginning with 22.5 miles of surveyed trail, this program expanded; in 2010 more than 200 volunteers throughout King County surveyed over 100 miles of trail. The EDDMapS is a system intended to provide the most complete information nationwide and has been operated by the Center for Invasive Species and Ecosystem Health at the University of Georgia since 2005. Unlike the reporting forms of trail incident observations that are rather diverse, the invasive species reporting forms are similar, requiring that information for any invasive plant be predefined although the KCWW and EDDMapS modified the reporting forms based on their specified needs.

The KCWW enables participants to monitor and control invasive plant species simultaneously. Like WA-DNR-P, an education and outreach coordinator is responsible for volunteer training and data handling. King County collaborates with environmental non-profit organizations and volunteers to maximize invasive plant control by using a list of focused plant species. In 2012, King County launched a web-based reporting and data search website that enables volunteers to report and search for monitored plant data and also provides an original paper and Excel reporting form (2006–2012). On the King County monitoring form, trails and lakes are used to summarize VGI because they contain spatial information for land managers to locate and reach easily.

The purpose of EDDMapS is to integrate all existing invasive species records into one system. The historical records were assembled from other databases and herbarium collections. In addition, multiple data collection methods (file-based, online forms, and mobile applications[1]) were developed to enable a broader range of volunteers to monitor invasive plants. Unlike the KCWW program, the records collected from EDDMapS include various spatial references. For example, previous records have used the county boundary as spatial references, whereas most data collections have used Global Positioning System (GPS) coordinates as spatial references.

10.3 Spatial Decision Unit and Structured Monitoring Forms

The primary observations from four cases pointed out two variations of VGI. First, a free-form (narrative-oriented) VGI used in LHRP is different from form-based VGI observed in rest three cases. Second, various spatial units have been used in different cases, such as GPS (EDDMapS), trails (KCWW, WA-DNR-P, LHRP), entire park (LHRP), or county boundaries (EDDMapS).

In this paper, the spatial reference and content of VGI will be adjusted and structured. The unit of spatial reference, which we call spatial decision unit (SDU), is defined as a physical and fundamental unit for management activities that volunteers can easily recognize. Having a common unit to connect project activities, associated budgets, and staff has been shown to be important for management (Kelly 1985). Specifically for monitoring, researchers have suggested that monitoring design should directly connect to management decision-making to track management performance (Gibbs et al. 1999) and improve future decisions (Williams et al. 2007). To transfer similar information cohesively during each stage of decision making, identical scales and fixed units are required (Lyons et al. 2008). This research used SDUs to organize VGI so that VGI can be used directly for decision-making and increase the value of VGI. In our cases, LHRP, WA-DNR-P, and KCWW use trails as SDUs to organize existing VGI. EDDMaps did not use any SDU to organize their VGI.

The second comparison involves different degrees of structured monitoring forms, which is derived from decision-making problems and the information needs of management tasks. Ideally, the form should be simple and clear for volunteers and sufficient for public environmental managers to use. Since trail incident monitoring and monitoring are less specialized than invasive species identification, it is found that a semi-structured form is chosen by LHRP for more descriptive detail and more interaction with volunteers. However, the WA-DNR-P would like to enhance their current free-form reports into more structured report form so more information can be collected and processed at the same time with preliminary understanding of incidents happening in managed areas. To WA-DNR-P, the ideal form would

[1]Because the mobile application targets general users, the form was designed to be simple and require minimal profession knowledge. Others forms require detailed information regarding habitat conditions and spreading estimations.

connect every possible management action with various selectable options instead of text fields. For example, volunteers can choose logs of various sizes (with photos) when reporting a dead tree blocking trails in the trail maintenance form. Each log size requires an associated action, and the reports assist the WA-DNR staff in arranging actions. In addition, this form enables volunteer contributions to be evaluated according to various quantitative criteria (e.g., hours, expense, survey miles, and expenditures). Based on various decision needs (e.g., trail maintenance, park enforcement, and environmental concern reporting), several forms were developed. As for invasive species control, the last section of this paper discusses how KCWW and EDDMapS adopted a general template created by the United States Department of Agriculture for invasive species reporting; and as such, both of their forms are highly structured.

10.4 Capacity Examination Through Case Comparison

This study conducted two assessments, capacity of SDUs and various degrees of structured forms, using four empirical cases in Washington State. The capacity of SDUs was assessed according to the SDUs' ability to connect VGI with management targets for decision-making. The capacity of structured forms was assessed according to how they change values of VGI to match management tasks with decisions made by land managers and impacts to volunteers.

The first capacity comparison compare how information can be arranged differently with and without SDUs based on a decision question within a specific area. Assessment elements include (a) time to search and find required information, (b) time required to retrieve and organize information for decision needs, and (c) abundance of information (percentage of content) retrieved were compared. This comparison compared (a) traditional meeting minutes[2] (LHRP) arranged according to trails SDUs, (b) invasive species monitoring records (KCWW) arranged according to trail or lake SDUs with GPS coordinates, and (c) invasive species monitoring records (EDDMapS) with GPS coordinates or county information.

The second comparison examined the capacities of three structured monitoring forms. Our main focus is to identify how different types of structured forms of VGI match with tasks involving land management decision making and how these information collection changes affect volunteers. Assessment elements include (a) integrity of VGI transformed into summarized numerical values, (b) the accuracy at which VGI were matched with the appropriate staff or managers, and (c) the efficiency at which VGI were connected to management activities or decisions were compared, and (d) what impact can be observed to volunteers. This research

[2]Generally, the meeting minutes and agenda are grouped according to priorities or themes based on personal preferences. The person who calls the meeting, designs the agenda, or writes the meeting minutes makes the final decision regarding how the gathered information is processed.

compared (a) LHRP non-structured VGI records arranged by spatial decision units, (b) a newly developed semi-structured form for LHRP, and (c) a highly-structured form for WA-DNR-P.

10.5 Results

The first case comparison examined the capacities of spatial decision units applied in four case situations. LHRP meeting records arranged by trails provides the first tier of search possibilities for the park management and is able to generate work orders when necessary. Because the meeting agenda and minutes (LHRP) were not organized using spatial decision units, all materials must be browsed in each data search. If the materials are not carefully archived or distributed, then the database provides incomplete information and has little value. VGI must be condensed and organized according to keyword tags supporting searches. Organizing every new search is time consuming. Finally, the abundance of information that can be visualized is limited. The meeting minutes do not provide spatial information as a visual aid that enables participants to achieve a unified understanding. Data search, retrieval, and use were estimated to require approximately one-half day to one-day in the LHRP case. This arrangement creates a direct connection between textual content and GIS data, enabling the incident-location relationship to be visualized. Once this table has been created, managers can add management-related tags such as the 'date' and 'management tasks'. Organizing all records simultaneously shortens the time required to search for information. However, when no systematic arrangement (such as a database) was established to enhance query efficiency, data search, retrieval, and use was estimated to require less than one-half day in the LHRP case.

Using a database and a searching interface, the KCWW program and EDDMapS were expected to exhibit higher query efficiency than the LHRP system. The only difference between KCWW and EDDMapS is the uses of SDU to organize VGI. The KCWW uses trail as their SDU to organize VGI. By contrast, EDDMapS uses GPS coordinates and county boundaries to categorize VGI, and SDU is used. Although the KCWW has a simple user interface (a textual filter) only that categorizes data according to trail, trail group, and land manager, users can locate the trail and land managers and submit a query in approximately 0.5–5 min, depending on the familiarity of the area. The retrieval results arranged according to query can be downloaded as an Excel spreadsheet. Because the records are attached to the closest trails, the content can be visualized based on trails. A search of the entire King County yielded 766 records, over one-half of which contained GPS coordinates for exact locations and 490 invasive species have been identified and reported. Finally, there are 33 volunteers who contributed to data collection. Data search, retrieval, and use were estimated to require approximately less than 1 h in the KCWW system.

The EDDMapS system offers a map-based interface that enables users to search for information easily. For the same search area, the system provided a list of 416 invasive species, but each invasive species had a different web page and information had to retrieve separately. Users had to read through web pages to download needed information. The advanced query tool from the EDDMapS can be found only after contacting the technician from the EDDMapS. The advanced query tool in EDDMapS is similar to the one in the KCWW system, thus enabling search functions to be shortened for frequent users. The advanced query tool can shorten the time of search. However, depends on user's familiarity with the interface, the time may vary. Since EDDMapS only uses invasive species to group their information, the most common spatial reference that can be used is any county boundary. Although some records have GPS coordinates; that information seems to lack organization. Finally, regarding the abundance of information, although EDDMapS contained 1207 records, of which 156 included GPS coordinates only 4% of the invasive species (20/455) can be located using GPS records. As to the volunteer who contributed the information, only 12 volunteers provided information. Nearly all GPS records (94.87%) were collected from one volunteer for the United States Geological Survey (USGS) Non-indigenous Invasive Species Database. Data search, retrieval, and use were estimated to require approximately a few hours to a few days in the EDDMapS system.

10.5.1 Using Monitoring Forms to Structure the Information Needs of Land Managers

Comparison of structured monitoring forms focused on articulating different aspects of decision-making. First non-structured, free form VGI was gathered from meeting minutes plus discussions and arranged by trails. It is the best representation of participants' concerns. However, free-form VGI does not enable managers (or other data users) to make decisions. The VGI that managers received must be assigned to responsible people to define necessary tasks and to denote specific needs. In addition to a long processing workflow, these reports might possibly be transferred easily to staffers not associated with the information. The free-form meeting minutes exhibit poor accuracy and efficiency in regards to information needed for decision-making. It is less likely that free-form VGI can be utilized and accepted as a main source of decision-making. The narrative nature of free-form VGI prevents them from being transformed into a summary of numerical values, although integrity can be maintained in transferring textual content.

The semi-structured form categorizes VGI based on management tasks, decisions, and responsible staff. These categories can typically be grouped into routine management, scheduled events, or status reports. The volunteer can select main categories, and because these categories have been connected to management tasks, they can enhance the efficiency and accuracy of transferring reports. Regarding the description of trail status, the LHRP semi-structured monitoring form provides textual space in which volunteers can describe observations. Because the semi-structured monitoring form must be read individually, it would not be expected to achieve efficient processing measurements. Consequently, reading reports individually is not an option when processing large amounts of VGI. Although the semi-structured monitoring form can be used to transfer information to staff members or managers and retain trail status descriptions, it fails to provide an efficient method for processing VGI.

Finally, the highly structured monitoring form (in the WA-DNR-R case) categorizes all possible decisions or management activities into options that volunteers can select. The highly structured monitoring form enables VGI to be transformed into numerical values directly because all possible reporting categories are selectable options, and enables VGI to be processed and transferred to appropriate staff members or managers efficiently and accurately. However, because all options are predefined, volunteers can only report their observations based on limited selections. When the status is complex as when a combined incident requires more than one selection, the structuring might create biases and affect the credibility of the VGI.

10.6 Discussion of Findings

The following subsections describe the findings about capacities of spatial decision units and structured monitoring forms. Findings about effects of public images on stakeholder reporting and the decision-making process and stakeholder knowledge, credibility, and structured monitoring forms in EDDMapS are included as well.

10.6.1 Capacity of Spatial Decision Units

Although the SDU concept has been applied in various fields, this study combined SDU with visualizing textual VGI through mapping. Using visualized spatial information is crucial for managing large areas. Trails were used as SDUs in park trail management and invasive species control to connect narrative VGI. Trails are easy for volunteers to identify and easy for managers to reference while conducting management activities. When trails are used as an index, VGI can be transferred to the land manager who is responsible for that area. Even within a single park (the LHRP case), categorized VGI that was based on trails assisted managers in filtering information, particularly when they were unfamiliar with the park details. Arranging VGI according to SDUs resulted in a substantial increase in data management efficiency, thereby creating an incentive for managers to use this information to improve communication with volunteers. Easy spatial reference promotes shared understanding, and fosters information use.

Beyond comparing the capacity of having SDUs, the comparison between the KCWW program and EDDMapS also examined whether other common spatial references (i.e., GPS coordinates and county boundaries) can perform a similar function as SDUs. The result showed that although GPS coordinates were embedded in newly generated EDDMapS reports, the lack of a common unit for organizing and summarizing reports still becomes a barrier for management use[3] (Gibbs et al. 1999; Lyons et al. 2008). In addition, the comparison revealed that the EDDMapS couldn't be used to integrate previous records seamlessly in the database because no GPS point is associated with these records, considerably reducing the usefulness of information.

In sum, our comparisons reveal that no matter which SDU is used to organize VGI, having a unified unit is important to improve the data-cleaning process before and after VGI collection. When multiple SDUs were used in the same collection practices, we found that the spatial relationships among multiple SDUs were well constructed. Using SDUs to summarize VGI by management targets or tasks provided a good foundation for decision-making. Our results also indicated that using SDUs enabled qualitative results to be connected to GIS, thereby enhancing visualization. Because SDUs are based on management tasks, they shorten data processing and enable information to be matched to associated management tasks.

10.6.2 Inevitable Tradeoffs with Structured Monitoring Forms

The second comparison revealed that tradeoffs must be made because each structured monitoring form features different functions. The free-form VGI provide great flexibility in enabling participants to describe their focus, regardless of land manager needs. However, reading and reprocessing free-form VGI is a time-consuming task, and there is a greater chance to transfer this information to responsible staff(s) who does not necessarily need it. The semi-structured LHRP enhanced form uses functional categories to clarify and identify possible incidents that occur on trails. Because the categories are predefined, this form reduces the risk of mis-transference because these categories are more specific to decision-making and management. However, this form did not save enough time for managers and thus attained limited effectiveness. Reading each report separately provides limited incentives for land managers to use the information.

The highly structured monitoring form developed for the WA-DNR-P provides the highest efficiency of data processing. Because all management activities are seamlessly connected to selectable options in a reporting form, this form connects management tasks with VGI. The managers can look up all information in one summarized file. To gain efficiency, a one-time, massive time investment of

[3]Gibbs et al. (1999) stated, "monitoring must be done to satisfy some desired state of an appropriate indicator that management is intended to meet."

characterizing management activities into highly structured forms with various selections is needed. The WA-DNR-P considers the semi-structured form to be applicable only if a volunteer can report information that matches management needs specifically (WA-DNR-P coordinator meeting minutes, 2012). However, processing large amount of VGI requires a tradeoff between data processing efficiency and the integrity of VGI. Various levels of structured forms might be needed to fit diverse purposes of VGI activities. In sum, our preliminary comparison showed that the free-form VGI might be good for communication and outreach due to the fact that volunteers can communicate their ideas without restrictions on the types of information is needed. Semi-structured forms provide a bridge between manager and volunteers when the collaboration between both parties is well established. Highly structured form is used to enhance the speed of data processing when crowdsourced information is needed. Structured monitoring forms help to reveal patterns of interactions between volunteers and land managers. Therefore, land managers need to select and/or craft structured forms to fit the purposes of their volunteer programs.

10.6.3 Potential Pressures Associated with Using Spatial Decision Units

In addition to bridging the gaps between VGI and information needs for decision-making, SDUs were associated with several potential pressures. Although SDUs facilitated shortening the data-cleaning process, they might exert a different public image than desired by the departments using them, and this depends on how the public interprets the image. The recreation managers mentioned that additional VGI pertaining to the area might imply poor trail maintenance practices and create a negative public image. However, for invasive species control case, a land manager who collects additional VGI on invasive species is proactive in eliminating this threat and attempts to gain an understanding of how invasive species can be controlled. VGI can be a positive public image for land managers in this case. Hence, creating SDUs should be considered part of a data processing task as well as a means of improving a manager's public image.

Second, SDUs may trigger complex administrative workflows pertaining to responses and liability. In the LHRP and WA-DNR-P cases, if the VGI cannot be properly managed, then the land manger may face unexpected lawsuits because managers fail to handle incidents occurring in parks. At a WA-DNR-P meeting on 2012-11-06, the volunteer coordinator and nature resource managers from the WA-DNR-P mentioned that, without careful planning regarding liability concerns, stakeholders reporting trail incidents may trigger unexpected risk. Another concern is authority sharing. Certain land managers do not treat volunteers as equal partners. One-way data collection may result in staff shortage (in the LHRP case) or scale limitations (nationwide EDDMapS); however, this circumstance limits communication between land managers and volunteers. Regarding the regional scales of cases (i.e., the KCWW Program and the WA-DNR-P Forest Watch Program), managers tend to share responsibility and authority with volunteers to ensure that the volunteers are motivated and willing to continue participating in activities.

Finally, for most cases, SDU information is incomplete or is not appropriately updated. When trails are used as SDUs, it is assumed that the existing GIS trail layer or map has been recently updated. However, this assumption is not practical in all cases. A complete trail layer may not be applicable in most cases (i.e., LHRP, the WA-DNR-P, and KCWW programs) because land managers do not want to be responsible for certain trails that they have not managed, thereby creating the jurisdictional barriers mentioned in Johnson and Sieber (2013). This concern should be addressed by encouraging a consortium GIS approach (i.e., a GIS virtual center in which all agencies collaborate) to maintain complete information about trails because trail clearing should be a one-time effort, regardless of the land manager and situation.

Acknowledgments We thank all of the volunteers, volunteer and education coordinators, and managers in LHRP, King County, and Puget Sound Region of WA-DNR's for providing access to their voluntary reporting system implementation processes to form the cases in this study. We also gratefully acknowledge Prof. Peter Schiess who provided generous efforts for critical review and many discussions.

References

Brewer GD, Stern PC (2005) Decision making for the environment: social and behavioral science research priorities. National Academies Press, Washington, DC

Brudney JL (1999) The effective use of volunteers: best practices for the public sector. Law Contemp Probl 62:219–256

Cash, D.W., W.N. Adger, F. Berkes, P. Garden, L. Lebel, P. Olsson, L. Pritchard, and O. Young. 2006. Scale and cross-scale dynamics: governance and information in a multi-level world. Ecol Soc 11(2): 8. [Internet]. [Cited 2014 Jun 20]. Retrieved from http://www.ecologyandsociety.org/vol11/iss2/art8/

Dunn CE (2007) Participatory GIS a people's GIS? Prog Hum Geogr 31(5)616–637

Elwood S, Goodchild MF, Sui DZ (2012) Researching volunteered geographic information: spatial data, geographic research, and new social practice. Ann Assoc Am Geogr 102(3)571–590

Gibbs JP, Snell HL, Causton CE (1999) Effective monitoring for adaptive wildlife management: lessons from the Galápagos Islands. J Wildl Manag 63(4)1055–1065

Goodchild MF (2007) Citizens as sensors: the world of volunteered geography. GeoJournal 69(4)211–221

Gregory RS, Keeney RL (2002) Making smarter environmental management decisions. J Am Water Resour Assoc 38(6)1601–1612

Gregory RS, Failing L, Harstone M, Long G, McDaniels T, Ohlson D (2012) Structured decision making: a practical guide to environmental management choices. Wiley-Blackwell, Chichester

Grira J, Bedard Y, Roche S (2010) Spatial data uncertainty in the VGI world: going from consumer to producer. Geomatica 64(1)61–72

Gunderson L, Peterson G, Holling CS (2008) Practicing adaptive management in complex social-ecological systems. In: Norberg J, Cumming GS (eds) Complexity theory for a sustainable future. Columbia University Press, New York, pp 223–245
Holling CS (1973) Resilience and stability of ecological systems. Annu Rev Ecol Syst 4(1)1–23
Holling CS (1978) Adaptive environmental assessment and management. Blackburn, Caldwell
Jankowski P, Nyerges T (2001) Geographic information systems for group decision making: toward a participatory geographic information science. CRC Press, Boca Raton
Johnson PA, Sieber RE (2013) Situating the adoption of VGI by government in crowdsourcing geographic knowledge. In: Sui D, Elwood S, Goodchild MF (eds) Crowdsourcing geographic knowledge: Volunteered Geographical Information (VGI) in theory and practice. Springer, London, pp 65–83
Kelly L (1985) Budgeting in nonprofit organizations. Drexel Libr Q 21(3)3–18
Lyons, J. E., M. C. Runge, H. P. Laskowski, and W. L. Kendall. 2008. Monitoring in the context of structured decision-making and adaptive management. J Wildl Manag 72 (8): 1683-1692.
McNie EC (2007) Reconciling the supply of scientific information with user demands: an analysis of the problem and review of the literature. Environ Sci Pol 10(1)17–38
National Research Council (NRC) (1983) Risk assessment in the federal government: managing the process. National Academy Press, Washington D.C.
Nyerges, T., M. Barndt, and K. Brooks. 1997. Public participation geographic information systems. In Auto-Carto 13 American Congress on Surveying and Mapping Proceedings, 224–233. Bethesda.
Obermeyer N (1998) The evolution of public participation GIS. Cartogr Geogr Inf Syst 25:65–66
Olsson P, Folke C, Berkes F (2004) Adaptive comanagement for building resilience in social-ecological systems. Environ Manag 34(1)75–90
Propst DB, Jackson DL, McDonough MH (2003) Public participation, volunteerism and resource-based recreation management in the U.S.: what do citizens expect? Soc Leisure 26(2)389–415
Reed MS (2008) Stakeholder participation for environmental management: a literature review. Biol Conserv 141(10)2417–2431
Schlossberg M, Shuford E (2005) Delineating "Public" and "Participation" in PPGIS. URISA J 16(2)15–26
Silvertown J (2009) A new dawn for citizen science. Trends Ecol Evol 24(9)467–471
Simon HA (1979) Rational decision making in business organizations. Am Econ Rev 69(4)493–513
Smith LG (1982) Alternative mechanisms for public participation in environmental policy-making. Environ 14(3)21–34
Steffen W, Sanderson A, Tyson PD, Jäger J, Matson PA, Moore BIII, Oldfield F, Richardson K, Schellnhuber HJ, Turner BL, others (2006) Global change and the earth system: a planet under pressure. Springer, Berlin
U.S.C.S. § 706(2)(A)
United States Fish and Wildlife Services. 2012. The cost of invasive species. [Internet]. [Cited 2016 Feb 20]. Retrieved from https://www.fws.gov/verobeach/PythonPDF/CostofInvasivesFactSheet.pdf
Uychiaoco AJ, Arceo HO, Green SJ, Cruz MT, Gaite PA, Aliño PM (2005) Monitoring and evaluation of Reef protected areas by local fishers in the Philippines: tightening the adaptive management cycle. Biodivers Conserv 14(11)2775–2794
Walters CJ (1986) Adaptive management of renewable resources. Macmillan, New York
Williams, B. K., R.C. Szaro, and C.D. Shapiro. 2007. Adaptive management: the U.S. department of the interior technical guide. Washington, D.C.: Adaptive Management Working Group, U.S. Department of the Interior.
Wyk E, Roux D, Drackner M, McCool S (2008) The impact of scientific information on ecosystem management: making sense of the contextual gap between information providers and decision makers. Environ Manag 41(5)779–791

Chapter 11
Volunteered Geographic Information for Building Territorial Governance in Mexico City: The Case of *The Roma* Neighborhood

Elvia Martínez-Viveros, Rodrigo Tapia-McClung, Yezmín Calvillo-Saldaña, and José Luis López-Gonzaga

Abstract A case study of Volunteered Geographic Information (VGI) for building territorial governance in a Mexico City neighborhood is presented. It is an ongoing project carried out since mid-2014 between a research center and a citizen organization of an urban neighborhood, namely the Consejo Vecinal Roma (CoVe). This project is aligned with research lines in VGI, Citizen Science and Humans as Sensors, and is supported by the development of a geospatial digital platform representing an instance of how the current way of making, reading, and using maps can be linked to citizens' interest to participate in decision-making processes that affect their daily life. The platform was designed and developed using free and open source software, and with the expectation of being easily used and adopted by citizens. The purpose of the project is to help consolidate CoVe as a policy community by means of integrating a territorial vision of *The Roma*'s problems and opportunities, as perceived by the citizens involved in the process. An additional purpose is to use this vision in order to support the construction of a citizen's agenda driven by CoVe with the intention to propose and organize courses of action with neighbors and citizens interested in the betterment of the neighborhood.

Keywords Geospatial web platforms • Volunteered geographic information • Citizen science • Governance

11.1 Introduction

Citizens hold an increasing interest to participate in the decisions that impinge upon their quality of life and work, and the well-being of the natural environment. They are witnessing a transition from a governmental management of public affairs,

---

E. Martínez-Viveros • R. Tapia-McClung (✉) • Y. Calvillo-Saldaña • J.L. López-Gonzaga
Centro de Investigación en Geografía y Geomática 'Ing. Jorge L. Tamayo' A.C. (CentroGeo), Contoy 137, Lomas de Padierna, Tlalpan, 14240, México, Distrito Federal, México
e-mail: emartinez@centrogeo.org.mx; rtapia@centrogeo.org.mx; ycalvillo@centrogeo.org.mx; jlopez@centrogeo.org.mx

M. Leitner, J. Jokar Arsanjani (eds.), *Citizen Empowered Mapping*, Geotechnologies and the Environment 18, DOI 10.1007/978-3-319-51629-5_11

underpinned in formal regulations and procedures, to emergent forms of governance in which non-governmental organizations (NGOs) and various stakeholders from civil society become involved in the design of policies and programs and in the decisions that have an impact on their daily life.

The power of the Internet is acknowledged as a tool for linking government and citizenship in a previously unthought-of dialogue, as well as for connecting people in social networks for sharing their common interests. Individuals, groups, and communities are generating and interchanging across distributed networks data and information relevant for decision making in domains of public interest. Also, cartography and citizenship are two terms that have started to be heard in ensemble. Digital maps have shown to be part of our everyday life and, more importantly, have also proved to be useful for the advancement of data collection. Phenomena like crowdsourcing or volunteered geographical information (VGI) are bringing people with the abilities to contribute with reliable and relevant data together. Finally, it is recognized that geospatial information on the web and the tools for its generation, visualization, and analyses are changing the vision of how people structure and investigate the world. People with spatial consciousness have the ability to interpret maps, detect functional relationships between processes happening in different places, highlight ongoing processes in the geographic space, and adopt a new scope for thinking about and solving problems.

In this framework, an ongoing project is presented. It has been carried out since the second quarter of 2014 and being spearheaded by a research team from a Mexican public research center and by a citizen organization of an urban neighborhood in Mexico City, namely the Consejo Vecinal Roma (CoVe – Roma Neighbors' Council). This project proposes a digital platform that relates research lines such as Volunteered Geographic Information, Citizen Science and Humans as Sensors, as proposed by Goodchild (2007). This digital platform is aligned with the aforementioned academic trends and represents an instance of how the current way of making, reading, and using maps can be linked to citizens' interest to participate in the decision-making processes that impact and affect their daily life.

The purpose of the project is to help consolidate CoVe as a policy community by means of integrating a territorial vision of the neighborhood's problems and opportunities, as perceived by the citizens involved in the process. This vision would be used to support the construction of a citizen's agenda for the period 2015–2018, conducted by CoVe in order to propose and organize courses of action with neighbors and citizens interested in the betterment of the neighborhood's public space, quality of life, and working conditions. The intention is to use both the digital platform and the agenda as mechanisms to support deliberations with governmental authorities –at both the county and city levels– and with different stakeholders able to aid in the agenda's implementation. An assumption is that these instruments will add to CoVe's abilities to build legitimate ways to participate in decisions that affect the neighborhood's life, thus empowering the citizens involved in the process. This can be framed in the development of a strategy aiming at the citizens' direct participation in shaping the scope and practice of policies, programs, and services rendered by the government.

This chapter presents the project as a case study of VGI for building territorial governance in Mexico City. The project started its planning phase in mid-2014, went through a developing stage in the second half of 2014, implemented a pilot study in early 2015, and was launched at full scale shortly after. It can be described from two interrelated perspectives, namely a technical one and a socio-organizational one. The technical perspective involves the development and implementation of the digital platform and the further adaptations and functionalities needed to adequately respond to the needs emerging from its implementation and use. The socio-organizational one involves the citizens' participation along the process: From the definition of the platform content and user requirements, to the organization of the data collection process, the recruitment of volunteers, the analysis of results, and the agenda's integration and use.

The meaning of governance in the context of this project pertains to the process that CoVe, as a community of policy, explicitly tries to build in participation with private and public stakeholders. It is in the context of this process that the research team embraced the challenge to develop a geospatial web platform to support a VGI process, and in order to provide evidence to help bridge perceptual gaps of social actors that could potentially be involved in the process. The first section of the chapter explores the issue of governance as a conceptual guide of the project's process. The second section deals with the definition of the information to be gathered in the VGI process and is followed by the description of the platform design and its technological development in section 3. Section 4 describes the process from collecting data to building the agenda. Finally, the discussion section addresses the VGI concepts involved in *The Roma* project, some considerations related to volunteered information, and a comparison of the project with other crowdsourcing platforms managed by NGOs that mediate between citizens' demands and government actions.

11.2 What Governance?

The Valley of Mexico Metropolitan Area is comprised by Mexico City together with an agglomeration of 59 municipalities of the State of Mexico and one of Hidalgo. It is the 10th largest urban agglomeration in the world, a metropolitan area without a metropolitan government. Mexico City's mayor is elected for a six-year term and so are the heads of each of the 16 municipalities that integrate its administrative territory. *The Roma* neighborhood is located in the Cuauhtémoc municipality, about 3.5 km southwest of the historical city center, covers approximately 3.8 km^2 and houses an intense economic activity coupled with dense movements of both vehicles and pedestrians (see Fig. 11.1). It was founded in the early 1900s as a high-class residential area, but the population growth and the gradual urban deterioration left their footprint in it, and so did the havoc left by the 8.1-degree magnitude earthquake of 1985. Since the 1960s, residential use has been shared with shops, offices, schools, and many varied businesses. During the last few years, *The Roma*

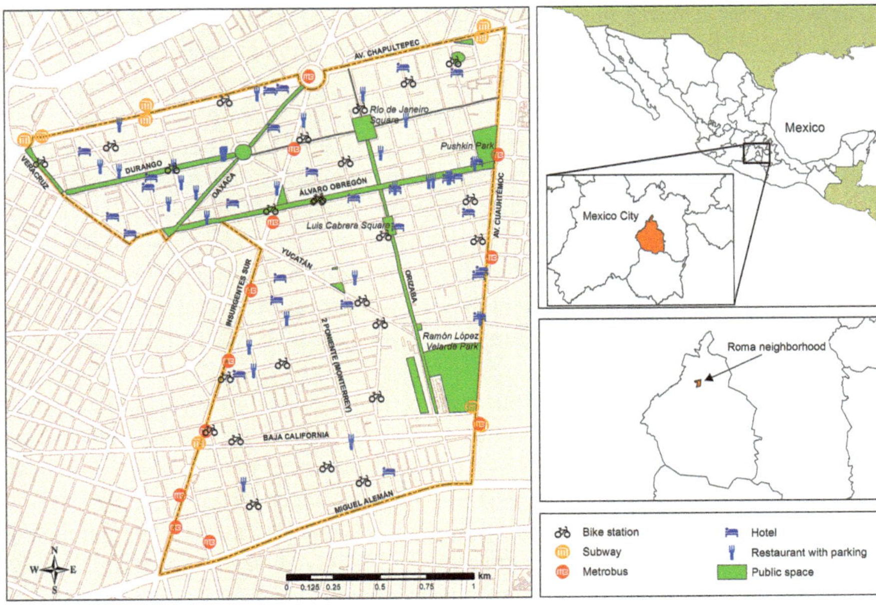

Fig. 11.1 Location of *The Roma* neighborhood in Mexico City

has experienced a gentrification process and has become one of the favorite places to establish new amenities and for young and educated settlers that share the neighborhood with families that have lived in it for decades. Gentrification gives life and dynamism to *The Roma*, but also carries with it tensions and risks derived from changes in land use, stressed public services, possible increases in insecurity, or the wear and tear of both architectural heritage and of public spaces. How to profit from the opportunities and how to reduce risks were the topics around which CoVe emerged.

CoVe is organized in a neighborhood council, led by a collegiate group of counselors and a coordinator. Since its creation, it has hold regular town hall meetings and has integrated a citizen space called *SUMATE* (JOIN US). CoVe defines itself as a citizen initiative to exert participative democracy and boost collaborative dialogs and actions with both local authorities and different sectors interacting together in the neighborhood. It proposes dialog as a process to encourage a shared sense and to align the diversity of purposes, interests, and values in order to create a learning environment. By doing so, it tries to focus the processes of reflection and deliberation to both problems and possibilities (CoVe (Consejo Vecinal Roma) 2015).

CoVe has become increasingly involved in citizen participation for public space improvement in *The Roma*. Such engagement has entailed novel forms of organization and communication, as well as the construction of cooperative partnerships with local entrepreneurs and local governmental authorities. It seeks a stake in the governance of its neighborhood and this project was formulated, from its inception, as an asset to consolidate citizen participation in the process. But what is the meaning of territorial governance in *The Roma* where, as in many Mexican towns, governmental powers and institutions of the three government levels –the federal, the city, and the municipal– overlap? Where the authorities in each of these levels most frequently belong to opposing political parties responding to partisan interests? Where the scheme for citizens' involvement was designed by the government in terms of heads of territorial sectors elected with little participation, in mostly opaque processes and which, in reality, function only as figureheads? How can a citizen organization change a form of governance of a corporate State that has historically relied on forms of patronage and bureaucracy and on an apathetic and non-participative citizenry?

The term governance takes on multiple meanings that vary in time and space. De Vries (2013, 4) points out that many of its definitions have in common the vision of a process related to the steering of societal developments. He also notes the trend in academic and international organizations of proposing dimensions or criteria for a 'good' governance, for instance, rule of law, citizens' voice, accountability, good regulation, corruption control, effectiveness, openness, coherence, and so on. The process would be good if the government's governance agenda paid attention to these dimensions (Grindle 2004). In this sense, the process is steered by a government whose representatives have been legitimized in the arena of democratic elections, and citizens have transferred them the responsibility and the authority for making decisions of public affairs.

But there are also many authors who see governance as a term that extends the approach from what formal government does to encompass the collective action practices that seek public purpose (for instance Raadschelders 2003; Dreschler 2013). This meaning had been gaining momentum in the face of neoliberal trends, such as the privatization of the delivery of public goods or market deregulations. Accordingly, governance is seen as an alternative to the hierarchical control of government, and entails a form of control that includes the active participation of multi-sectorial networks of stakeholders sharing the responsibility for policies and programs that consider the public interest.

This project addresses a governance process which, according to Healy (2006) and Davoudi and Strange (2009), refers to collective action promoted to achieve public purposes. An activity which is neither left to the market forces nor to the government's decision, but is instead guided by the complex interactions between the spheres of the State, the economy and the civil society, and in a modality in which the objectives, strategies, and ideology of public policy have feedback with the knowledge and meanings developed in the civil society's policy communities. These interactions underpin interventions in the form of regulations, programs or actions oriented towards shared goals related to distributive justice, quality of life, environmental well-being, and/or economic vitality (Healy 2006).

This meaning of governance, framed in values of participative democracy (as opposed to representative democracy), is the one behind the emergence of CoVe as a community of policy. It is in this framework that the challenges to connect citizenry, government, and different stakeholders emerge, and it is in this context that the research group saw a social base able to profit from a VGI process to gather information about problems and opportunities in this urban area, which may provide the evidence needed to bridge deliberations about *The Roma*'s public affairs among the parties involved.

The working hypothesis was that spatial knowledge of *The Roma*'s problems and opportunities could serve as an asset with the power to create a synergy between citizen participation and influence the actions of local government. This knowledge could be able to brake sectorial silos, recognizing the coexistence of different problems in a same place, and to gear actions that take into account the spatial organization, the qualities, and the identity of the place. *The Roma* experience is a case trying to show the potential power of place, space, and citizenship in building capacities to be heard in the policy arena.

The research team supported the building of this vision by developing a collaborative web mapping application that aids in the visualization of problems and opportunities that may help the construction of a governmental-civil society working agenda. An agenda based on the local knowledge of the space they experience on a daily basis, and above all, an agenda that may have an influence in discussions to include the citizens' voice in the governance process.

11.3 What Information?

CoVe's members perceived many problems across *The Roma*'s territory but they lacked the tools which enabled them to show their magnitude and specific location. They were aware of the need to have such information in order to sustain with evidence the proposals they were articulating to those who made decisions. Some of its members tried, with little success, to depict these problems on a paper map. The idea to capture information derived from citizen's perceptions on a web mapping application emerged at an informal meeting.

The first step to devise a plan of action for this project involved the definition of the data that were going to be gathered. Initial interviews with CoVe's members gave overall ideas about the general issues at stake, but the research team conducted and facilitated several workshops with members of CoVe aimed at the definition of the set of variables that were to be collected. Concerns such as how to profit from certain opportunities, or how to reduce some risks, were topics that guided these workshops in which participants identified what information to collect for their neighborhood. This included objects and situations that are perceived as threats to their security, obstacles to their mobility, deviations from legal and established land use, as well as activities and practices that deteriorate the environment and the quality of public spaces.

Although the research team facilitated and conducted workshops, the neighbor committee members were responsible for the definition of variables. Local author-

Table 11.1 Six categories and corresponding sub-categories for field data collection

Businesses	Services	Real estate	Security	Mobility	Waste
Specialized business	School	Empty lot	Dark place	Sidewalks with appropriate ramps	Waste accumulation
Self-service	Health services	Unoccupied or invaded house	Nonworking streetlight	Pothole	Large dumpster
Restaurant	Bank	House or building under construction	Formal security	Obstacle relevant for mobility	Small dumpster
Bar	Hotel	House or apartment for sale	Drugs or alcohol consumed	Crosswalk	Illegal dump
Gas station	Religious center		Vandalized place	Bike station	Place where truck separates waste
Mall	Cultural, recreational or sports complex		Loud noise emission	Taxi stand	Place where businesses leave waste or oil
Fixed street vendors	Government office		Visibility obstacle	Public transport stop	
Non-fixed street vendors	Public parking		Urban equipment		

ities of the municipality participated as observers in some of these workshops. Through successive iterations, six categories were defined to be collected by human sensors, including businesses, services, real estate, security, mobility, and waste. These, in turn, encompassed a series of sub-categories, with a total of 41 different variables to be collected that are shown in Table 11.1.

The research team's opinion was that data collection for businesses and services categories was redundant, given the existence of official data sources for the majority of these variables, provided by the National Statistical Directory of Economic Entities (INEGI (Instituto Nacional de Estadística y Geografía) 2015). However, citizens strongly defended the need to collect them, arguing that the life cycle of many of these establishments is too variable to be reflected in the official records, that official information was outdated, and that by using the official classification it was not possible to observe the variables they desired, such as street vendors (both fixed and non-fixed).

There were several reasons that led neighbors to define these categories and variables. For instance, in the case of businesses, the motivation stemmed from topics such as finding out which restaurants offer parking space or valet parking service, an issue related to traffic congestions in the neighborhood. In the case of real estate, they were interested in finding out if there were catalogued heritage buildings that were being modified, detect the possibility of illegal changes in land use, or the need to quantify and locate property occupied by illegal tenants. Also, they wanted to detect the degree of deterioration of such buildings and their spatial relation to other issues in the neighborhood, such as waste or insecurity. Waste was another topic that drew a lot of attention. It is one that affects the whole city and even though it relates to the efficiency and effectiveness of the local government public services, it is intertwined with the need to ameliorate civic and environmental culture among the city population. The perception of insecure places was deemed central to foster improvements of the urban space, encourage the appropriation of public spaces for leisure activities, and enhance the cultural ambiance of the neighborhood in general. Mobility was also of much interest because, in spite of the increasing cost of land in the neighborhood in the last few years, the conditions of sidewalks, curbs, and accessibility ramps represent obstacles, mainly for disabled people, but also for strollers, bicycles, and other vehicles as well.

Variables chosen by neighbors do not necessarily hold a one-to-one correspondence with the way in which official agencies decide and make operational concepts and variables collected via censuses, surveys, or administrative records. The variables in this study are not adjusted to a formally established conceptual framework, but rather are the product of the knowledge the citizenry derives from everyday experience and from the way in which they use and appropriate themselves of the space they live in, go through, work, amuse, or relate to. Data that are socially integrated and that differ from official statistics emerge from the local knowledge people have about their neighborhood. These official statistics tend to be homogeneous for the whole urban space in the city and are collected in accordance to rigorously regulated and supervised procedures. For instance, reports on waste collection inform about the amount of tons of garbage collected, but do not mention what was not collected, the problems that this action generates, or locations where waste is accumulated. Variables defined by neighbors obey to certain particularities that they want to highlight in their neighborhood, are collected by means of a constant upgradable process and are derived from both their experience and perception. Therefore, it makes sense to crowdsource these data.

Additionally, a participatory diagnosis in order to characterize problems of 19 public spaces located in *The Roma* was undertaken. Each one holds its own value for citizens. These spaces include ten squares and public parks and gardens, three roundabouts, sports pits, and five emblematic walkways. A diagnosis of these spaces contemplates the recovery of their history, records of previous uses, and the evaluation of their current state, problems, and possibilities.

The variables' definition constituted the content of the citizens' map, an alternative to official cartography that, using the words of Mattioli (2014, 151), [could reveal] "the complexity of stories and practices taking place in the 'blanks' of the map". Stories that are captured in maps depicting places where itinerant trading takes place, garbage piles up, or passersby feel insecure. Stories that cannot be captured by hard and structured data, formally and officially collected in accordance with accepted standards. Maps that emerge as collective constructions cannot be assessed for the accuracy and robustness of their data, but they express local knowledge of the people involved. They contain citizens' perceptions and, in this sense, reflect the space in which they live in, their sensations, emotions, and the meanings they have in their everyday life. Maps created in such a way may put forward arguments with respect to a certain issue, derive their usefulness from their capacity to evoke it, as well as to invite to reflect on it. Also, these maps become active participants in the search and finding of a solution to the issue. Goodchild (2007, 220) arrives at the conclusion that "the most important value of VGI may lie in what it can tell about local activities in various geographic locations that go unnoticed by the world's media, and about life at a local level".

11.4 The Geospatial Platform

Parallel to workshops that were taking place, the research team was developing a geospatial web platform in which to host citizen's perceptions. The design of such a platform was adjusted to the specific needs of CoVe. However, a review of different platforms and VGI exercises reported in the literature accompanied the platform design. This review allowed the research team to detect experiences relevant to different aspects of the project. From the governance point of view, a very interesting experience is a platform for connecting groups of citizens with local authorities of Cali, Colombia. It is managed by an NGO interested in fostering social development by means of citizen participation campaigns. It crowdsources information about: (1) Problems, solutions' proposals, and projects in specific zones for 169 items in 23 categories. For example, conviviality, corruption, culture, human rights, education, entrepreneurship, vulnerable groups, urban infrastructure, or public works; and (2) georeferenced warnings that include reports on issues like damages or maintenance problems on urban infrastructure, mobility obstacles, crimes and insecurity problems, landslides, or abuse of power. Problems, proposals, and reports can be displayed and filtered by category on a geospatial platform based on Google Maps. Geotags are used in order to describe users introducing data, number of people affected, and date of publication, among other elements. Users can record opinions supporting initiatives and local governmental authorities can respond. The pulse of initiatives is measured and those which raise major interest of citizens and supporters are reported. The platform is directly linked to Cali's governmental institutions and received the Open Territorial Government prize granted by the Organization of American States and the excellence prize for e-government 2013 (Activo Group 2013).

In the same line of thought, another NGO implemented a web platform connecting citizens and authorities in the metropolitan area of Monterrey, Mexico. This platform crowdsources citizen reports on diverse incidents in the public space, such as accidents, traffic jams, obstacles in the streets or sidewalks, vehicle thefts, burglaries, passers robberies, and suspects. When last consulted, it contained about

150,000 reports. CIC's (Centro de Integración Ciudadana) staff monitors reports, forwards them to the authority in charge of its attention, and records the progress on the platform. The platform is based on Google Maps and its reports can be filtered by six categories and 24 items, time of reports, distance between reports, municipality, and reports details. Displays facilitating analysis include graphs showing timely and seasonal historical trends, reports at the municipal level, a scorecard comparing data between municipalities, heat maps, and animated maps showing space-time patterns. It is worth mentioning that, although these tools seem very complete, its reading and handling is counterintuitive and cumbersome. The platform is web-based and offers mobile apps (CIC (Centro de Integración Ciudadana) 2015).

The aforementioned platforms, as compelling as they are, have been on the web for over five years and have had time to become relatively stable and well-known. They are supported by formal NGOs with access to different forms of funding, which mediate between citizen demands and local authorities' actions. Citizens act as human sensors and crowdsource data on platforms designed under the NGO's purposes. In many ways, both are platforms that share purposes with the one presented here. However, CoVe relies only upon citizens' small donations and lacks a professional and full time staff devoted to manage, canalize, and monitor each report. Hence, during the first stage of the project, the decision was made to develop the platform tools essential for gathering and analyzing citizens' data needed for building CoVe's agenda and to implement tools for structuring a dialogue between authorities and citizens. Additional functionality could be added during a future stage when, and if, protocols for response and monitoring could be agreed upon with different public and private stakeholders, and when CoVe would be able to become financially and professionally stable.

Other reporting platforms have been developed for different interests and applications. The general goal of these volunteer-driven platforms is to have a great number of contributors sharing their observations on specific topics. For example, data can be collected about observations of the natural world using iNaturalist (California Academy of Sciences 2015). In terms of creating tree inventories useful for urban forestry and ecosystem management OpenTreeMap (2015) is very helpful. Ushahidi (2015) is a well-known platform for crisis management mapping. GeoCitizen (Atzmanstorfer et al. 2014) was developed as a platform for community-based spatial planning. GeoKey (University College London, 2015) is a more general platform that allows the creation of customized projects. All these platforms are available on the web and most of them provide their own mobile apps. GeoKey, however, can be configured to use the EpiCollect+ mobile app (Imperial College London, 2015) to collect data.

As noted by Elwood (2008), most crowdsourcing and VGI platforms are supported on web 2.0 services, handheld devices with GPS, broadband internet connectivity, and geotagging. One disadvantage for this project is that most of these platforms are in English and it is a cumbersome process to localize all frontends to Spanish. It is convenient to provide users with applications in their own language. Otherwise, one runs into the risk of the platform not being properly adopted. Ushahidi is the exception that already provides some translations.

From a technological perspective, some platforms are interesting and in retrospect, could be useful for the goals of this project. Based on the experience of the platforms mentioned above, it was not suitable to undertake a very complex development for this project at first. Only when platforms are adopted it becomes convenient to elaborate on different components and modules that users can access. The platform was first conceived to provide support for data collection. It was deemed easier and faster to implement a simple online map, rather than using a full blown platform (which also required a dedicated team of designers, administrators, and programmers). As the project grew and once it became necessary, additional functionality was added to support specific user requirements.

Based on time and budget restrictions, the decision was made to have the platform available as a web app and not as native smartphone or tablet apps. Three additional reasons played a role in this decision. First, even though technological advancements and achievements are within reach of an ever increasing percentage of the population, having signed-up volunteers to participate in the data collection process, it was not deemed reasonable to make them spend their data plans while on the go. Not everybody has unlimited data plans on their mobile devices and, for instance, just loading a base map can consume up to a few megabytes. It is even worse for users that use prepaid services as data connection fees are usually higher. Second, even though mobile data-enabled devices are more common everyday (according to the World Bank (2015), in Mexico 83 out of 100 persons are subscribed to a public mobile telephone service), not everybody has one. Making the platform only available through apps would eliminate potential users. Last, staff constraints made it impossible to develop for both desktop and mobile devices, so efforts were focused on having a robust platform that was accessible via web browsers on desktop computers, laptops, tablets, and smartphones. A native mobile app is expected to be released for a subsequent stage of the project that will not be data-hungry on the users' data plan and that will allow data collection locally on the device without the need of being permanently online in constant communication with the remote server to store and retrieve data.

The geospatial platform was designed and developed using free and open source software and tools, with the expectation of being easily used and adopted by citizens. It is accessible at http://roma.dev.centrogeo.org.mx/.

Map layers are displayed using Leaflet JS, a JavaScript library for interactive maps (http://leaflet.com). The backend for data collection and storage is PostgreSQL, a powerful and advanced open source database supporting spatial data (http://www.postgresql.org). The frontend was developed using the Bootstrap (http://getbootstrap.com) and jQuery (https://jquery.com) JavaScript libraries with FontAwesome (http://fortawesome.github.io/Font-Awesome/) and Maki (https://www.mapbox.com/maki/) icons. Server-client communication is done with PHP (https://www.php.net) to retrieve data from the backend and dynamically display it on the frontend.

The initial view for users during the data collection exercise was a web map interface. The neighborhood region was masked in order to avoid users from

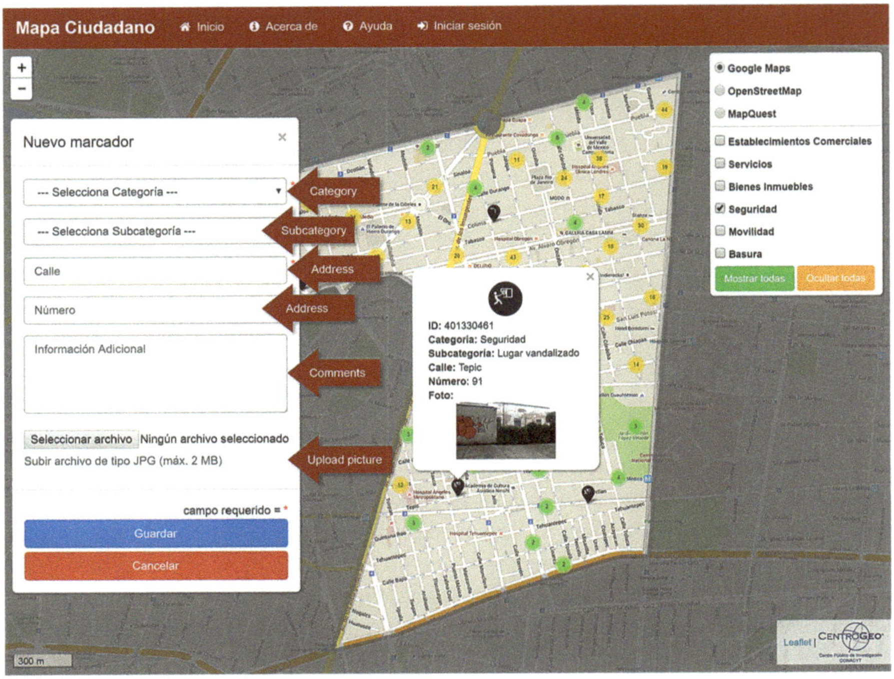

Fig. 11.2 An example of the web app showing the masked neighborhood with clusters of observations, the list of categories included in the study, and the options for the base map layers. Inset: user form to add a report to the database

entering data outside of the study area. The map shows a clustered view of the different categories. Each category is retrieved from the database and a clustering algorithm is applied to each category layer before being added as a map layer. This aids in the user experience avoiding lags, without hindering data display, as during the period of data collection there were more than 6000 points in a relatively small area (3.8 km^2). The viewer can toggle which categories to display on the map and switch the base layer from Google Maps, OpenStreetMap, and MapQuest tile layers, to aid in the identification of different features in the field, since they typically provide complementary bits of information (see Fig. 11.2).

Any visitor could navigate the map, pan, zoom, and click on existing data points to read reported incidences. Citizens wanting to add points to the map needed to authenticate. The reason for users authenticating before adding data to the map is that there were strong concerns from the policy community about the possibility of collecting significant amounts of false data. Even though many of them have used social networks and profit from crowdsourced products (such as Wikipedia, traffic apps, etc.) and granted that this was their very first hands-on experience, it was interesting to observe their reaction to actually participating in crowdsourced data collection procedures. For this stage of the project, an access-control mechanism was requested and thus the user authentication was implemented. User registration and authentication was also present in order to help in the validation of volunteered geographic information.

Once a user was authenticated, and after panning and zooming to an adequate view, clicking on the map added a new data point with its corresponding geographical coordinates. Then, users are prompted to select a category and sub-category, as shown in Table 11.1, and enter additional information for the report (see inset in Fig. 11.2).

The platform also included functionality to support data analysis with displays and interactivity, aligned with geospatial visual analytics (Andrienko et al. 2011). Although, the ones developed here are not very sophisticated, they are based on friendly and intuitive forms of interactivity and support the user's understanding of the data, synthesizing information, and translating map readings into more effective proposals for decision-making.

When accessing the data collection map, citizen reports are displayed with a simple clustering strategy for each of the categories that changes when zooming in or out. This helps users to get a general idea of the distribution of certain types of observations in the study area. Users can also create heat maps for different categories or variables and overlap them, which is useful for identifying initial possible correlations between the categories of collected data. This data visualization uses the citizen observations' database, so when users add points to the map, heat maps can be updated to reflect the live database.

Finally, it is worth mentioning that the platform functionality enables detecting places where problems accumulate and overlap with clusters of other problems, hence signaling the need to opt for more integral strategies for the re-designing or the rehabilitation of these places. That is, the platform does not only deal with georeferenced data but guides the analysis in order to detect what is needed and where it is needed. This knowledge is a decisive factor for the citizens' empowerment.

11.5 From Gathering Data Towards Building the Citizen's Agenda

Field work was organized and managed by CoVe. They were in charge of recruiting neighbors and students, train them in the process of field data collection and adding data to the web app, and track their progress. The study area was comprised of 320 blocks that were further subdivided into 16 zones. Data collection in each of these zones was under the responsibility of a volunteer neighbor member of CoVe, who was also responsible for checking data integrity and quality of their area as well as coordinating a team of volunteer surveyors. The supervisory group was formed by individuals actively involved and committed to the project. Each team of citizen surveyors was trained on how and what to observe while walking their study area, to pour their perception about the specified categories of each block first into data collection sheets and later into the web map. CoVe recruited 190 volunteers in order to carry out the data collection process. This group was mainly comprised of neighbors, with the significant addition of Graphics Design students from the Universidad Iberoamericana together with a group of young and enthusiastic junior high school students from the local school Decroly that teamed up with proactive teachers and parents.

Finally, CoVe established a situation room to oversee the supervisory territorial allocation as well as the advance in volunteer training. This was located in an art gallery that was lent at no cost by the owner (which happened to be a neighbor) to organize the data collection exercise. Volunteers were given t-shirts with the project's logo on them to help identify themselves, donated by a local business person. Additionally, local authorities issued an official letter stating their knowledge and consent about the exercise that volunteer surveyors were carrying out. This was useful to protect citizens against possible mistreatments or aggressions.

These volunteers formed what Goodchild calls a network of human sensors, "each equipped with some working subset of the five senses and with the intelligence to compile and interpret what they sense, and each free to rove the surface of the planet" (2007, 218). They observed the space in *The Roma* neighborhood and captured information related to the aforementioned variables and finally transferred that information to the digital platform developed by the research team.

What was observed, when it was observed, and how derived observations were introduced as data in the platform, was a process preceded by a coaching and training work on how and when to observe different categories and how to capture their observations. Coordinators and some students were trained by the research team and this exercise was cascaded down to the rest of the volunteers. Training on the functionality of the digital platform followed shortly after a brief pilot process. In this sense, the network of human sensors was given protocols and recommendations to standardize data collection in as much as possible.

As this was a pilot project, it was necessary to complete a first stage of data collection in order to assess and evaluate preliminary results. For this first phase, citizens went out in the field for about two months. Initially, data collection was expected to be complete within a month, but delays due to several reasons forced CoVe to allow more time for citizens to map their neighborhood. One reason for this delay was that due to the nature of the different variables to be collected, some had to be collected both at day and night, some others only throughout the day or the night. This meant that volunteers had to revisit their study areas more than once. The progress of the data collection process was tracked on a web map (see Fig. 11.3).

At first, it was expected for users to access the web app while out in the field and efforts were made in order to guarantee proper functionality for all kinds of mobile devices. After a test run in the field, and given the fact that the system was on a trial stage, an executive decision was made not to collect directly on the web app. Instead, and taking into consideration what was mentioned earlier about volunteers with or without unlimited data plans, data collection was carried out with pen and paper. Citizens were provided with sheets of paper with tables of categories to fill in with

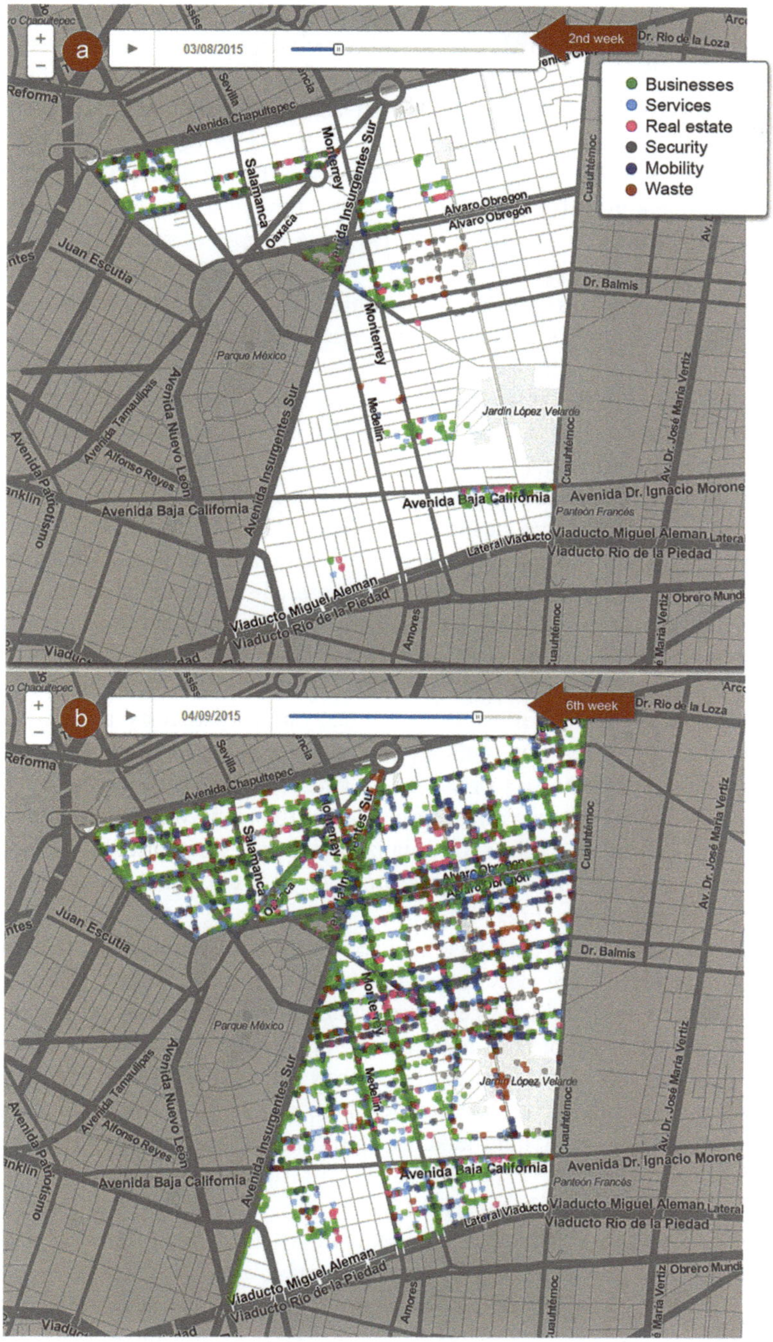

Fig. 11.3 Data collection tracking for (**a**) the second week of field work, (**b**) the sixth week of field work

the information collected in the field. Afterwards, this locational and tabular data were "translated" into a spatial domain in the web application. This also served to double-check the information that was being uploaded to the server, as mistakes and corrections could be made before points were actually committed to the database.

It is important to stress the fact that, as in every volunteered geographic information exercise, obtained data represent participants' perceptions and reflect a variable moment in their observation, such as a time of day or a day of the week. Collected data are by no means exhaustive for a couple of reasons. First, a few blocks were not visited and have no available data and second, it is quite possible that citizens may have missed the observation of some variables at certain locations. Nonetheless, the richness in the data shows a clear tendency on those topics that are of special interest to neighbors. On the one hand, they pinpoint specific locations of particular problems that need attention. On the other, they provide a first glimpse into a territorial image that will evolve as the map is continuously updated with new observations and reports, and in the best case scenario, with news about exercised solutions from both local authorities and neighbors as well.

Data collection took place during the first quarter of 2015, after which there was a brief period to carry out both statistical and geographical analysis. Data extracted from the database were used to produce several distribution maps of different variables. Kernel Density Estimator surfaces were used to produce comparable maps about the density of incidences in the study area. These maps were produced using the same window and adjusting the density color to be within an appropriate range for all different categories. This provided one way to compare densities for different variables. In addition to these, interactive web-based heat maps were also implemented in which the viewer can select categories/subcategories to display. It is also possible to overlay two different heat maps to provide basic exploratory data analysis capabilities. This allows to visualize interesting relationships, for instance, the one between restaurants and waste accumulation, or the one between dark places and vandalized places (see Fig. 11.4).

As part of the data analysis stage, and recognizing that public spaces play both an important role in the society and is of utter importance to CoVe, intersections of citizen reports and public spaces were obtained and maps were produced. This was a helpful exercise to provide additional information about citizen's perceptions on public spaces. As stated earlier, the approach to public spaces was different to the rest of the study area. However, results obtained from spatial intersections provided further insights about the situation of public spaces in the neighborhood (see Fig. 11.5).

The platform was publicly presented on June 18, 2015. The venue for the event was provided by the Universidad de la Comunicación, an institution located in *The Roma* (which also houses the SUMATE space). The event was crowded with local neighbors, business people, and academics, including the head of the National Council for Science and Technology, who remarked the fact that he had very few opportunities to witness instances where, in Mexico, science and technology were put at the service of active and participative citizens.

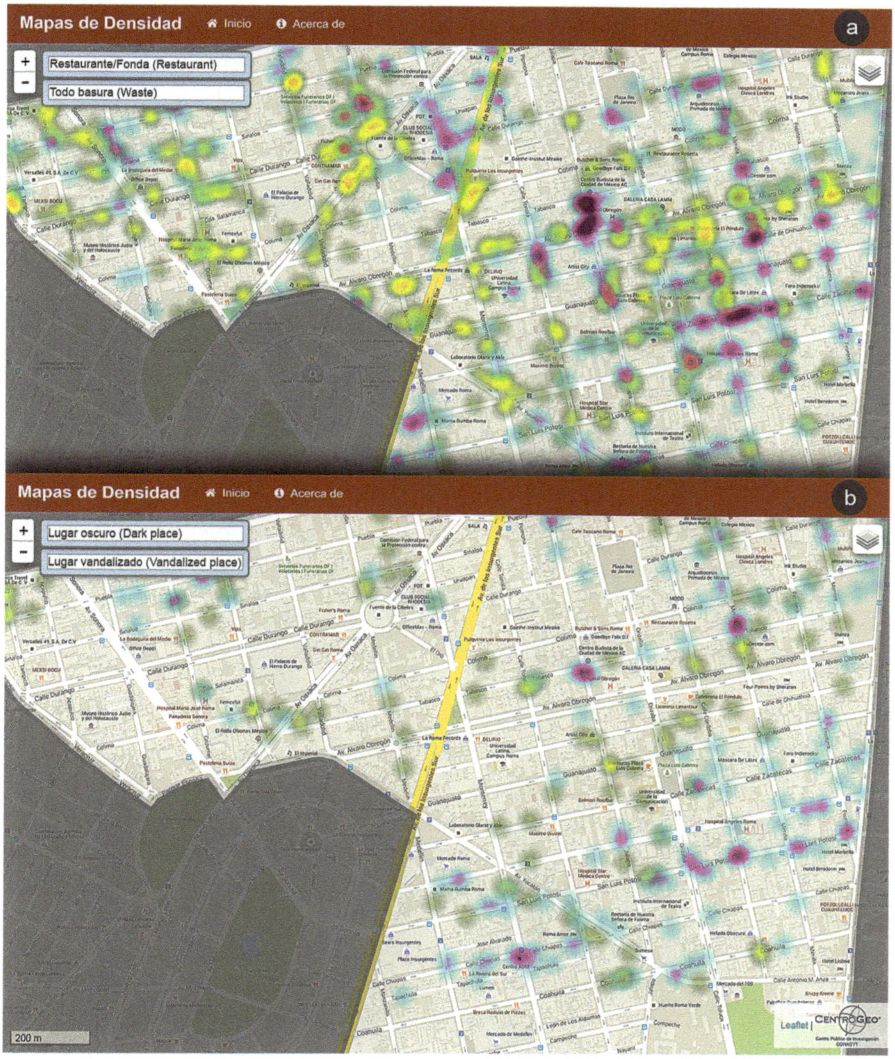

Fig. 11.4 (**a**) Heat map of intersections for restaurants and waste accumulation, (**b**) Heat map intersections for dark places and vandalized places

The research team explicitly reminded the audience that these results showed a tendency in the topics of interest to the neighbors, they pointed problems at specific locations, and gave a first territorial image that would surely evolve as long as the map was kept alive and fed with new observations. In the best case scenario, the map will be updated with solutions from local authorities, or even the neighbors themselves. At the same time, the announcement was made that the platform would remain open to receive new observations, with CoVe being in charge

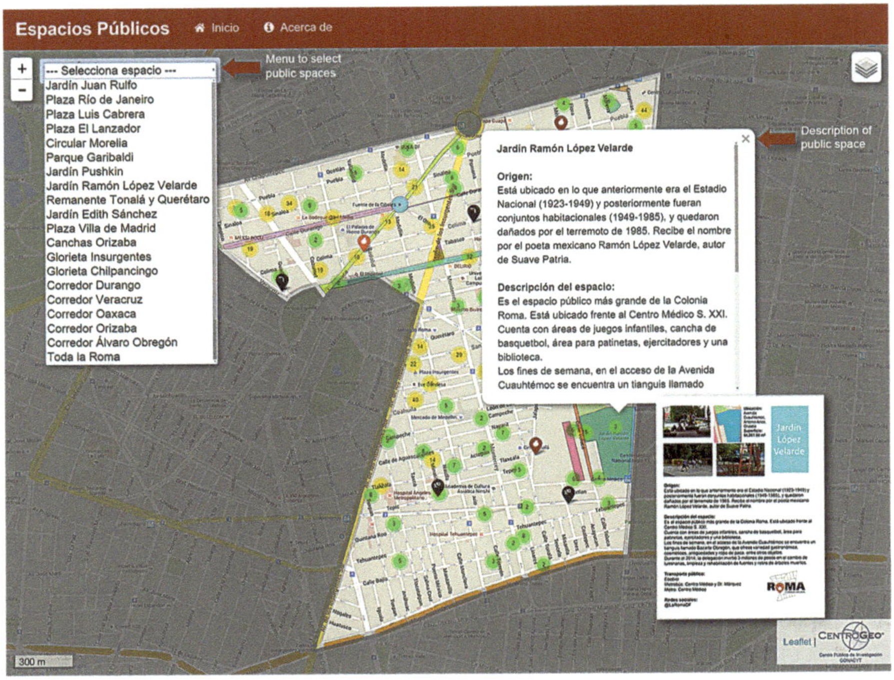

Fig. 11.5 Map of public spaces in *The Roma* with an example of additional information and citizen observations

of user management and new data analyses that stemmed from new incoming data. Additionally, the research team will continue to host the digital platform as long as CoVe deems it appropriate, and the research team will remain an observer of the evolution of the process and its results.

During this public presentation, six working groups were formed –one per category– and coordinated by CoVe counselors. Their intention was to convene participants to adhere to one or more of them in order to get involved in a particular topic and help build the citizen's agenda.

A month later, a CoVe meeting took place and using an approach based on appreciative enquiry (Cooperrider and Srivastva 1987), groups of participants were divided by categories and used the maps to structure messages they derived from this process. They went on imagining what *The Roma* could be, if problems they detected were solved and identifying stakeholders that may help detonate a change process directed towards these visions. The research team participated in this meeting only as an observer.

These groups continued to meet regularly and individually in different places and times in order to articulate suggestions for action that would help put forward some proposals to provoke change. Some of these explicitly articulate the spatial dimension of the problem (for instance, when a priority area of the neighborhood is planned to be targeted for insecurity prevention), while others have a less explicit spatial link (like one that proposes a campaign for responsible consumption in order to reduce waste generation). In any case, the platform will be useful in following the impact of their proposals and updating the different issues they tackle. An interesting topic will be how the use of the platform is sustained over time as this will be a key indicator of its appropriation by citizens. As far as the agenda is concerned, CoVe's coordinator still considers it to be in a building stage.

11.6 Discussion

Using the classification scheme for different kinds of VGI proposed by Deparday (2010), it can first be noted that data do not emerge from a digital footprint, but from participants' intention to geocode their observations in the platform. They are human sensors, in the sense Goodchild (2007, 217–218) assigns to this notion, who proactively observe space to capture information. Their observations are expressed as points on the map with a close approximation to a postal address. These points are associated to both non-structured attributes (such as photographs or comments in the text) and to attributes that can be classified as structured because they obey to an operationalization in the categories. But, this structure originates from a frame of reference that stems from the local knowledge of the citizens involved in the design workshops.

Participants' observations were based on perceptions that can be classified into two types, including ones that are more subjective (such as noise perception) and others that are more objective (such as detecting a broken streetlight). The degree of commitment of volunteer citizens was generally high. Although, there were some slight variations between residents and non-resident students, since the latter were somewhat required to participate and they did not show as much interest as the former, maybe because they do not live there. Also, participation quality had variants that are reflected in the map as areas with very low densities of points. They are derived from the lack of coverage or deficiencies in data collection, and may bias results. The digital geospatial platform was fed by the local knowledge of citizens, which is primarily qualitative and is related to opinions and perceptions.

Elwood (2008, 175) points out that several authors note that the characterization of information as 'volunteered' "implies an intentionality or altruism that may not be present" and informs that "efforts to theorize why individuals volunteer information note that socially- and politically-grounded motivations for volunteering or withholding will shape the dynamics of inclusion and exclusion in VGI development and affect data content." (Elwood 2008, 177). In this framework and besides acknowledging CoVe's strengths, it must be recognized that, as with any organization, it counts among its ranks people with very different visions with respect to the most pressing issues in the neighborhood, namely what and how they should be tackled. These differences were noticeable in the different workshops that took place in order to define both variables and agenda and were expressed, for instance, in the neighbors' solid sense of appropriation and protection of the area and their strong interest in defending their community. Residents think about the presence of imaginary boundaries that cannot exist in an urban area embedded in the central life of the city, subject to high density crossings of both people and vehicles and of commercial activity that overlap social and cultural processes. Additionally, a sense of *otherness* from different social and economic actors that converge in the neighborhood with respect to its inhabitants was perceived in some of the workshop's participants. Nevertheless, it was possible to overcome these feelings and values of tolerance and inclusion prevailed throughout the workshop's discussions.

These observations acknowledge the heterogeneity of points of view involved in governance processes. Citizens by no means pursue the same vision, they rather are a melting pot of divergent and even conflicting interests and so are other stakeholders involved. CoVe's desire to build an agenda can be seen as a strategy to build agreements among neighbors that have proved to engage in participative processes and are recognized as a sort of informal local leaders. They are a robust coalition with shared purposes to enter deliberations and negotiations with local government and other stakeholders in order to pursue actions beneficial to their neighborhood. The meanings that CoVe's members may derive from *The Roma* citizens' maps provide feedback of the projects and purposes of the agenda in the context of a process of participative democracy. In this sense, this case illustrates the kind of social and political practices for creating, sharing, and using geospatial knowledge that Elwood (2008, 176–177) asserts are emerging around VGI and that may support governance processes but that this author alerts that they may have impacts upon "participation, power relations, and existing inequalities in access to spatial data and technologies". These are impacts already documented in the mid-1990s participative-GIS research.

Deparday (2010) points out that VGI may facilitate community discussions in order to identify conflicting areas and to generate agreements. Haklay (2013) proposes a new form of participation called "extreme citizen science" in which users "can choose their level of engagement and can be potentially involved in the analysis and publication or utilization of results". *The Roma* is emblematic of this type of citizen science because, as previously mentioned, citizen participation encompassed both the platform design and the data collection process and went on in meetings discussing results and building courses of action. Analysis was facilitated by the platform's functionalities that allow the generation of density maps of different variables and enables to visualize spatial intersections of these densities, which in turn gives the user the possibility to venture in an exploratory data analysis exercise. Besides, the ongoing involvement of CoVe's members in deliberative processes placed the analyses of these results in the citizens' local knowledge schemes, which were in turn translated into broad proposals structured to enhance the neighborhood's quality of life. The VGI process may allow tracking both the evolution of problems and the possible impacts of actions that may crystallize.

Finally, it is worth mentioning that online instances of VGI processes induced by governments in order to crowdsource data about citizens' demands, such as the ones quoted for Cali and Monterrey, need to be structured in terms of response protocols embedded in procedures of agencies involved. In this sense, the focus is on the specific manifestation of problems that can be solved with an action or a service rendered. The VGI exercise in *The Roma* followed a slightly different approach. Although the points collected in the map indicated specific manifestations of problems (for example, a corner where garbage was accumulated), the purpose was to derive a more integral reading from the map. A reading that may facilitate the inclusion in the agenda of strategic lines of action addressing the factors that cause the problem and guide participative processes involving the neighbors with authorities and different stakeholders. These lines of action may have an effect on the problem detected at the *(x, y)* coordinate, but they may also shield the place for its repeated manifestation. This poses the need to adapt the platform and set it in the midst of the main actors involved in governance. The rationale is to induce local authorities and different stakeholders to introduce data about issues like the solution of specific problems at a point in space, projects to solve problems in the area, or opportunities detected by different stakeholders. This idea is somewhat in the line of the Cali's platform, though narrowing the categories and items to subjects of public interest agreed and negotiated between public and private stakeholders participating in building a governance process. New functionalities have to be developed in the platform in order to support dialogues among social actors involved and a more complete set of social base needs to be constructed to embrace governance in a more inclusive way.

Of the utmost interest is the amount of issues to include in this kind of platforms. It is impressive to see the many categories and items collected and managed by the Cali platform and to a lesser extent by the Monterrey one. To process the volume of information generated for all variables into effective action is highly problematic, because this endeavor involves a great variety of agencies and procedures and as Herbert Simon rightly expressed back in 1971, "... a wealth of information creates a poverty of attention and a need to allocate that attention efficiently among the overabundance of information sources that might consume it" (Simon 1971, 40–41). Throughout workshops organized in *The Roma*'s project, the research team tried to limit the number of categories and items to collect. But for CoVe, the purpose was not to manage solutions point by point, but to use them as ways to identify problematic areas and to use this information to support participatively built projects following a methodology they sought suitable.

A key factor that has been recognized for the success of this project is the linkage between the academic knowledge and technical capacities of the research team with those of a civil organization. Such capacities include previous civil participation, problems analyses, proposals for solutions, and the desire and will to take action in topics of public interest that affect citizen's everyday life. CoVe has an assertive leadership with high convening power and is a promoter of legitimate ways to participate in the decisions affecting their community.

Other factors that played a key role in the successful implementation of this project were the use of both technology and open-source software. As mentioned above, it was desirable to deploy a native mobile app for smartphones and tablets but it was not possible. Nonetheless, the web app was entirely developed using open source software and standards. This is important in the sense that the platform is extensible and adaptable for uses other than the particular one presented here, so new potential users can adopt the platform without licensing costs related to the software pieces that comprise it. It is expected to release an application for mobile platforms in the near future. This will need to incorporate a few alternatives for its users such as off-line data collection and the capability to synchronize on demand or via Wi-Fi only, just to name a few.

11.7 Conclusions and Future Work

A case study of VGI with a citizen community to support its voice in a territorial governance processes in Mexico City was reported. The project uses a digital platform that allows capturing citizen observations in six different categories with a total of 41 variables inside a neighborhood in Mexico City. It allows the dynamic generation of heat maps from these variables as well as the cross-relation of these maps to detect overlaps in the accumulation of issues in different locations.

The platform was fed with data generated by 190 volunteers coordinated by a neighborhood association with no political affiliation, committed to participative democracy. This platform is already being an instrument in several key factors for the neighborhood. It is helping in the construction of a common idea of some striking problems in the area, it is helping in the generation of agreements about these issues and their priority locations in order to create a vision towards the future and to build social relations with local authorities, business persons, more neighbors, and other interested groups that may trigger innovative processes of territorial governance.

The public presentation of the platform was very welcome by the audience. Its diffusion with the media motivated social groups from other neighborhoods and public officers from different public institutions that approached the research team with interest in using this type of platform for their own needs to foster citizen dialogs. Glimpsing into the near future, the continuous updating of the data contained in the platform will be followed with great attention, as well as uses that can be derived by inhabitants of *The Roma* that can result in the improvement of their well-being.

Actions proposed by citizens will be posted online in the platform and, when available, these could be consulted there. The geospatial platform remains open to the public and the authorities. The administration and management process is in the hands of CoVe for further updates and modifications, with the research team providing guidance and support as well as acting as an observer of how the system evolves and what results can be obtained from it. It will be interesting to assess how much the platform is being used in order to have an indicator of how citizens are appropriating it.

Regardless of the course that *The Roma* process follows, future plans for the research team can be summarized as follows. On the platform side some improvements will be made in order to (1) remove entries from the data base and/or comment on them. This is an issue that was not contemplated at first, but it is certainly relevant in regards to detecting the reasons why a certain observation should be removed; (2) manage similar reports in close proximity as a way to up- or down-vote reports; (3) allow users to choose in an interactive and friendly way the urban space they wish to work on, along with categories and variables they want to capture; (4) integrate the platform with a database that supports observation tracking through time, and (5) include graphical visualizations for summary statistics from within the platform. These improvements will certainly come in handy when the platform is used beyond the scope of this study, in order to create a generic platform that could be used for many heterogeneous cases.

On the user side, the research group has already been approached by citizens representing another neighborhood and with specific claims to crowdsource geographic information. Their goal is to work with a network of local residents in order to directly negotiate with local authorities the attention in a set of more specific issues, based on a reduced and somewhat slightly altered set of variables used in *The Roma* study, such as nonworking streetlights, buildings with more than three levels of construction, detection of drainage problems that lead to waterlogging and flooding during the rainy season, and locations of schools and businesses.

Also, the platform has been seen with great interest by the new local authorities of Tlalpan, one of the 16 municipalities that form Mexico City and by a group of stakeholders interested in launching a platform for the city of León, Guanajuato, a large metropolitan area located in the center of the country. There is a good chance for the research group to work in close relation with them. Setting up the platform as part of a dialog initiated by local authorities will allow to encompass the geographical space of several neighborhoods, those contained in the municipality in question, but most importantly, it will allow to compare the citizens' participation dynamics when efforts are implemented in a top-down fashion.

Acknowledgments The authors would like to thank Laura Sarvide and the CoVe team for their invitation to participate in this project. Also, thanks to all volunteers who embarked on the data collection process and played a very important role in this exercise, especially students, teachers, and parents from Decroly and students from the Universidad Iberoamericana. Finally, thanks to the Universidad de la Comunicación for participating so willingly and allowing the use of their space for the public presentation of the results. Thanks to Rafael García for helping with the design of the website and figures, and Gabriela López with finishing the figures. The authors thank the anonymous reviewers who provided constructive feedback to improve this paper.

References

Activo Group (2013) *Ciudadanos Activos*.http://www.ciudadanosactivos.com/. Accessed 10 Mar 2016

Andrienko, Gennady, Natalia Andrienko, Daniel Keim, Alan M. MacEachren, and Stefan Wrobel (2011) "Challenging problems of geospatial visual analytics." J Vis Lang Comput 22 (4). Elsevier: 251–256. doi:10.1016/j.jvlc.2011.04.001.

Atzmanstorfer, Karl, Richard Resl, Anton Eitzinger, and Xiomara Izurieta (2014) "The GeoCitizen-approach: community-based spatial planning – an Ecuadorian case study." Cartogr Geogr Inf Sci 41 (3). Taylor & Francis: 1–12. doi:10.1080/15230406.2014.890546.
California Academy of Sciences (2015) *iNaturalist*.http://www.inaturalist.org/. Accessed 9 Nov 2015
CIC (Centro de Integración Ciudadana) (2015) Tehuan.http://tehuan.cic.mx. Accessed 9 Sept 2015
Cooperrider DL, Srivastva S (1987) Appreciative inquiry in organizational life. In: Woodman RW, Pasmore WA (eds) Research in organizational change and development, vol 1. JAI Press, Stamford, pp 129–169
CoVe (Consejo Vecinal Roma) (2015) Cartografía Participativa. Ciudadanía en Acción. http://roma.dev.centrogeo.org.mx/cove.html. Accessed 10 Sept 2015
Davoudi S, Strange I (2009) Conceptions of space and place in strategic spatial planning. Routledge, London. doi:10.4324/9780203886502
de Vries, M (2013) The challenge of good governance. The Innovation Journal: The Public Sector Innovation Journal 18 (1): article 2.
Deparday V (2010) Enhancing volunteered geographical information (VGI) visualization with open source web-based software. University of Waterloo, Waterloo
Dreschler W (2013) Islamic public administration: the missing dimension in NISPAcee Public Administration Research? In: Vintar M, Rosenbaum A, Jenei G, Dreschler W (eds) The past, present and future of public administration in central and Eastern Europe: Twenty Years of NISPAcee, 1992–2012. NISPAcee, Bratislava, pp 57–76
Elwood S (2008) Volunteered geographic information: future research directions motivated by critical, participatory, and feminist GIS. GeoJournal. doi:10.1007/s10708-008-9186-0
Goodchild MF (2007) Citizens as sensors: the world of volunteered geography. GeoJournal 69(4):211–221. doi:10.1007/s10708-007-9111-y
Grindle, Merilee S (2004) "Good enough governance: poverty reduction and reform in developing countries." Governance 17 (4). Blackwell Publishing Ltd.: 525–548. doi:10.1111/j.0952-1895.2004.00256.x
Haklay M (2013) Citizen science and volunteered geographic information: overview and typology of participation. In: Sui D, Sarah E, Goodchild M (eds) Crowdsourcing geographic knowledge: volunteered geographic information (VGI) in theory and practice. Springer, pp 105–122. doi:10.1007/978-94-007-4587-2_7
Healy P (2006) Urban complexity and spatial strategies: iowards a relational planning of our times. Routledge, London
INEGI (Instituto Nacional de Estadística y Geografía) (2015) Directorio Estadístico Nacional de Unidades Económicas. http://www.inegi.org.mx/est/contenidos/proyectos/denue/presentacion.aspx. Accessed 10 Sept 2015
Mattioli, C (2014) "Crowd sourced maps: cognitive instruments for urban planning and tools to enhance citizens' participation." In Innovative technologies in urban mapping, edited by Antonella Contin, Paolo Paolini, and Rossella Salerno, 10:145–156. Sxi — Springer per l'Innovazione/Sxi — Springer for Innovation. Springer International Publishing. doi:10.1007/978-3-319-03798-1
OpenTreeMap (2015) OpenTreeMap. https://www.opentreemap.org/. Accessed 9 Nov 2015
Raadschelders, J C N. 2003. Government, a public administration perspective. M.E. Sharpe.
Simon HA (1971) Designing organizations for an information-rich world. Computers, Communications, and the Public Interest 72:37
Usahidi (2015) Usahidi.https://www.ushahidi.com/. Accessed 9 Sept 2015
World Bank (2015) Mobile cellular subscriptions (per 100 people).http://data.worldbank.org/indicator/IT.CEL.SETS.P2. Accessed 10 Sept 2015

Chapter 12
Crowdsourcing of Environmental Health Quality Perceptions: A Pilot Study of Kroměříž, Czech Republic

Jiří Pánek, Lenka Mařincová, Lenka Putalová, Jiří Hájek, and Lukáš Marek

Abstract Public participation is an inherent part of urban planning and the Local Agenda 21 initiative (LA21). Nevertheless, opportunities to involve citizens in the process of creating, using, and evaluating public spaces are still limited. The research presented in the paper involves the integration of Gould-style mental maps with Participatory GIS (PGIS) technologies. This paper describes the testing and implementation of the web-based crowdsourcing tool PocitoveMapy.cz, used for the collection and visualisation on maps of people's perceptions of environmental deprivation. In the case study the tool was used in the city of Kroměříž with 99 users (n=99) collecting 1257 perception-based shapefiles on eight different topics. The application is based on Leaflet library, an open-source JavaScript library for mobile-friendly interactive maps. It allows developers without a GIS background to very easily display tiled web maps hosted on a public server, with optional tiled overlays. The collected vector data were analysed as a hexagonal grid with sides of 25 metres, where each hexagon was assigned a value based on the values of points, lines, and polygons spatially overlapping it. Visual analysis showed a similar spatial distribution for some environmental topics, and the non-parametric Kendall's rank correlation supported the visual observations.

Keywords Emotional maps • Environmental perception • Mapping • Kroměříž • Participatory planning • Crowdsourcing

---

J. Pánek (✉) • L. Mařincová • L. Putalová • J. Hájek
Department of Development Studies, Palacky University in Olomouc, 17, listopadu 12, Olomouc 771 46, Czech Republic
e-mail: JirkaPanek@gmail.com

L. Marek
Department of Geography, University of Canterbury, Private Bag 4800, Christchurch, New Zealand

M. Leitner, J. Jokar Arsanjani (eds.), *Citizen Empowered Mapping*, Geotechnologies and the Environment 18, DOI 10.1007/978-3-319-51629-5_12

12.1 Introduction

With the democratisation of cartography (Rød et al. 2001) and GIS (Butler 2006), and with the era of social media, the Internet, and crowdsourcing; city planners and decision makers have new tools and methods that can include both qualitative and quantitative data about cities, their dynamics, and the people living there (Kloeckl et al. 2011). Most geospatial applications rely on *objective* data only, although there can be a discussion concerning the extent to which GIS data are objective, as there is always a level of generalisation, uncertainty, and authorial bias (Pickles 1995). The call for a more humanised and participatory approach to geospatial information and technologies has been heard since the publication of *The Ground Truth* in 1995 (Pickles 1995).

Public participation has become an integral part of the urban planning process in Western countries and it involves the entire community in the strategic and management processes of urban planning; or, community-level planning processes, urban and rural; and it is often considered to be part of community development (Lefèvre et al. 2001; McTague and Jakubowski 2013). Within the Czech Republic the rise of cooperation, partnership, and participation in local administration is in its inception (Čermák 2001). Community mapping as a tool for linking participation and urban planning has gained popularity, especially after the development of the Local Agenda 21 Planning Guide created during the United Nations Rio Conference on the Environment in 1992, where it was identified as a best practice for locally-based sustainability planning (IDRC 1996). Since the Rio conference many scholars have been engaged in both the theory and the practice of community mapping (Chambers 2003, 2006; Perkins 2007; Pánek and Vlok 2013; Forrester and Cinderby 2012; Elwood 2002; Craig and Elwood 1998). Nevertheless, it was only recently that a *subjective layer* (Huang et al. 2014) and the concept of *qualitative GIS* (Elwood and Cope 2009) were introduced.

The authors collected and analysed subjective data formed from people's perceptions of the quality of the environment and its effect on public health in the city of Kroměříž in the Czech Republic. The data were collected in the form of emotional maps and the outcomes can be seen as a version of a Gould-style mental map (Gould 1986). It is possible to argue that emotional mapping is not the correct term as it is not emotions that are mapped but merely people's perceptions or experiences of places. Nevertheless, the authors decided to keep the emotional mapping term mainly based on the argument of Perkins (2009, p. 130), who states that *emotional maps chart human feelings onto a cartographical landscape ... and allow users to devise and customize their own emotional landscape, choosing what kinds of thoughts or experiences, feelings or passions, to map*. Emotions and spaces are connected because every location can evoke an emotion (Mody et al. 2009) and places can be perceived as attractive, boring, dangerous, or scary, among other qualities (Korpela 2002). In the past 10 years several projects have dealt with georeferenced emotions and methods used to gather emotional data can be divided into three groups: (1) Biometric measurements (Nold 2009; Bergner et al. 2011); (2) extraction from user generated content such as Twitter, Flickr, Facebook, etc.

(Bollen et al. 2011; Mislove et al. 2010; Biever 2010); and (3) surveys (Mody et al. 2009; Huang et al. 2014; MacKerron and Mourato 2010). The authors´ approach can be considered a survey.

Emotions strongly influence how an environment is perceived and emotions have an effect on the spatial layout of people's perceptions (Zadra and Clore 2011). Griffin and Mcquoid (2012) distinguished three categories when talking about maps and emotions/perceptions. These categories are (1) maps of emotions, (2) using maps to collect emotional data, and (3) emotions while using maps. The tool described in this paper is a combination of the first two categories. Maps were used to collect the information and also to visualise the data. Emotions are one of the defining characteristics of every human being and yet their presence on maps and in spatial data is uncommon (Griffin and Mcquoid 2012). Although historically cartography mainly focused on treating that which was visible or could be mapped (including air temperature and wind speed) (Wilson 2011), critical cartographers always advocated mapping a space as people experience it, with emotions included (Pearce 2008). Hauthal and Burghardt (2016, p. 2) argued that ... *mappers of georeferenced emotions are almost exclusively researchers*... using emotional maps in various fields such as tourism (Mody et al. 2009), navigation (Huang et al. 2014; Gartner 2012), urban safety (Pánek et al. 2017) and city planning (Raslan et al. 2014).

12.2 From Global to Local – Health and Sustainable Development

The concept of sustainable development has its roots in the 1970s. These years are considered by many to be a crucial moment in changing people's minds about the connections between the environment, the economy, and social well-being (International Institute for Sustainable Development 2015). After its conceptual development it became one of the most famous approaches and formed the international agenda for global and local development. Sustainable development is, according to the Brundtland's report, development which meets the needs of current generations without compromising the ability of future generations to meet their own needs (UNWCED 1987).

Health is one of the most important and inseparable parts of sustainable development. The clear argument about the role of health in sustainable development was mentioned by Von Schirnding and Mulholland (2002, p. 3): *"The goals of sustainable development cannot be achieved when there is a high prevalence of debilitating illnesses, and population health cannot be maintained without ecologically sustainable development."* The global environment faces many anthropogenic threats and challenges with the main impact being on the quality of life, health, and well-being of people. The global community has initiated many global action plans and strategies, but as we can observe, the most effective and useful way is to address these problems through local actions. This is according to the golden rule of sustainable development, namely think globally, act locally.

Agenda 21 (IDRC 1996) serves, among other things, as a cornerstone for the global implementation of sustainable development. This 24 year old document which was adopted by 178 governments at the United Nations Conference on Environment and Development in Rio de Janeiro as a voluntary initiative, follows the main dimensions of sustainable development and its implementation on a global, national, and local level (Norton 2014). It seems that its variation for local and community strategy action (Local Agenda 21 or LA21), which emphasizes the central role of local authorities and calls upon them to develop local strategies for sustainable development, as quoted by Dooris (1999), can be progressive, inspiring, and perceived more positively by people. The main reason for this is the added value of the emphasis on communication, participation, and cooperation between various entities at the local level (Kveton et al. 2014).

Concurrently with the elaboration of concept LA21, the concept of the Healthy Cities Programme was also developed and established by The World Health Organisation in 1988. The programme is a long-term initiative with its main aims being to place health high on the agenda of decision makers and to promote comprehensive local strategies for health protection and sustainable development (WHO 2015a). It tries to bring the technical language of the Health for All strategy into the twenty-first century (WHO Europe 1998)) and translate the principles of the Ottawa Charter for Health Promotion (WHO 2015b) into tangible action (Dooris 1999). Here we can see an important overlap and the common drivers of two concepts, which come from mutual values and in practice have complemented each other. The Network of Healthy Cities of the Czech Republic (HCCZ) was created in 1994 and its mission, which goes hand in hand with the above mentioned premises, is to goad Czech municipalities into stipulating in their statutes that they will consistently work towards sustainable development, health, and the quality of life in cities, municipalities, and regions of the Czech Republic.

12.3 Participatory Environmental Perceptions Research

Recently, a new development has come from planning practitioners, geographers, GIScientists, and Citizen Science enthusiasts, and it has led to the deployment of Participatory Planning Support Systems (PPSS) such as *softGIS*, *Geo-Questionnaire*, *SprayCan,* and many others. All these tools are examples of Internet-based Participatory GIS methods, and they allow residents to communicate local spatial knowledge to the administration body (Kahila-Tani et al. 2015). The first support for public participation came through European Union initiatives which promoted public involvement in local governance, including the 1998 Aarhus Convention and subsequently, the 2007 Leipzig Charter on Sustainable European Cities. In the Czech Republic participation is granted by the Constitution of the Czech Republic as well as the Act of Parliament 128/2000 – Act concerning Municipalities. Emotional mapping allows the display of subjective, qualitative, and bottom-up spatial information about the environment in highly hierarchical, quantitative, and top-down GIS settings.

The authors see the main research gap in the current use of a combination of analogue and digital mapping approaches. A relevant amount of research has been done in the area of internet-based public participation GIS (Hemmersam et al. 2016; Kahila-Tani et al. 2015; Kytta et al. 2015; Kahila and Kyttä 2009; Huck et al. 2014; Jankowski et al. 2015). Nevertheless there is still the digital divide between generations that refers to the gap between those demographics and regions that do have access to modern information and communications technology, and those that do not or only have restricted access. For example the authors are unable to contact elderly people through Internet-based mapping tools, hence the use of a combination of paper-based and Internet-based questionnaires is one issue to explore further. Generally, in Central Europe, there is a lack of both practice as well as research in the areas of (1) participatory planning and (2) community mapping, therefore this paper presents two topics which are under-represented in Central European geographical research discourse.

The authors acknowledge that the practical aim of the research is to identify the quality of the living environment at the level of a town. This paper addresses one main research question: What is the community perception of environmental health issues within the Kroměříž urban area? The practical aim of the paper was to deploy and test a tool and a methodology for crowdsourcing people's perceptions of urban public spaces, in order to promote urban planning within the Central European area.

12.4 Methodology

During the development of the tool and its deployment, the authors took part in the process as action researchers, with the aim of achieving practical benefits and generating practical information from their research (Brydon-Miller et al. 2003). The methodology involved sketchable maps with questions (Jankowski et al. 2015) in both an analogue as well as a digital interface. The combination of collection methods was selected because in ordinary participatory methods face-to-face meetings combined with time and space commitments narrow down the number of participants and lead to the exclusivity of a single opinion (Kahila-Tani et al. 2015). The involvement of Computer-Assisted Web Interviews (CAWI) is in alignment with the concept of a participatory planning support system as defined by Kahila and Kyttä (2009). The method used by the authors helps to learn about perceptions and preferences of city residents regarding specific environmental determinants.

The preferred environmental determinants of health were selected and adopted from the HELEN (Health, Life-style and Environment) study and included pollution of public spaces, air pollution, day-time noise levels, night-time noise disturbances, smell, criminality, quality of swimming water, and motor traffic. Selected social and environmental determinants of health were inspired by the large survey HELEN carried out by the National Institute of Public Health in 3 years (1998, 2002 and 2010) (National Institute of Public Health 2015). The evaluation scale was kept in its original state using a Likert scale which ranged from 1 (*not bothersome)* to 6 (*extremely bothersome*). Each question about a respondent's perception of a selected determinant was extended by a Stamen designed toner (black-and-white) version of the OpenStreetMap (Stamen Design 2015) on a scale of 1: 35,000 with the middle of the city centre transformed to the .jpg format (130 × 190 mm). Respondents were asked to draw points, lines, and polygons for each topic and to number the level of bothersomeness for every topic. They had an unlimited number of attempts and were able to skip questions and the map. Trained interviewers randomly approached each respondent with the main task being to establish contact with him/her and convince them to participate in the study. Subsequently, interviewers conducted quick data control and took some additional notes to help increase data analysis.

The second option for collecting the data, as opposed to classical paper-based mapping, was the crowdsourcing online tool, called PocitoveMapy.cz. The tool is designed as a web-application based on Leaflet library. Similar to other web-based tools for crowdsourced mapping, it allows users to collect spatial data on a slippery map background. Unlike Ushahidi, Umap, ArcGIS Online, and many others, PocitoveMapy.cz does not require the registration or installation of any specific software, plug-in, or virtual server. Responders had the opportunity to toggle between maps, as can be seen in Fig. 12.1. The default basemap for the application is OpenStreetMap, but users are free to choose between OpenStreetMap, OpenStreetMap toner, Esri topographic map, National aerial map, and a National map of the Czech Republic (State Administration of Land Surveying and Cadastre 2014). The skills required to use the tool effectively are similar to those of the average Internet user.

Users of the online tool had the same eight topics as the paper-based questionnaire, with options to use point, line, polygon-freehand, and polygon-clicking. Once

Fig. 12.1 Printscreen of PocitoveMapy.cz online crowdsourcing tool with examples of points and polygons already collected

the feature was created, a pop-up window appeared which asked users to mark the level of their answers. At the end of the questionnaire users were asked about their gender, age group, and relationship to the city, with the options *I live here*, *I commute to this place for services*, *I work here*, *I study here*, *I am a tourist,* and *Other*. The additional question, *how long* appeared when they selected the option *I live here*. Similarly, the additional question, *how often* appeared when they selected the option *I commute to this place for services*.

12.5 Pilot Study

12.5.1 Study Area

Kroměříž is the second largest city in the Zlín region of the Czech Republic, with a population of 29,035 (Czech Statistical Office 2015) and it is also a strong natural regional centre. Kroměříž has been important not only as an administrative centre, but also as a centre of culture (the gardens and château were added to the UNESCO World Cultural and Heritage List in 1998), religion, history, and education. It lies in the southern part of Hornomoravský uval (Province of Western Carpathians) at an altitude of 201 m above sea level on a broad alluvial plain of the Morava River. The cadastral area of the municipality of Kroměříž is currently 55.6 square kilometres and the city is built on 17.7 square kilometres. The city of Kroměříž (excluding suburban areas) occupies 11% of the cadastral area and a significant portion of its surface has urban greenery which, combined with the historical and modern buildings, helps to create a valuable, purpose-built, architecturally aesthetic town with a balanced environment. Nevertheless, according to the inhabitants' perception of their surroundings, Kroměříž faces some challenges and finding solutions to these challenges could enhance the living environment (City of Kroměříž 2015).

The main reasons for selecting Kroměříž for this research were (1) Kroměříž is a proactive member of HCCZ, (2) Kroměříž is implementing the LA 21 criteria, and (3) Kroměříž participated in the national survey HELEN (Health, Life-style and Environment) carried out by The National Institute of Public Health in 1998, 2002, and 2010 (National Institute of Public Health 2015). According to the report by Kubinova et al. (2006, p. 448–449) *the survey (HELEN) was focussed on the supplementation of demographic and health statistics with selected indicators of health to estimate the prevalence of important chronic non-infectious diseases and their risk factors in the urban population*. Moreover, part of the survey also focussed on the respondent's view of the quality of the surrounding environment in connection with their health status. During the first period 25 cities took part in the survey, 4 years later there were 27, and in the last phase only 19 cities participated. Kroměříž was included in all three phases of the HELEN research.

12.5.2 Collection of Data

In July and August 2015 the authors visited the city of Kroměříž several times in order to conduct the paper-based mapping questionnaires. In the same time period, the online version of the questionnaire (Fig. 12.1) was distributed via the local municipality webpage, the university Facebook, and through the snowball effect. For all paper-based questionnaires the online tool was later used to transform/digitise the analogue answers into the digital database. A total of 99 respondents completed the questionnaires, 27 used the online tool and 72 used the paper-based tool. The potential of the crowdsourcing tool lies in the better propagation of the survey, nevertheless it was also useful for digitising and analysing the paper-based questionnaires. The authors used the *sample size calculator* (Raosoft 2004) with a margin of error of 10%, a confidence level of 95%, a population size of 29,000 inhabitants, and a response distribution of 50% (set as default) to calculate the size of population needed for the case study to be representative. The sample size was then calculated to be 96 respondents. The group of respondents consisted of 55 females, 40 males, and four users who did not indicate their gender. Two thirds of the group were residents of the city (as can be seen on Fig. 12.2), another 15% were people who either work in the city, study in the city, or at least often visit the city for services (hospital, shopping, social care, etc.). Regarding the age distribution of respondents (Fig. 12.3), the majority were in their productive age (21–60 years) and had lived in the city for an average of 31 years.

Crowdsourced data were saved in the GeoJSON format, which was later exported to shapefile, therefore the data were analysed as a vector. The original dataset included points, lines, and polygons, which were later merged by the *spatial join* (Fig. 12.4) tool with a hexagonal grid (created by Repeating Shapes for ArcGIS toolbox (Jenness 2006)), where each hexagon had sides of 25 m, a diameter of 50 m, and an area of 1624 square metres. The orientation offset of the hexagons, that refers to the general orientation of the array in comparison with perfectly vertical and horizontal rows and columns, was 30°. A hexagonal grid was selected based on the research of Burian et al. (2014), where hexagons were identified as an optimal

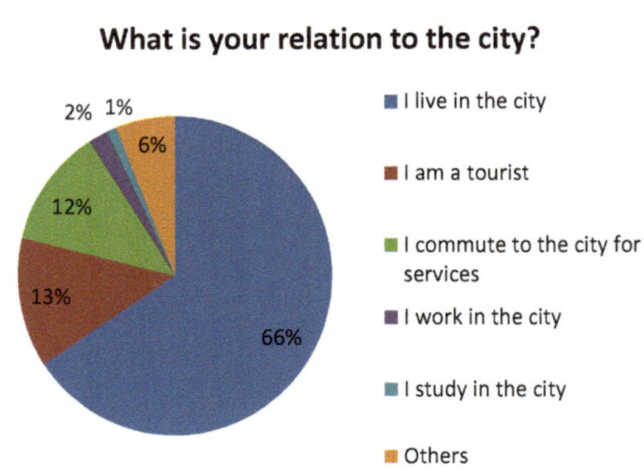

Fig. 12.2 Survey answers about the question: What is your relation to the city?

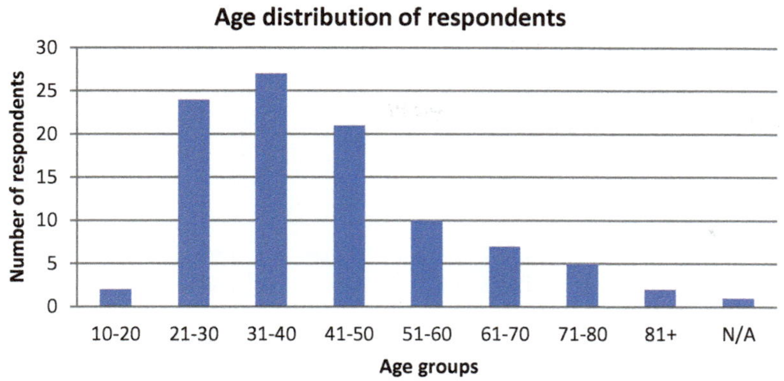

Fig. 12.3 The age distribution of survey respondents

Fig. 12.4 Example of the *Spatial join* function in the hexagonal grid with point and line feature

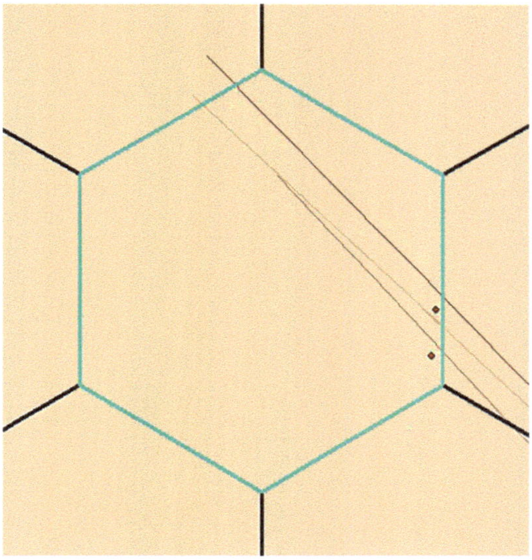

distribution grid for urban spaces. The visualisation of results was done via a white-red colour ramp as a simple choropleth map. Before the spatial join was carried out, the authors removed all features with the value *1*. It was understood from interviews with users that they assigned the value *1* to features where they were satisfied with the environmental conditions.

Table 12.1 describes the number of features that were collected for each variable. Each variable also lists the number of points, lines, and polygons collected. The second part of Table 12.1 explores some basic descriptive statistics for each variable. The variable *Low quality of bathing places* has the highest range of values as well as the highest standard deviation.

Table 12.1 Basic statistical variables of the crowdsourced dataset

	Pollution of public spaces	Air pollution	Noise (daytime)	Noise (night-time)	Bad smell	Criminality	Low quality of bathing places	Motor traffic
Number of features	245	139	177	99	146	151	120	180
Points	171	104	107	69	114	92	87	105
Lines	23	18	48	17	14	22	2	52
Polygons	51	17	22	13	18	37	31	23
Number of hexagons with assigned values	1805	3109	4862	2046	4774	2983	337	1550
Average hexagon value	5.39	5.36	9.18	4.87	4.52	6.49	37.74	11.25
Minimum hexagon value	2	2	2	1	2	2	2	1
Maximum hexagon value	33	49	123	33	33	62	126	116
Standard deviation	3.14	5.87	13.58	3.58	2.70	4.68	46.33	18.28

12.6 Results of Pilot Study

The main result of this research is a dataset of answers (n=99) and data points (n=1257) based on emotions and perceptions related to the quality of the environment and its effect on public health in the city of Kroměříž. During the research process the authors also tested the crowdsourcing application entitled PocitoveMapy.cz (meaning EmotionalMaps in Czech), as the aim of the research was to deploy and test a tool and a methodology for crowdsourcing people's perceptions of urban public spaces. The tool was developed as a crowdsourcing application and results were presented on eight maps which reflect perceptions of specific environmental determinants of health.

The pollution of public spaces (Fig. 12.5) has its hot-spots mainly in the city centre, city parks, along the River Morava, and in two suburban settlements on the south-western part of the city. There is also a smaller hot-spot in the area of the main train station in the eastern part of the city. Air pollution and day-time noise levels are related to the same issue, namely to the traffic in the city. Kroměříž does not have a bypass road and all the traffic is dependent on the main roads that run through the city. According to the latest official noise map published by the Ministry of Health of the Czech Republic (2007), the average noise on the northern tip of the main road exceeds 75 dB. Nevertheless, noise is only measured on selected main roads outside city areas, therefore the authors do not have the official noise data for the whole city to compare people's perceptions with actual noise levels. The noise (night-time) perception is mainly highlighted in the historical city centre and surrounding streets containing pubs and restaurants. The main hot-spot for the perception of bad smell (Fig. 12.6) was outside Kroměříž in the nearby village of Těšnovice (south-east of Kroměříž), where the largest piggery in the region is situated. Within the city, participants mainly mentioned the area around the main train station. Additionally, the visualisation clearly indicates one of the main roads in Kroměříž, so the perception of smell is also related to traffic. The increased perception of criminality mainly has three hot-spots, including the main train station, the town centre, and the park called Bezruč in the city centre. The area around the River Morava is also indicated as a place where, according to one participant, homeless people and young delinquents spend their time. Similar to bad smell, the low quality of bathing places has only one main hot-spot, namely the natural pond called Hraza. However, it is questionable to evaluate this issue because in Kroměříž only the Hraza pond could be assessed on this matter. The perception of motor traffic aligns with the data for a typical traffic on a weekday on Google Maps (2015). Additionally, the city centre was also frequently marked on this matter, but mostly because of a lack of parking places.

Fig. 12.5 The visualisation of perceptions of environmental deprivation of surveyed topics, including (1) Pollution of public spaces, (2) Air pollution, (3) Noise (daytime), and (4) Noise (night-time)

12.7 Spatial Statistics

Hexagonal grids of aggregated values were entered into the evaluation using the local Spearman's ρ. Each hexagonal cell contained information about people's average perception of the place in relation to a certain topic, which was calculated as a mean value of point/line/polygonal evaluation of the place as gained from the public through the spatial questionnaire. The second order neighbourhood of individual grid cells was used for the evaluation of the association between citizens' feelings about a place, which meant that every grid cell was bordered by 18 (see Fig. 12.7) surrounding grid cells.

The computation of the correlation coefficient is the most common method of exploring and enumerating the association between two (or more) characteristics. The correlation is stated as the measure of the symmetric statistical linear dependence between two events, characteristics, or variables. Its value ranges from -1 to $+1$, where a value equal to -1 expresses a perfect negative association, a value equal to $+1$ expresses a perfect positive association, and a correlation equal to 0 expresses linear independency. However, correlation does not determine the mutual causality of the investigated factors. Pearson's correlation, Spearman's rank correlation, and Kendall's rank correlation are widely used methods for the correlation calculation (Reimann et al. 2008). The correlation per se usually provides a global overview of

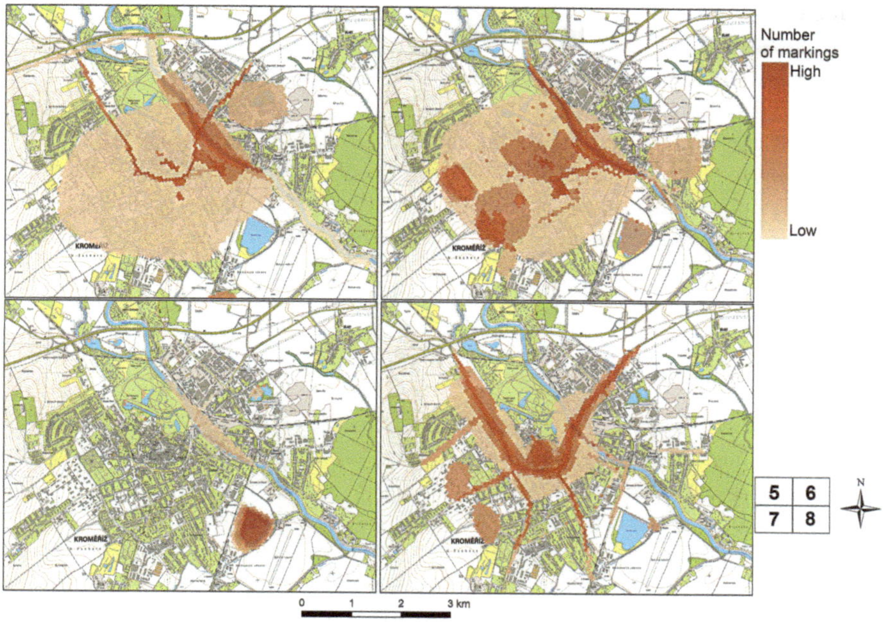

Fig. 12.6 The visualisation of perceptions of environmental deprivation of surveyed topics, including (5) Bad smell, (6) Criminality, (7) Low quality of bathing places, and (8) Motor traffic

Fig. 12.7 Second order neighbourhood for one hexagonal grid cell

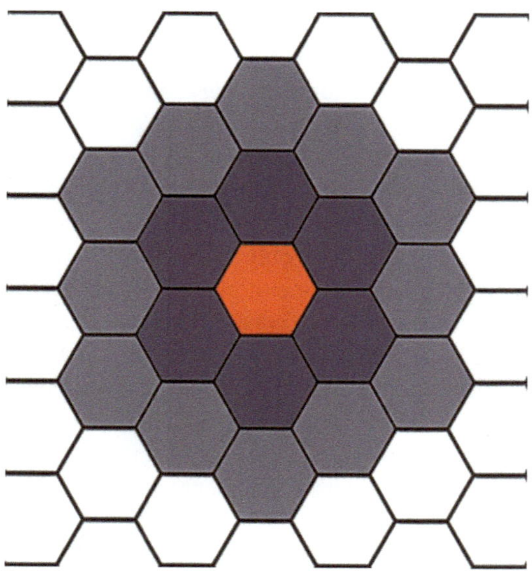

the association described, which may appear to be an inconvenient method in the case of local studies.

A non-parametric Spearman's rank correlation (Spearman's ρ) was used to analyse associations among people's cognition of the urban environment. Spearman's correlation can explore nonlinear relationships of characteristics, and this was the main reason for using it, because the character of the analysed data does not allow us to assume a normal probability distribution. Measured values are substituted for their rank in the calculation of the Spearman's correlation.

Formula 1:

$$\rho = 1 - \frac{6\sum_{i=1}^{n}(p_i - q_i)^2}{n\,(n^2 - 1)}$$

Formula 1 describes the non-parametric Spearman's rank correlation. The expression (pi-qi) means the difference in the rank of values corresponding to the measured characteristics and n is the number of pairs. In the study, the authors applied the local version of the Spearman's ρ, which calculated the correlation between observations neighbouring in geographical space (second order neighbourhood). The local Spearman's ρ (including statistical significance) was computed using the R language, utilizing an adapted function based on the package lctools (Kalogirou 2011, 2015).

The visualisation of the Spearman's ρ for all pairs of investigated factors is depicted in Fig. 12.8. Positive associations are shown in red, while negative ones are shown in blue. The lighter the colour of an area, the weaker the associations between feelings that exist in the area. In contrast, the darker the colour of the place, the stronger the associations among characteristics is present in the neighbourhood. The areas with a correlation of around zero depict places where associations among characteristics are weak, i.e. where public opinion concerning the characteristics in question is mixed, or where no answers were recorded in the map. It is important to note that positive associations not only occur among places people mentioned more often on both topics (represented by more records), but also among places that were mentioned only rarely. This means that correlations express the (local) topical agreement among places. The correlation also depicts a mutual/two-sided relationship. Some associations between investigated characteristics are usually perceived naturally by the public (e.g. air pollution nearby and noise associated with the traffic). But this is rather general statement that is hard to prove without additional analysis. The spatial correlation allows quantifying this kind of statements in a finer spatial scale, while subsequent geovisualisation of results provides their factual representation in easy to understand form.

The maps in Fig. 12.8 support the findings depicted by the maps in Figs. 12.5 and 12.6. The strongest positive correlations between investigated factors were heavily associated with the presence of main roads within the city. This is why people agreed on mentioning the level of air pollution, noise during day and night, bad smell and traffic, with traffic being the main cause for all of this. It is interesting that the sense of pollution of public spaces in relation to noise changes its pattern during the day

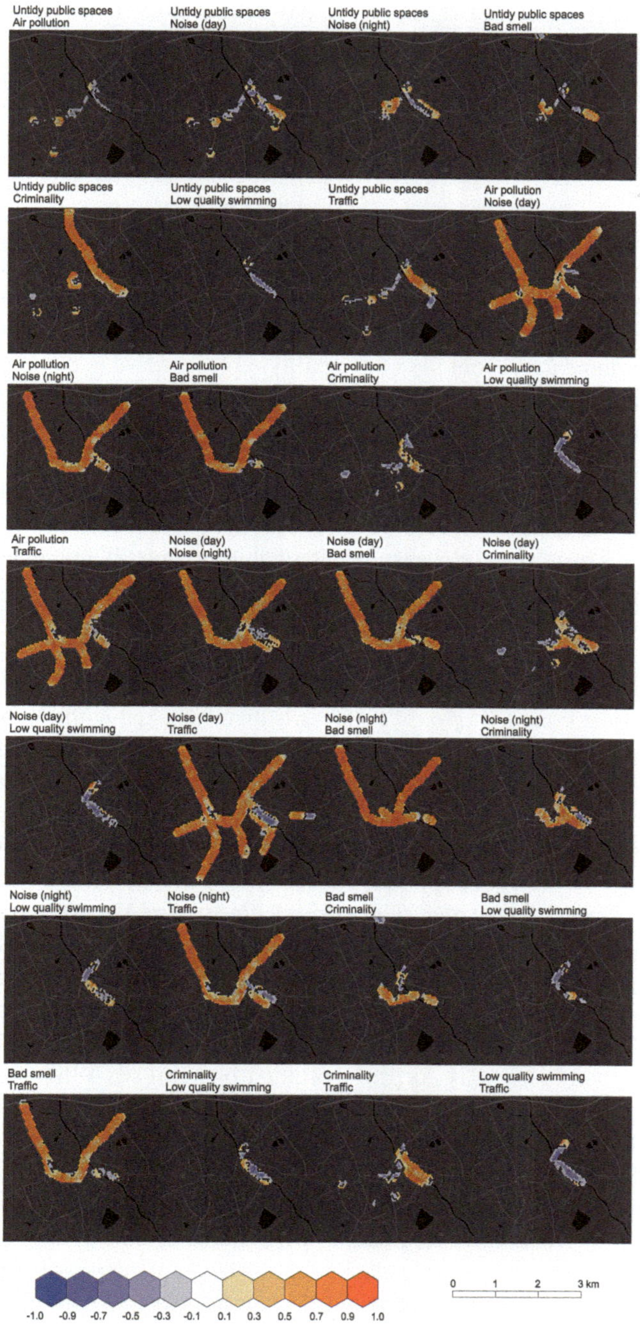

Fig. 12.8 The visualisation of Spearman's ρ for all pairs of investigated variables

and at night, when it relocates closer to the city centre. Noisy places during the night are also evaluated as being related to crime and to bad smell.

12.8 Discussion

Based on the experience gathered during the implementation of the project described in this paper, the authors would like to comment mainly on three issues encountered. As action researchers, the authors view it as important of delivering the results back to the community, since that is a crucial part of all participatory action research. Hence, the commentary from the municipality representative is the first part of the discussion. Following this local expert knowledge, the second section deals with the national response from The Network of Healthy Cities of the Czech Republic, with the evaluation of a possible bias in the data collection forms and its varying results. The third part of the discussion presents obstacles faced from the methodological point of view in relation to data analysis and data visualisation. It presents our "failure" to use heat-maps in the study, although they are still a popular, yet in this case not so useful, visualisation method.

The authors consulted a municipality representative (Viliam Staněk, coordinator of Local Agenda 21 in Kroměříž) about the results and he confirmed the veracity of most of the results in the survey. Based on his expertise and local spatial knowledge he stated that the *pollution of public spaces* is mainly linked with vandalism issues in the northern part of the city and the neglected condition of the city's parks. *Air pollution*, *noise (daytime)*, and *motor traffic* correspond with the volume and extent of traffic in the city and the perception of *noise (night-time)* can be, according to Staněk, identified with the opening times of restaurants and, partly, with traffic. The Local Agenda 21 coordinator also agreed that there was only one hot-spot for *bad smell* and one for *low quality of bathing places*, and he also agreed with the main hot-spots for the perception of *crime occurrences*. In general, Viliam Staněk evaluated the results from the case study in Kroměříž as *suitable*, *matching with reality* and *appropriate*. Regarding the future use of the results from the presented case study, Viliam Staněk confirmed that the results may be used in a general development plan (especially for transportation planning) and the results may be released to the public on the GIS platform of the city.

The results of the case study were also endorsed by representatives of The Network of Healthy Cities of the Czech Republic, who agreed on further deployments of emotional maps in other cities that are part of the network. These would focus on various topics ranging from citizens' perceptions of safety to housing estates revitalisation programmes. The inclusion of subjective perceptions from the emotional mapping activities may be one of the input sources into community planning action plans. The authors would suggest keeping the combination of a paper-based questionnaire with a web-based crowdsourcing tool, as it helps to address different stakeholders and various target groups. Nevertheless, the combination of results from the two sources can be a source of data bias. Despite the possible impact of data comparability, the authors are confident in the usability of both sources and their data richness for further studies, especially when involving elderly people (paper-based maps) and local youth (web-based questionnaires).

Testing the methodology was also a crucial aspect of the case study. The most challenging part was the visualisation of 1257 data points. The initial idea had been to create heat maps, but these maps mostly covered vast areas of Kroměříž and did not provide specific information. The most important findings were merged and in some cases created continuous areas that did not reflect the specific and unique findings. Therefore, another approach was adopted concerning the visualisation of findings and the authors decided to use a hexagonal grid. This improved visualisation insofar as it offered an increasingly clearer representation of the perception results (Figs. 12.5 and 12.6).

12.9 Conclusion

The authors presented the testing and implementation of their own web-based crowdsourcing tool, called PocitoveMapy.cz and a case study of the collection and visualisation of perceptions of environmental deprivation on maps. The case study (n=99) analysed in this paper was localised in the city of Kroměříž. The authors collected a total of 1257 perception-based shapefiles on eight different environmental topics. The collected data (points, lines, and polygons) were transformed into a hexagonal grid with a side of 25 metres and each hexagon was assigned a value based on spatially overlapping features from the survey. Representatives of the city of Kroměříž confirmed the findings of the case study in relation to the deprived areas. Furthermore, the visual analysis showed similar spatial distribution of some environmental variables (such as Air pollution, Noise (night-time) and Motor traffic) and the non-parametric Kendall's rank correlation supported the visual observations. The authors see the main novelty of the research presented in this paper in the fact that it is the first deployment of a geoparticipatory online tool for the collection of subjective perceptions about the quality of the environment in the Czech Republic. The tool is designed as an easy in-browser application that is scalable and adjustable for various themes and spatial scenarios. The future outlook of the research will include the exploration of a mobile app development and the better availability for users of interaction with the map and the survey. The authors see potential in the participatory planning support tools, mainly due to their connectivity to GIS-based urban planning applications and their ability to speak the *language of city administrators*. The use of geoparticipatory applications in public consultancies and community planning has been much tested and used in Finland and other western-European countries. However, their deployment in the post-soviet countries of central and eastern Europe is in its beginnings.

References

Bergner BS, Zeile P, Papastefanou G (2011) Emotional barrier-GIS – A new approach to integrate barrier-free planning in urban planning processes. In Proceedings REAL CORP, 247–257
Biever C (2010) Twitter mood maps reveal emotional states of America. New Sci 207 (2771). Elsevier: 14.
Bollen J, Mao H, Zeng X (2011) Twitter mood predicts the stock market. J Computational Sci 2(1). Elsevier: 1–8
Brydon-Miller M, Greenwood D, Maguire P (2003) Why action research? Action Res 1(1):9–28. doi:10.1177/14767503030011002
Burian J, Pászto V, Langrová B (2014) Possibilities of the definition of city boundaries in GIS – the case study of a medium-sized city. In 14th SGEM GeoConference on Informatics, Geoinformatics and Remote Sens 3:777–784. doi:10.5593/SGEM2014/B23/S11.099
Butler D (2006) Virtual globes: the web-wide world. Nature 439(7078):776–778. doi:10.1038/439776a
Čermák J (2001) Universum: Všeobecná Encyklopedie, Díl 6. Euromedia Group - Odeon, Praha
Chambers R (2003) Whose reality counts? Putting the first last. ITDG Publ, London
Chambers R (2006) Participatory mapping and geographic information systems: whose map? who is empowered and who disempowered? who gains and who loses? Electron J Inf Syst Dev Ctries 25(2):1–11
City of Kroměříž (2015) Geografie – Město Kroměříž. http://www.mesto-kromeriz.cz/fakta-o-meste/demografie-mapy-a-statistiky/geografie/
Craig WJ, Elwood S (1998) How and why community groups use maps and geographic information. Cartogr Geogr Inf Sci 25(2):95–104. doi:10.1559/152304098782594616
Czech Statistical Office (2015) Population of municipalities of the Czech Republic, 1 January 2015. https://www.czso.cz/documents/10180/20556287/1300721503.pdf/33e4d70e-e75f-4596-930c-63406c9068d0?version=1.1
Design S (2015) Toner.. http://maps.stamen.com/#toner
Dooris M (1999) Healthy cities and local Agenda 21: The UK experience – challenges for the new millennium. Health Promot Int 14(4):365–375. doi:10.1093/heapro/14.4.365
Elwood S (2002) GIS use in community planning: a multidimensional analysis of empowerment. Environ Plan A Abstract 34(5):905–922
Elwood S, Cope M (2009) Qualitative GIS: a mixed methods approach. SAGE, Los Angeles
Forrester J, Cinderby S (2012) A guide to using community mapping and participatory-GIS. http://www.tweedforum.org/research/Borderlands_Community_Mapping_Guide_.pdf
Gartner G (2012) Putting emotions in maps–the wayfinding example. Mountaincartography.org, 61–65
Google Maps (2015) Traffic in Kroměříž. https://www.google.com/maps/@49.2983205,17.3953272,15.25z/data=!5m1!1e1
Gould P (1986) Mental maps. Taylor & Francis, London
Griffin AL, Mcquoid J (2012) At the intersection of maps and emotion : the challenge of spatially representing experience. Kartographische Nachtrichten 62(6):291–299
Hauthal E, Burghardt D (2016) Mapping space-related emotions out of user-generated photo metadata considering grammatical issues. Cartogr J, February. Taylor & Francis.
Hemmersam P, Martin N, Westvang E, Aspen J, Morrison A (2016) Exploring urban data visualization and public participation in planning. J Urban Technol, February. Routledge, 1–20. doi:10.1080/10630732.2015.1073898
Huang H, Gartner G, Klettner S, Schmidt M (2014) Considering affective responses towards environments for enhancing location based services. ISPRS-International Arch Photogramm Remote Sens Spat Inf Sci 1:93–96
Huck J, Whyatt D, Coulton P (2014) Spraycan: a PPGIS for capturing imprecise notions of place. Appl Geogr 55 (December): 229–37. doi:10.1016/j.apgeog.2014.09.007

IDRC (1996) The local agenda 21 planning guide. The International Council for Local Environmental Initiatives (ICLEI), The International Development Research Centre (IDRC), The United Nations Environment Programme (UNEP)

International Institute for Sustainable Development (2015) What is sustainable development? https://www.iisd.org/sd/#one

Jankowski P, Czepkiewicz M, Młodkowski M, Zwoliński Z (2015) Geo-questionnaire: a method and tool for public preference elicitation in land use planning. Trans GIS, December. doi:10.1111/tgis.12191

Jenness J (2006) Repeating shapes for ArcGIS. Jenness Enterprises. http://www.jennessent.com/arcgis/repeat_shapes.htm

Kahila M, Kyttä M (2009) SoftGIS as a bridge-builder in collaborative urban planning. In: Stan Geertman and John Stillwell (eds) Planning support systems best practice and new methods. 95:389–411. The GeoJournal Library. Springer Netherlands, Dordrecht. doi:10.1007/978-1-4020-8952-7

Kahila-Tani M, Broberg A, Kyttä M, Tyger T (2015) Let the citizens map—public participation GIS as a planning support system in the Helsinki Master Plan process. Plan Pract Res, December. Routledge, 1–20. doi:10.1080/02697459.2015.1104203.

Kalogirou S (2011) Testing local versions of correlation coefficients. Jahrb Reg 32(1):45–61. doi:10.1007/s10037-011-0061-y

Kalogirou S (2015) Local correlation, spatial inequalities and other tools | Lctools. http://rpackages.ianhowson.com/cran/lctools/man/lctools-package.html

Kloeckl K, Senn O, Di Lorenzo G, Ratti C (2011) Live Singapore!-an urban platform for real-time data to program the city. In Computers in Urban Planning and Urban Management, CUPUM. Vol. 4

Korpela K (2002) Children's environment. Handbook of environmental psychology. Wiley, New York, pp 363–373

Kubinova R, Zejglicova K, Kratenova J, Maly M, Volf J (2006) Monitoring population health status in the Czech Republic. Epidemiology 17(6):S448–S449

Kveton V, Louda J, Slavik J, Pelucha M (2014) Contribution of local agenda 21 to practical implementation of sustainable development: the case of the Czech Republic. Eur Plan Stud 22(3):515–536. doi:10.1080/09654313.2012.753994

Kytta M, Broberg A, Haybatollahi M, Schmidt-Thome K (2015) Urban happiness: context-sensitive study of the social sustainability of urban settings. Environ Plann B: Plann Des 47:1–24. doi:10.1177/0265813515600121

Lefèvre, P., Kolsteren P, De Wael MP, Byekwaso F, Beghin I (2001) Comprehensive Participatory Planning and Evaluation (CPPE). Antwerp. http://dspace.itg.be/handle/10390/1553

MacKerron G, Mourato S (2010) Mappiness. http://www.mappiness.org.uk/

McTague C, Jakubowski S (2013) Marching to the beat of a silent drum: wasted consensus-building and failed neighborhood participatory planning. Appl Geogr 44 (October): 182–91. doi:10.1016/j.apgeog.2013.07.019

Ministry of Health of the Czech Republic (2007) Noise maps. http://hlukovemapy.mzcr.cz/image.aspx?obr=Mapy/Silnice/ZL_Ldvn/ZL 12.png

Mislove A, Lehmann S, Yong-Yeol Ahn, Jukka-Pekka Onnela, Niels Rosenquist J (2010) Pulse of the nation: U.S. mood throughout the day inferred from twitter. http://www.ccs.neu.edu/home/amislove/twittermood/?utm_campaign=Facebook+Page&utm_content=Pulse+of+the+Nation+US+Mood+Throughout+the+Day+inferred+from+Twitter&utm_medium=postitat&utm_source=facebook

Mody RN, Willis KS, Kerstein R. (2009). WiMo: location-based emotion tagging. In Proceedings of the 8th International Conference on Mobile and Ubiquitous Multimedia, 14

National Institute of Public Health (2015) Studie HELEN (Health, life style and environment). http://www.szu.cz/publikace/studie-helen?lang=1

Nold C (2009) Emotional cartography: technologies of the self. http://emotionalcartography.net/EmotionalCartography.pdf

Norton R (2014) Agenda 21 and its discontents: is sustainable development a global imperative or globalizing conspiracy? Urban Lawyer 46(2):325–360

Pánek, Jiří, and Chris Vlok. (2013). Participatory mapping as a tool for community empowerment – a case study of community engagement in Koffiekraal, South Africa. In 26th International Cartographic Conference, edited by Manfred F. Buchroithner, 26. Dresden. http://icaci.org/files/documents/ICC_proceedings/ICC2013/_extendedAbstract/969_abstract.pdf

Pánek J, Pászto V, Marek L (2017) Mapping emotions: spatial distribution of safety perception in the city of Olomouc. In: Igor I, Singleton A, Horák J, Inspektor J (eds) Lecture notes in geoinformation and cartography: the rise of big spatial data. Springer, Ostrava, pp 211–224. doi:10.1007/978-3-319-45123-7

Pearce MW (2008) Framing the days: place and narrative in cartography. Cartogr Geogr Inf Sci 35 (1). Taylor & Francis: 17–32.

Perkins C (2007) Community mapping. Cartogr J 44(2):127–137. doi:10.1179/000870407X213440

Perkins C (2009) Performative and embodied mapping. International Encyclopedia of Human Geography, Oxford: Elsevier, 126–132.

Pickles J (1995) Ground truth: the social implications of geographic information systems. 1st ed. New York: The Guilford Press.

Raosoft (2004) Sample size calculator by Raosoft. http://www.raosoft.com/samplesize.html

Raslan R, Al-hagla K, Bakr A (2014) Integration of emotional behavioural layer 'EmoBeL' in city planning. In Real Corp 2014 8:309–317

Reimann C, Filzmoser P, Garrett RG, Dutter R (2008) Statistical data analysis explained: applied environmental statistics with R. John Wiley & Sons, Ltd.

Rød, Jan Ketil, Ferjan Ormeling, Corné Van Elzakker (2001) An agenda for democratising cartographic visualisation. Norsk Geografisk Tidsskrift 55 (1). Taylor & Francis: 38–41

State Administration of Land Surveying and Cadastre (2014) Fundamental base of geographic data of the Czech Republic. http://geoportal.cuzk.cz/(S(z2w1esrjen1dfactx0vyepfg))/Default.aspx?mode=TextMeta&text=dSady_zabaged&side=zabaged&menu=24

UNWCED (1987) Our common future. Oxford University Press, Oxford

Von Schirnding, Yasmin ER, and Mulholland C (2002) Health and sustainable development: key health trends. Geneva, Switzerland. http://apps.who.int/iris/handle/10665/68755

WHO (2015a) Types of healthy settings. http://www.who.int/healthy_settings/types/cities/en/

WHO (2015b) The Ottawa charter for health promotion. http://www.who.int/healthpromotion/conferences/previous/ottawa/en/

WHO Europe (1998) HEALTH21 – Health for all in the 21st century. http://www.euro.who.int/en/publications/policy-documents/health21-health-for-all-in-the-21st-century

Wilson MW (2011) 'Training the eye': formation of the geocoding subject. Soc Cult Geogr 12 (04). Taylor & Francis: 357–76

Zadra JR, Clore GL (2011) Emotion and perception: the role of affective information. Wiley interdisciplinary reviews. Cogn Sci 2(6):676–685. doi:10.1002/wcs.147

Outlook

The recent proliferation of location aware devices that are small and inexpensive together with the opportunities provided via web 2.0 have greatly empowered citizens to easily collect, share and use geographical information in a diverse range of applications. These technological advances as well as a growing desire for openness and collaboration have had a revolutionary effect across a broad spectrum of activity including a suite of issues connected with mapping. For example, resources such as Google Earth, Bing maps and citizen derived maps such as those generated via collaborative projects like OpenStreetMap (OSM) are now very widely and routinely used by diverse communities, both amateur and professional.

This book has explored a series of key issues in the subject of citizen empowered mapping. It has focused on fundamental issues covering activity from the collection of data through topics linked to data quality and usability through to topical applications of the acquired data. The book provides examples that highlight the considerable potential and role of citizens in mapping. The considerable value of citizens as large, geographically distributed and inexpensive source of valuable geographical information has been stressed. It has been shown, for example, that citizens can acquire useful data and do so to such a standard that they even represent an important source of data for professional bodies such as the national mapping agencies who until recently were effectively the sole source of many data sets and map products. This situation highlights that not only are citizens able to generate data, the quality of the data, while naturally regarded with a degree of suspicion by professional mappers trained to work to exacting standards, is also often high, and higher than many users may need. This then makes the data suitable for a range of uses. A number of example applications are presented in the book with an emphasis on environmental monitoring but also on perceptions. The latter is also a topic on which citizen sensors may be particularly well-suited to act as data source. As such the book represents an important point of reference in this rapidly growing subject.

As highlighted in the book the subject is growing and developing. This creates opportunities but also challenges. Amongst the latter are issues highlighted in the

M. Leitner, J. Jokar Arsanjani (eds.), *Citizen Empowered Mapping*,
Geotechnologies and the Environment 18, DOI 10.1007/978-3-319-51629-5

book such as the need for enhanced means to deal with the vast and growing amount of often heterogeneous data. Indeed, the topic of big geospatial data is likely to become more important as data sources expand. There will also be a desire to make greater use of citizen sensor data plus those from other traditional data sources as well as rapidly growing sensor networks and continued growth in data collection from remote sensing systems on platforms operated by communities that range from citizens (e.g. using cameras mounted on drones) to major space agencies (e.g. the expanding array of Earth-orientated satellite remote sensing systems). Associated with the growth in data collection, subjects such as the Internet of Things and Data Analytics are likely to become increasingly important in how the subject of citizen empowered mapping develops. In addition, it is likely that future developments will require an increased consideration of the citizen as well as the end use of the data generated, especially if data are open to re-use and/or harvested from, for example, wearable technology or social media sites. The latter issues links to a set of legal and ethical issues that may require attention and as these have the potential to aid or hinder citizen sensing will need to be approached with care recognising that there are often conflicting demands and tensions linked to citizen activity. How the subject develops from the snapshot presented in this book is difficult to predict but its current status points to an exciting and impactful period of growth for broad benefit. The tremendous power of citizens is recognised widely, as perhaps evident in the following two quotes from international leaders with very different standpoints:

> The human animal cannot be trusted for anything good except en masse. The combined thought and action of the whole people of any race, creed or nationality, will always point in the right direction. (Harry S. Truman, President United States of America, 1884–1972)

> I have witnessed the tremendous energy of the masses. On this foundation it is possible to accomplish any task whatsoever. (Mao Zedong, Chairman of the Communist Party of China, 1893–1976)

There is clearly considerable scope for citizens to further advance mapping and this should be for the benefit of all.

School of Geography, University of Nottingham Giles Foody
Nottingham, UK

Biographies of Editors and Book Chapter Contributors

Editors

Michael Leitner received a Master Degree at the Department of Geography and Regional Research, University of Vienna, Austria (1990) and a second Master (1993) and a Doctoral Degree in GISc (1997) at the Department of Geography, State University of New York at Buffalo, US. Since 2013, he is a Professor of Geography in the Department of Geography and Anthropology, Louisiana State University, US and a faculty member in the Doctoral College "GIScience" at the University of Salzburg, Austria. He was the recipient of a Fulbright Scholarship as a student (1990–1992) and has received a Fulbright Specialist Program grant for short-term visits to the Jagiellonian University in Krakow, Poland in 2016 and 2017. He was also the recipient of the 2007 Meredith F. Burrill Award from the Association of American Geographers. His research interests are in GISc and their applications to spatial crime analysis, medical geography, and geospatial privacy. He has published four books, five co-edited journal volumes, and 50+ refereed journal articles and book chapters. He has secured grants as PI or Co-PI of over $4.5 million and was the editor of the *Cartography and Geographic Information Science* (*CaGIS*) journal from 2008–2014.

Jamal Jokar Arsanjani received his doctoral degree in Geographic Information Science (GISc) from the Department of Geography and Regional Research, University of Vienna, Austria. He is currently an assistant professor for Geoinformatics at Aalbrog University Copenhagen, Denmark. He was an Alexander von Humboldt Fellow at the Institute of Geography, Heidelberg University, Germany within 2013–2015 and a senior research fellow in 2016. His interdisciplinary research interests are in volunteered geographic information and crowdsourcing, geocomputation, remote sensing of the environment, and disaster management. He has published articles in leading international journals of his discipline, including International

M. Leitner, J. Jokar Arsanjani (eds.), *Citizen Empowered Mapping*,
Geotechnologies and the Environment 18, DOI 10.1007/978-3-319-51629-5

Journal of Applied Earth Observation and Geoinformation, International Journal of Digital Earth, Transactions in GIS, Cities and single-authored a book and co-edited two books with Springer.

Contributors

Pierre Aumond is a postdoctoral research associate in acoustics at the Environmental Acoustics Laboratory, IFSTTAR, France. He carried out his PhD from 2008–2011 at the University of Maine (Le Mans, France) focusing on simulation of acoustics propagation. After his PhD, he was appointed as R&D acoustics engineer (2011–2014) in an engineering company, Santiago de Chile. Thereafter, he took up a postdoctoral position at MRTE laboratory, Cergy-Pontoise University, France, to participate in the research project GRAFIC (2015).

Matthieu Baley was a master student at the Engineering School for Land Surveying and Topography (ESGT) in Le Mans, France, and made a 6-month internship with the COGIT team in 2013, where he tested the first ideas presented in this book chapter.

Jessica G. Benner is a Teaching Fellow in the Information Culture and Data Stewardship Department and Researcher in the Geoinformatics Laboratory in the School of Computing and Informatics at The University of Pittsburgh. Jessica is interested in wayfinding and mobility in the real world and how crowdsourced applications and practices can be used to increase the availability of information about the physical accessibility of the environment for all people, especially people with disabilities. She is also interested in the use of GIS and spatial analysis tools in libraries particularly how these tools are utilized by different disciplines.

Mu-Ning Wang Brandeis has a PhD from the School of Environmental and Forest Science, University of Washington (2008–2015) with specialty of Participatory Geographic Information System (PGIS), Volunteered Geographic Information (VGI), and Citizen Science in natural resource and environmental applications. Her dissertation, Volunteered Geographic Information Reporting System: A Cross-Case Comparison, aims to develop a methodology to compare practical VGI, PGIS, and Citizen Science cases and discover the best available practices for future research. Her recent three publications are focused on the case comparison with different analytical methods and applications. Before starting her PhD at the University of Washington, her research focused on community conservation projects in Taiwan (2005–2008). She received her master degree from the National Taiwan University. She has always been passionate about participatory applications in natural resource and environmental management.

Yezmín Calvillo-Saldaña is a junior researcher at CentroGeo, where she also teaches Cybercartography. In 2013, she obtained her master's degree in Geomatics from CentroGeo. Her current research efforts focus on the distribution of human capital, crowdsourcing, and spatial analysis of crime. She has served on the International Map Year Committee in Mexico, organized by the International Cartographic Association (ICA) and supported by the United Nations (UN).

Kari J. Craun is the Director of the U.S. Geological Survey, National Geospatial Technical Operations Center (NGTOC). This Center performs a wide range of functions in support of maintaining a seamless, current, nationally consistent coverage of base geospatial data for the United States, including development of digital and graphic products such as U.S. Geological Survey topographic maps. Craun is a Past-President of the American Society for Photogrammetry and Remote Sensing (ASPRS) and the Cartography and Geographic Information Society. She received a B.S. degree in Geology from the University of Missouri-Kansas City in 1984, an M.S. degree in Photogrammetry from Purdue University in 1987, and a Master of Science degree in Geospatial Information Science through Northwest Missouri State University in 2014.

Catherine Domingues is a researcher at the French mapping agency, the Institut national de l'information géographique et forestière. Her research interest is in the cartographic representation of geographic data. Her research activities are the automatic detection and the analysis of geolocated information (geographical and thematic objects, named entities) in texts with natural language processing tools, in order to offer personalized cartographic representations.

Saúl Gomez is a former research engineer at the French mapping agency, the Institut national de l'information géographique et forestière. He started his carrier at the urban laboratory of civil engineering of the city of Paris, where he worked in the urban resilience department. His research focuses on geographic data exploitation for urban issues applications.

William A. Gough has been a climatologist at the University of Toronto Scarborough since 1993. He received his Master degree in Atmospheric Physics at the University of Toronto and his doctoral degree in Atmospheric and Oceanic Sciences at McGill University in 1991. Gough's research interests include the nature of climate and climate change in the Eastern Arctic, Hudson Bay, and southern Ontario. More recently his research has focused on climate change impacts and opportunities. He is currently serving as the Vice-Principal (Academic) & Dean at the University of Toronto Scarborough.

Weihe Wendy Guan is the Executive Director of the Center for Geographic Analysis at Harvard University in Cambridge, MA. She has been managing the Center's operation since 2006, which develops and applies geospatial technologies in support of research and teaching across Harvard. She also teaches GIS at the Harvard Extension School. She has a PhD in ecology from the University of Georgia, an MA in Canadian heritage and development studies from Trent University, an MS in geography from Beijing University, and a BS in biology from South China Normal University. She was an assistant professor in urban ecology at Qinghua University; a postdoctoral research associate in GIS for environmental modeling at the State University of New York, Buffalo; and an adjunct professor in geography at Florida Atlantic University. Prior to working at Harvard, she managed GIS professional services at Marshall, a GIS consulting firm, headed the geospatial information technology department for Weyerhaeuser, a multinational forestry corporation, and supervised GIS teams in the South Florida Water Management District. Her research interest is in GIS applications for natural resource management and environmental studies. She also has extensive experience in enterprise GIS implementation.

Jiří Hájek graduated from the International Development Studies at the Palacky University, Olomouc, Czech Republic and is now a PhD candidate there. His professional interest is in sustainable consumption with focus on spatial distribution of sustainable materials used in constructions. He also works as a manager of international development cooperation projects mainly aimed on awareness rising.

Ming-Chih Hung is a professor of Geography and GIScience in the Department of Humanities and Social Sciences, Northwest Missouri State University. He earned his B.S. in Geography from National Taiwan University in 1993, M.S. and Ph.D. in Geography from the University of Utah in 1996 and 2003, respectively. After his graduation, he began his teaching career at Northwest Missouri State University. His teaching responsibilities and research interests include remote sensing, GIS, GPS, cartography, geovisualization, and applications of such techniques in urban areas and environmental issues.

I. Kapouranis holds a degree in Computer Science from the School of Informatics, Faculty of Sciences, Aristotle University of Thessaloniki, Greece, since 2014. He was involved in many initiatives, including start-ups (PhotoCityView, MyReception). He currently participates in two European projects. He is a software developer expert (C/C++/Java) in computer science fields, such as graphs, databases, parallel and distributed systems, as well as mobile application development.

Hassan A. Karimi is a Professor and the Director of the Geoinformatics Laboratory in the School of Information Sciences at the University of Pittsburgh. Karimi's research interests include computational geometry, geospatial big data, distributed/parallel computing, mobile computing, navigation, and location-based services. He has published the following books: Indoor Wayfinding and Navigation (sole editor), published by Taylor and Francis (2015); Big Data: Techniques and Technologies in Geoinformatics (sole editor), published by Taylor and Francis (2014); Advanced Location-Based Technologies and Services (sole editor), published by Taylor and Francis (2013); Universal Navigation on Smartphones (sole author), published by Springer (2011); CAD and GIS Integration (lead editor), published by Taylor and Francis (2010); Handbook of Research on Geoinformatics (sole editor), published by IGI (2009); and Telegeoinformatics: Location-Based Computing and Services (lead editor), published by Taylor and Francis (2004).

Erin Korris is a Geographer at the US Geological Survey. She works within the National Geospatial Technical Operations Center (NGTOC), which is responsible for the creation and maintenance of The National Map and the US Topo map series. Erin currently focuses full time on the volunteer mapping program, The National Map Corps. Erin began working at the USGS in 2010 while finishing up her B.A. in Geography with a concentration in Environmental Studies at the University of Colorado Denver (CU Denver). She also holds a certificate in Geographic Information Science (GIS) from CU Denver.

Catherine Lavandier is professor in the ETIS Laboratory (Information Processing and System Research Lab) at the University of Cergy-Pontoise, France. She teaches building acoustics at the civil engineering department of the Technological Institute. The topic she is involved in focusses on the sound quality of the environment. Catherine Lavandier has been involved in international networks (European COST action "Soundscape of European Cities and Landscapes", or ISO/TC43/SC1/WG54 standards) and is presently the vice president of the French Acoustical Society (SFA).

Andrew C. W. Leung is a PhD student in environmental science at the University of Toronto Scarborough. His PhD research focuses on climate change impacts on northern Canadian airports. He received best paper awards for his research in the master (2012) and doctoral (2015) student categories from the Canadian Association of Geographers – Ontario Division (CAGONT). He is also a member of the Canadian Meteorological and Oceanographic Society since 2014.

Benjamin G. Lewis is the Geospatial Technology Manager for the Harvard Center for Geographic Analysis and system architect/project manager for the WorldMap and HHypermap platforms, open source infrastructures to extend collaborative research centered around geospatial information. Before joining Harvard, Ben was a project manager with Advanced Technology Solutions of Pennsylvania, where he led the company in adopting platform independent approaches to GIS development. Ben studied Chinese at the University of Wisconsin and has a Master in Planning from the University of Pennsylvania. After Penn, Ben helped start the GIS Lab at U.C. Berkeley, founded the GIS group for transportation engineering firm McCormick Taylor, and coordinated the Land Acquisition Mapping System for the South Florida Water Management District. Ben is especially interested in technologies that lower the barrier to spatial technology access.

José Luis López-Gonzaga is a biologist with a specialization in Geomatics and has more than 23 years of experience using GIS in both the private and public sectors. He has ample experience analyzing the quality of the information contained in spatial databases, geoprocessing, and carrying out spatial analysis with them. He has worked on studies related to weather, risk and vulnerability, ecological ordering, natural resources, human capital, discrimination indices, crowdsourcing, and public safety.

Lukáš Marek is currently employed as a postdoctoral fellow/geospatial researcher at the Department of Geography, University of Canterbury (New Zealand). He received a Ph.D. from the Department of Geoinformatics, Palacký University in Olomouc, Czech Republic. His main research interests are focused on geovisualisation and spatial analyses of health data using spatial and spatio-temporal modelling, spatial statistics, and smart technologies. He is part of a team that aims to create a smart system based on (near) real-time environmental monitoring and individual exposures of patients with breathing issues in order to provide the tool that can inform the public about impaired conditions that may affect their health conditions. Besides his research activities, he is co-founder and editor of the web-based journal GISportal.cz.

Lenka Mařincová graduated from the International Development Studies at the Palacky University, Olomouc, Czech Republic and is now a PhD candidate there. Her main professional interests are in global health issues, health impact assessment, health literacy, community participation, and other public health challenges. During her studies she has also spent some time abroad, namely attending projects, summer schools, and internships in Vietnam, Laos, the United States, Switzerland, Finland, and the UK.

Elvia Martínez-Viveros holds a PhD in Social Systems Science from the University of Pennsylvania and an MSc degree in Operations Research from the London School of Economics and Political Sciences. After 25 years of experience as a public consultant in planning and projects' evaluation, she has been a researcher at CentroGeo, a public research institution specialized in Geomatics, since 2003. She has coordinated several research projects, dealing with issues related to the spatial distribution of innovation capabilities, vulnerability to disasters, crime, and disorder, among others; and led the Roma project presented in this publication. She teaches in the Master in Geomatics program in this research center and is the academic designer of a new Master in Spatial Planning program that has recently been listed by the National Council of Science and Technology (CONACYT) in the National Quality Graduate Program.

Elizabeth McCartney is a Cartographer at the U.S. Geological Survey (USGS) National Geospatial Technical Operations Center. She currently serves as the Team Leader for the USGS Volunteered Geographic Information (VGI) project known as The National Map Corps which supports The National Map and US Topo maps.

Ms. McCartney graduated from Jacksonville State University (AL) with a dual undergraduate degree in Biology and Geography in 1996, and a Master's degree in Biology in 1998.

Lily Niknami is a cartography intern at the US Geological Survey. She began working at the USGS with The National Map Corps in 2015, focusing efforts on volunteer engagement and data quality assessment. She is currently pursuing a M.S. in Environmental Sciences at the University of Colorado Denver. Lily received her B.S. in Soil and Crop Sciences with an emphasis in Applied Information Technology from Colorado State University in 2013.

Timothy L. Nyerges is Professor of Geography and Director of the Master of Geographic Information Systems for Sustainability Management Program at the University of Washington. Dr. Nyerges specializes in teaching and research related to participatory geographic information systems (GIS) focusing on integrated studies about land use, transportation, and water resources using a sustainability management perspective. He received his Ph.D. from the Ohio State University in 1980 specializing in database management languages for GIS. For the past couple of decades he has undertaken research projects funded by the National Science Foundation, the National Oceanic and Atmospheric Administration, and the Department of Energy to explore development and evaluation of networked GIS, particularly as supported by cyberinfrastructure technology, for enabling stakeholder participation in decision support. Currently, his research focuses on extending geographic information science into a sustainability information science with particular application to watershed development resilience. He is a Fellow of the University Consortium for Geographic Information Science, a former appointee to the US National Geospatial Advisory Committee, and a recipient of the Distinguished Career Award from the American Association of Geographers' Geographic Information Science and Systems Specialty Group.

Gaëtan Palka is a researcher in Spatial Planning at the University of Tours, France. While doing his PhD, he was lecturer in GIScience at Polytech Tours, Department of Spatial Planning and Environment. His research focuses on mapping improvement based on an end-user centered approach and on spatial modeling for urban development.

Jiří Pánek holds a B.Sc. and a M.Sc. in Geography and GIS from the Department of Geoinformatics and a PhD from the Department of Development Studies, both Palacky University in Olomouc, Czech Republic. His research is focussed on GIS in development cooperation and humanitarian aid, with the main focus on Participatory GIS (PGIS/PPGIS). He has experienced mapping in Kenya and South Africa, studied GIS in India and currently he is co-developing a participatory platform for the collection of perceptions about the urban environment called PocitoveMapy.cz. Besides his research activities, he is the co-founder and editor of the web based journal GISportal.cz.

Alenka Poplin is an assistant professor of Geoinformation Science and GeoDesign at Iowa State University and a founder of the GeoGames Lab. Her research interests intersect geospatial modelling, interactive virtual geo-environments, game-based modelling and design, and interaction with online mapping systems. Her main application areas include civic engagement, public participation in urban planning, energy modelling, and smart cities. She holds a PhD in Geoinformation Science from Vienna University of Technology, a Master of Business Administration (MBA) from Clemson University, SC, and a Master in Surveying and Spatial Planning from the Technical University of Ljubljana, Slovenia. Prior to this position she was an associate professor at HafenCity University Hamburg, where she worked between 2007 and 2014. Alenka recently published in several journals including *Journal of Urban Technology; Environment and Planning B: Planning and Design; Computers, Environment and Urban Systems; The Cartographic Journal; Transactions in GIS;* and *Cartography and Geographic Information Science.* She is one of the co-editors of the forthcoming edited book on "*The Virtual and The Real: Perspectives, Practices and Applications for The Built Environment*", published by Routledge.

Lenka Putalová is a student in the master program of International Development Studies at Palacký University Olomouc, Czech Republic. In her bachelor thesis she analysed mental maps of Africa rendered by grammar school children in Slovakia and Czech Republic. She is currently doing research for her master thesis that deals with teachers' needs in incorporating global development education topics. She is volunteering as an administrator for the Slovak Governance Institute's project called Odkaz pre starostu that aims for more active civic engagement and participation.

Armanda Rodrigues is an Assistant Professor at the Departamento de Informática, Universidade NOVA de Lisboa, Portugal, where she teaches courses and develops research in Computer Science and Geographic Information Technology. She is an integrated member of the Multimodal Systems Group of NOVA LINCS and is interested in Web-GIS, Emergency Management, and Geo-collaboration. Armanda has been involved in several international and national research projects related with Geographic Information Systems, Simulation, Web-GIS, and Geo-Collaborative Systems with case studies in the areas of Emergency Management, Digital Heritage, and Agronomy. She is the author and co-author of several GI Science and Computer Science peer-reviewed publications. She has also reviewed for peer-reviewed journals and served on the program committee of various GI national and international conferences.

André Sabino is a PhD student at the Departamento de Informática, Universidade NOVA de Lisboa, Portugal, where he conducts research in Computer Science. His main focus are geographic information systems, cooperative networks, and machine learning. He has been involved in several research projects in the emergency management domain.

Kamal Serrhini is a lecturer and full research member of the laboratory "Unité Mixte de Recherche 7324 Cités, Territoires, Environnement et Sociétés" of the University of Tours. He took part in various international and national research projects (ESPON, Era-Net Crue, ANR). He focuses on both structural (mass evacuation plan) and nonstructural (hazard and vulnerability mapping) measures to manage natural risks in the urban context (flood, earthquake).

Monir H. Sharker is currently a Researcher in the Geoinformatics Laboratory of Information Science and Technology Program, University of Pittsburgh, US and an Assistant Professor in the Department of Computer Science and Engineering, Jahangirnagar University, Bangladesh. The focus of his PhD research is spatiotemporal big data analytics. His area of research interests include machine learning, data mining, high performance computing (HPC), and algorithm analysis and design. Mr. Sharker completed his B.Sc. (Honors) degree in Computer Science and Engineering, a MS Engg. in Information and Communication Technology, and a MS degree in Computational Mathematics. He served as Academic Head in APTECH Worldwide and ASSET International. He also worked as a Graduate Researcher in the Real-Time Outbreak and Disease Surveillance (RODS) Laboratory at the University of Pittsburgh, US.

Yehong Shi is a Master of Environmental Science candidate at the University of Toronto. He has years of experience in environmental science and geology data analysis. He also has a strong background in GIS application covering ArcGIS and automation map production.

E. Stylianidis studied Surveying Engineering at the School of Rural and Surveying Engineering, Faculty of Engineering, Aristotle University of Thessaloniki, Greece. From the same university he received his PhD in the area of Photogrammetry. Today, he is an Assistant Professor at the School of Spatial Planning and Development, Faculty of Engineering, Aristotle University of Thessaloniki, Greece. He is teaching Photogrammetry and Geoinformatics, as these are also his main research fields, including ICT. For the period 2015–2018, Dr. Stylianidis is acting as CIPA Heritage Documentation Secretary General. He is also active in various international organisations, such as ISPRS and ICOMOS. He published two dissertations and more than 60 scientific publications in journals, conference proceedings, and chapters in books. He is also the editor and co-editor of three books. He has participated in more than 40 research projects (H2020, FP7, EUROSTARS, AAL, LLP), in eight as project coordinator.

Rodrigo Tapia-McClung is an associated researcher at CentroGeo. He is also a PhD student working in crowdsourcing and geovisual analytics. He holds a Master in Geography and Environmental Studies from Wilfrid Laurier University, Canada, in which he focused mostly on the use of spatial statistics to work with environmental datasets. At CentroGeo he has participated in a wide variety of projects including public safety in Mexico City, spatial distribution of talent, migration of highly qualified human resources, and others that include visualizations of spatial data and crowdsourcing. He has recently taught geoinformatics in the Master in Geomatics program and is also part of the newly created Spatial Planning program.

Guillaume Touya is a senior researcher in the geovisualisation research group, COGIT team at IGN France, the French national mapping agency. From 2008–2011, he did his PhD in geographical information science at the University Paris-Est. His research interests are map generalisation, automatic cartography, and volunteered geographic information. He currently leads the MapMuxing project (ANR-14-CE24-0011-01) on the multiplexing of cartographic representations to improve the navigation smoothness in geovisualisation applications.

E. Valari received her degree in Computer Science from the School of Informatics, Faculty of Sciences, Aristotle University of Thessaloniki, Greece in 2008. In 2010, she received her Master in Information Systems, School of Informatics, Faculty of Sciences, Aristotle University of Thessaloniki, Greece. Currently, she is a PhD student and working as a research consultant and developer in ICT projects. She participates in more than five research projects. She is an expert in software development in various programming languages (JAVA, C++, etc.), as well as in web development. Elena is an expert in graph mining and data stream management systems, while she published five papers in conference proceedings and chapters in books.

Index

M. Leitner, J. Jokar Arsanjani (eds.), *Citizen Empowered Mapping*, Geotechnologies and the Environment 18, DOI 10.1007/978-3-319-51629-5

D

FSC
www.fsc.org
MIX
Papier | Fördert
gute Waldnutzung
FSC® C083411

Kolibri 360 GmbH
Ferdinand-Jühlke-Straße 7
99095 Erfurt, Deutschland
produktsicherheit@kolibri360.de